PHYSICS OF LIFE

T0073700

Committee on Biological Physics/Physics of Living Systems:
A Decadal Survey

Board on Physics and Astronomy

Division on Engineering and Physical Sciences

Board on Life Sciences

Division on Earth and Life Studies

A Consensus Study Report of

The National Academies of
SCIENCES · ENGINEERING · MEDICINE

THE NATIONAL ACADEMIES PRESS
Washington, DC
www.nap.edu

THE NATIONAL ACADEMIES PRESS   500 Fifth Street, NW   Washington, DC 20001

This study is based on work supported by Grant 1760032 with the National Science Foundation. Any opinions, findings, conclusions, or recommendations expressed in this publication do not necessarily reflect the views of any agency or organization that provided support for the project.

International Standard Book Number-13: 978-0-309-27400-5
International Standard Book Number-10: 0-309-27400-1
Digital Object Identifier: https://doi.org/10.17226/26403

Copies of this publication are available free of charge from

Board on Physics and Astronomy
National Academies of Sciences, Engineering, and Medicine
500 Fifth Street, NW
Washington, DC 20001

This publication is available from the National Academies Press, 500 Fifth Street, NW, Keck 360, Washington, DC 20001; (800) 624-6242 or (202) 334-3313; http://www.nap.edu.

Suggested citation: National Academies of Sciences, Engineering, and Medicine. 2022. *Physics of Life*. Washington, DC: The National Academies Press. https://doi.org/10.17226/26403.

The National Academies of
SCIENCES · ENGINEERING · MEDICINE

The **National Academy of Sciences** was established in 1863 by an Act of Congress, signed by President Lincoln, as a private, nongovernmental institution to advise the nation on issues related to science and technology. Members are elected by their peers for outstanding contributions to research. Dr. Marcia McNutt is president.

The **National Academy of Engineering** was established in 1964 under the charter of the National Academy of Sciences to bring the practices of engineering to advising the nation. Members are elected by their peers for extraordinary contributions to engineering. Dr. John L. Anderson is president.

The **National Academy of Medicine** (formerly the Institute of Medicine) was established in 1970 under the charter of the National Academy of Sciences to advise the nation on medical and health issues. Members are elected by their peers for distinguished contributions to medicine and health. Dr. Victor J. Dzau is president.

The three Academies work together as the **National Academies of Sciences, Engineering, and Medicine** to provide independent, objective analysis and advice to the nation and conduct other activities to solve complex problems and inform public policy decisions. The National Academies also encourage education and research, recognize outstanding contributions to knowledge, and increase public understanding in matters of science, engineering, and medicine.

Learn more about the National Academies of Sciences, Engineering, and Medicine at **www.nationalacademies.org**.

The National Academies of
SCIENCES · ENGINEERING · MEDICINE

Consensus Study Reports published by the National Academies of Sciences, Engineering, and Medicine document the evidence-based consensus on the study's statement of task by an authoring committee of experts. Reports typically include findings, conclusions, and recommendations based on information gathered by the committee and the committee's deliberations. Each report has been subjected to a rigorous and independent peer-review process and it represents the position of the National Academies on the statement of task.

Proceedings published by the National Academies of Sciences, Engineering, and Medicine chronicle the presentations and discussions at a workshop, symposium, or other event convened by the National Academies. The statements and opinions contained in proceedings are those of the participants and are not endorsed by other participants, the planning committee, or the National Academies.

For information about other products and activities of the National Academies, please visit www.national academies.org/about/whatwedo.

COMMITTEE ON BIOLOGICAL PHYSICS/PHYSICS OF LIVING SYSTEMS: A DECADAL SURVEY

WILLIAM BIALEK, NAS,[1] Princeton University, *Chair*
BRIDGET CARRAGHER, Columbia University
IBRAHIM I. CISSÉ, Max Planck Institute of Immunobiology and Epigenetics
MICHAEL M. DESAI, Harvard University
OLGA K. DUDKO, University of California, San Diego
DANIEL I. GOLDMAN, Georgia Institute of Technology
JANÉ KONDEV, Brandeis University
PETER B. LITTLEWOOD, University of Chicago
ANDREA J. LIU, NAS, University of Pennsylvania
MARY E. MAXON, Lawrence Berkeley National Laboratory
JOSÉ N. ONUCHIC, NAS, Rice University
MARK J. SCHNITZER, Stanford University
CLARE M. WATERMAN, NAS, National Institutes of Health

Staff

CHRISTOPHER J. JONES, Senior Program Officer, Study Director
STEVEN M. MOSS, Senior Program Officer
NEERAJ P. GORKHALY, Associate Program Officer
AMISHA JINANDRA, Associate Program Officer
RADAKA LIGHTFOOT, Senior Financial Assistant
LINDA WALKER, Program Coordinator
JAMES C. LANCASTER, Director, Board on Physics and Astronomy (until April 2021)
COLLEEN N. HARTMAN, Director, Board on Physics and Astronomy (since April 2022)

---

[1] Member, National Academy of Sciences.

---

[1] Member, National Academy of Sciences.

SABINA VADNAIS, Research Associate
HUANG-NAM VU, Program Assistant
DAISHA WALSTON, Program Assistant (until August 2022)

Preface

Every 10 years, the National Academies of Sciences, Engineering, and Medicine (the National Academies) survey the state of physics research in the United States. To make the task manageable, it is broken down by field, so that there are separate reports on astronomy and astrophysics; atomic, molecular, and optical physics; condensed matter and materials physics; elementary particle physics; nuclear physics; and plasma physics. In this cycle, for the first time, biological physics, or the physics of living systems, stands alongside these fields as part of physics. For many of us, this is a moment to celebrate.

It has been nearly 20 years since the National Science Foundation (NSF) launched a small program to support biological physics within its Physics Division. That effort has evolved into the Physics of Living Systems program. NSF is the sponsor of this survey, charging the committee to take a broad look at the field, its connections to other fields, and the challenges that the community faces in realizing the field's potential (see Appendix A). Our work began at a meeting on February 6–7, 2020, with presentations from NSF staff, followed by a lively discussion. It is a pleasure to thank Krastan Blagoev, Denise Caldwell, and their colleagues for their support of the project and for their candor. We could not know, of course, that this would also be our last gathering, as the COVID-19 pandemic soon swept through the country and around the world.

The February meeting was followed by a second meeting online, April 1–3, 2020. At these two events we heard from a number of colleagues—chosen to complement the expertise of the committee—about the state of science and education in the field: Cliff Brangwynne, Lucy Colwell, Catherine Crouch, Jeff Hasty,

Ted Hodapp, Sarah Keller, Chandralekha Singh, Xiaoliang Sunney Xie, and Xiaowei Zhuang. In addition we heard from representatives of other organizations and federal funding agencies beyond NSF, and about the federal budget process more generally, from Linda Horton (Department of Energy Office of Science), Matt Hourihan (American Association for the Advancement of Science), Peter Preusch (National Institute of General Medical Sciences, National Institutes of Health), and Elizabeth Strychalski (National Institute of Standards and Technology). Further insights into an important segment of our audience were provided by Mary Guenther and Alexis Rudd (Senate Committee on Commerce, Science, and Transportation) and by Dahlia Sokolov (House Committee on Science, Space, and Technology). All of these presentations led to significant discussion in the committee, and we appreciate the time and care taken by all of our speakers.

A crucial part of the decadal survey process is community input. We solicited written input through the survey website, and held two town hall meetings—one in person at the Biophysical Society Meeting in San Diego (February 16, 2020) and one online through the Division of Biological Physics of the American Physical Society (April 16, 2020). It was wonderful to hear from so many members of the community, speaking from many different perspectives—from undergraduates attending their first scientific conferences and from senior faculty; from researchers in industry, research institutes, and medical schools; and from faculty at community colleges, primarily undergraduate institutions, and research universities. As described in the report, it was exciting both to hear such a wide range of voices and to see the emergence of common themes.

The work of the committee was done in the two meetings described above and a third (July 27–29, 2020, also online), along with a series of 28 video conferences; there were numerous meetings of subgroups involved in generating first drafts of different chapters or addressing gaps in subsequent drafts. None of this would have been possible, especially in this challenging time, without the support of the National Academies staff, including Neeraj Gorkhaly, Amisha Jinandra, Steven Moss, Fran Sharples, and Linda Walker. Kim DeRose and Anne Johnson provided guidance on writing style and process. Throughout the project, Radaka Lightfoot managed the budget as plans were continually revised in response to the pandemic. At crucial moments James Lancaster came with excellent advice in his role as Director of the Board on Physics and Astronomy; late in the project he was succeeded by Colleen N. Hartman, who brought fresh eyes, insights, and enthusiasm. Very special thanks to Christopher Jones, who led this effort and provided both wise counsel and gentle reminders of the passing months.

This report argues that breadth is an essential part of the excitement in biological physics. The physicist's approach to understanding the phenomena of life is yielding fascinating results in contexts ranging from the folding of proteins to the flocking of birds, from the internal mechanics of cells to the collective dynamics of neurons in the brain, and more. We see glimpses of the sorts of unifying ideas that we hope for in physics, cutting across this huge range of scales. At the same time, each of these problems also connects to a larger community of biologists, sometimes reaching as far as applications in medicine and technology. Surveying all of this required assembling a committee that represents a broad range of interests and expertise, and even so each of us had to stretch to be sure that we could, together, provide our readers with a sense for all of what is exciting in the field. As expected, it has been wonderful fun for all of us on the committee to talk about all these scientific developments. Less expected, perhaps, has been the pleasure of participating in the emergence of consensus on how our excitement about the science translates into recommendations about policy on matters of great concern to the community.

The collective effort involved in producing this report has been substantial, especially in the context of the disruptions that all of us have experienced during the pandemic. I think I can speak for the whole committee in expressing thanks to the numerous colleagues, coauthors, friends, and family members who were patient and understanding with us during this time.

Finally, let me exercise the chair's prerogative and add a personal perspective. When I was a student, physicists who became interested in the phenomena of life were perceived as becoming biologists. Physicists and biologists agreed that there were productive applications of physics to biology, but the idea that living systems posed real challenges to our understanding of physics itself was not popular. I do not think that these views were fair to the history of the field, but they were widely held. Today, much has changed, both in the substance of what has been accomplished and in the perception of these accomplishments, especially by the physics community. The search for the physics of life now is a research program rather than a fantasy, and biological physics has emerged as a branch of physics. This happened not in one dramatic moment, but through decades of progress and gradual realizations. The result is nothing less than a redrawing of the intellectual landscape, the consequences of which continue to unfold in beautiful and sometimes surprising ways. I hope that we have done justice to these remarkable developments.

William Bialek, *Chair*
Committee on Biological Physics/Physics of Living Systems:
A Decadal Survey

Acknowledgment of Reviewers

This Consensus Study Report was reviewed in draft form by individuals chosen for their diverse perspectives and technical expertise. The purpose of this independent review is to provide candid and critical comments that will assist the National Academies of Sciences, Engineering, and Medicine in making each published report as sound as possible and to ensure that it meets the institutional standards for quality, objectivity, evidence, and responsiveness to the study charge. The review comments and draft manuscript remain confidential to protect the integrity of the deliberative process.

We thank the following individuals for their review of this report:

Catherine Crouch, Swarthmore College,
Robert Full, University of California, Berkeley,
Margaret Gardel, University of Chicago,
James C. Gumbart, Georgia Institute of Technology,
Judith Klinman, NAS,[1] University of California, Berkeley,
Robert Phillips, California Institute of Technology,
Stephen Quake, NAS/NAE[2]/NAM,[3] Stanford University,
Elizabeth Strychalski, National Institute of Standards and Technology,
Kandice Tanner, National Cancer Institute, and
Yuhai Tu, IBM T.J. Watson Research Center.

---

[1] Member, National Academy of Sciences.
[2] Member, National Academy of Engineering.
[3] Member, National Academy of Medicine.

Although the reviewers listed above provided many constructive comments and suggestions, they were not asked to endorse the conclusions or recommendations of this report nor did they see the final draft before its release. The review of this report was overseen by Thomas F. Budinger, NAE/NAM, Lawrence Berkeley National Laboratory, and Herbert Levine, NAS, Northeastern University. They were responsible for making certain that an independent examination of this report was carried out in accordance with the standards of the National Academies and that all review comments were carefully considered. Responsibility for the final content rests entirely with the authoring committee and the National Academies.

Contents

Executive Summary

Biological physics, or the physics of living systems, brings the physicist's style of inquiry to bear on the beautiful phenomena of life. The enormous range of phenomena encountered in living systems—phenomena that often have no analog or precedent in the inanimate world—means that the intellectual agenda of biological physics is exceptionally broad, even by the ambitious standards of physics.

For more than a century, the contrast between the complexity of life and the simplicity of physical laws has been a creative tension, driving extraordinarily productive interactions between physics and biology. From the double helical structure of DNA to magnetic resonance images of our brain in action, results of this collaboration between physics and biology are central to the modern understanding of life, and these results have had profound impacts on medicine, technology, and industry. Until recently, however, these successes were codified as parts of biology, not physics.

As the 20th century drew to a close, this began to change: Members of the physics community came to see the phenomena of life as challenges to our understanding of physics itself, challenges that are as profound and revolutionary as those posed by phenomena of the inanimate world. This wide range of explorations is united by the search for underlying physical principles, leading to the major conclusion of this study: A new field has emerged.

Conclusion: Biological physics now has emerged fully as a field of physics, alongside more traditional fields of astrophysics and cosmology, atomic, molecular and optical physics, condensed matter physics, nuclear physics, particle physics, and plasma physics.

Reacting to the emergence of this new discipline, the Committee on Biological Physics/Physics of Living Systems: A Decadal Survey of the National Academies of Sciences, Engineering, and Medicine was appointed to carry out the first decadal survey of this field, as part of the broader decadal survey of physics. Hundreds of scientific community members from a wide range of institutions and career stages provided valuable input to the committee, in addition to the funding agencies themselves. This report aims to help federal agencies, policymakers, and academic leadership understand the importance of biological physics research and make informed decisions about funding, workforce, and research directions.

Although the field is intellectually broad, unifying conceptual questions, in the physics tradition, help to organize exploration of the field.

- *What are the physics problems that organisms need to solve?*
  To survive in the world, living organisms must convert energy from one form to another, sense their environment, and move through the world. Exploration of these functions has led to surprising new physics, from the interplay of classical and quantum dynamics in photosynthesis to hidden symmetries in the dynamics of macroscopic animal behavior.

- *How do living systems represent and process information?*
  Understanding the physics of living systems requires us to understand how information flows across many scales, from single molecules to groups of organisms. From harnessing energy dissipation for more reliable information transmission on the molecular scale to using novel network dynamics as a neural code in the brain, life has found unexpected realizations of the physics of information.

- *How do macroscopic functions of life emerge from interactions among many microscopic constituents?*
  From the ordered structure of a folded protein to the ordered flight paths of birds in a flock, much of what fascinates us about life involves collective behavior of many smaller units. Theory and experiment have combined to show how these collective behaviors can be described in the language of statistical physics, while pointing to new kinds of order that have no analog in the inanimate world.

- *How do living systems navigate parameter space?*
  The numbers describing the mechanisms of life change in time through the processes of adaptation, learning, and evolution. The biological physics community has brought new perspective to these problems, envisioning life's mechanisms as drawn from an ensemble of possibilities. The char-

acterization of these ensembles—from the repertoire of antibodies in the immune system to the range of synaptic connections that are consistent with brain function—provides new physics problems.

These general physics questions are illustrated by many different biological systems, and answers will have concrete consequences. Details on each of these "Big Questions" appear in Part I of the report.

No healthy scientific field exists in isolation. Biological physics has drawn ideas and methods from neighboring fields of physics, but also has been a source of inspiration for new problems in these fields. Historical connections to many different areas of biology and chemistry continue to be productive, and results from the physics of living systems reach further into medicine and technology. The biological physics community has provided new tools for scientific discoveries, new instruments for medical diagnosis, new ideas for systems biology with applications in synthetic biology, new methods and theories for exploring the brain, and new algorithms for artificial intelligence. Results and methods from the biological physics community have been central in the world's response to the COVID-19 pandemic. More on these connections can be found in Part II.

Finally, the report addresses what must be done to realize the promise of biological physics as a field. Building a new scientific field is a multigenerational project, and the emergence of biological physics prompts rethinking of how we teach physics, biology, and science more generally. Funding structures, currently fragmented, need revision to respond to the full breadth and coherence of activity in the biological physics community. Fully realizing the potential of the field requires welcoming aspiring scientists from all over the world, and from all segments of our society, managing resources in ways that are both effective and just. The committee's general and specific recommendations in response to challenges in education, funding, and the human dimensions of science can be found in Part III, and are summarized in Appendix B.

The biological physics community is developing new experimental methods that expand our ability to explore the living world, and new theories that expand the conceptual framework of physics. These developments are redrawing the intellectual landscape of science and driving new technology. Ultimately, a mature physics of life will change our view of ourselves as humans.

Introduction and Overview

Physics is about things that we can hold in our hands, but also about the history of the universe as a whole. The discipline is defined not by the objects being studied, which have changed over time, but rather by the kinds of questions being asked and the kinds of answers being sought. The physics community takes seriously the Galilean dictum that "the book of Nature is written in the language of mathematics," and searches for an understanding that is expressible in a compact and compelling mathematical structure. This understanding emerges through an intricate dialogue between theory and experiment. Physicists build new instruments, extending humanity's ability to observe the world; these instruments test the predictions of our theories and enable discovery in places where our theories are silent. Physicists explore new theories, both as explanations of puzzling observations and as worlds unto themselves, sometimes connecting back to the world of our experience only after decades of work. Victory is declared when theory and experiment agree in quantitative detail, but the community also prizes approximate reasoning, allowing for less precise predictions with proportionally less effort. The history of physics teaches us that particular numerical facts about the world are explained by reference to more general principles, and that the most striking phenomena will connect to the deepest concepts. Biological physics, or the physics of living systems, brings this physicist's style of inquiry to bear on the beautiful and complex phenomena of life.

READER'S GUIDE

This report addresses three broad questions, in the three major parts of the report:

1. What is biological physics? (Part I)
2. How is biological physics connected to other parts of physics, to biology, and to applications of direct relevance to society? (Part II)
3. What challenges do we face in realizing the promise of the field? (Part III)

As will become clear, the intellectual agenda of biological physics is extremely broad, touching phenomena on scales ranging from molecules to ecosystems. The committee has organized the exploration of this agenda, in Part I, around four major conceptual questions, each of which is illustrated by multiple examples. Each major question is the subject of a separate chapter, each of these chapters begins with a general introduction that defines the question, and each example concludes with a perspective about the new physics that has been discovered. This organization is not intended to be canonical, but rather to give a feeling for the breadth and depth of the field as a branch of physics, and for its unifying themes, in the physics tradition.

Where Part I emphasized the internal coherence of biological physics, Part II emphasizes its strong connection to other fields of science and technology. There are deep and sometimes surprising connections to other areas of physics (Chapter 5), as well as interactions with disparate parts of biology, chemistry, and even psychology, sometimes extending over a century (Chapter 6). As with all areas of physics, progress in biological physics has practical consequences, and we see these consequences in contexts ranging from the doctor's office and the diagnostic lab to robots and artificial intelligence, with many stops in between (Chapter 7).

Realizing the ambitious agenda of biological physics will require addressing many challenges, as explained in Part III. New science is fueled by new young people entering the field, and it matters how they are educated. Integration of biological physics into the physics curriculum, at all levels, can be synergistic with a broader modernization of physics teaching and the nurturing of a more quantitative biology (Chapter 8). Progress also is fueled by funding, and here the major challenge is to align the funding mechanisms to the structure of the field, rather than fragmenting the field along the lines defined by different grant programs (Chapter 9). Finally, as with all areas of physics and science more generally, biological physics faces challenges in welcoming talent from all segments of our society, including those who have been the targets of historic and continuing injustice (Chapter 10). The bulk of the report's recommendations are found in Part III, addressing these challenges.

The present chapter is meant both as an introduction to the report and as a self-contained overview. In that spirit, we recapitulate the major "parts" of the report in the following sections. Because this is the first time that biological physics is being surveyed as a part of physics, the committee has taken the liberty of providing more than the usual historical background for the field, followed by a brief description of how this report fits into the larger context of the decadal surveys of physics. Finally, the introduction concludes with a summary of the committee's findings, conclusions, and recommendations.

DEFINING THE FIELD

The phenomena of life are startling. A single cell can make an almost exact copy of itself, with just a handful of errors along the millions or even billions of bases of DNA sequence that define its identity. In some species it takes just 24 hours for a single cell to develop into a multicellular organism that crawls away from its discarded eggshell and engages in a wide range of behaviors. The swimming of bacteria is driven by a rotary engine comparable in size to a single transistor on a modern computer chip. When we humans sit quietly on a dark night, we can see when just a few quanta of light arrive at our retina, and we remember these sensory experiences years or even decades later, when most of the molecules in our brain have been replaced. Physicists have been fascinated by all of these phenomena, and much more.

In defining a scientific field, it is important to remember that phenomena in nature do not come labeled as belonging to particular disciplines. Thus, the behavior of electrons in solids is of interest to chemists, engineers, and materials scientists, while at the same time being a core topic in condensed matter physics. Similarly, the phenomena of life have attracted the attention of biologists, chemists, engineers, and psychologists, as well as the growing community of physicists whose work is explored in this report. Faced with the same phenomena, scientists from different disciplines ask different questions and search for different kinds of answers. Answers to questions coming from one discipline often lead to new questions or even whole research programs in other disciplines, and this has been especially true at the borders of physics, chemistry, and biology. These rich interactions among scientific cultures are a source of excitement, but complicate any effort to define the boundaries of different fields.

In this first attempt to survey biological physics as a part of physics, the committee has taken a broad view: Biological physics is the effort to understand the phenomena of life in ways that parallel the physicists' understanding of the inanimate world, prizing the search for new physics that does not have an obvious analog outside the living world. Even though physicists ask different questions, the biological physics community often builds on foundations laid by generations of biologists. Similarly, the answers to questions posed by the biological physics community, and the tools developed in answering these questions, often have substantial impact on the mainstream of biology. Beyond the traditional confines of biology, the construction of analogs to living systems is an important way of connecting biological physics to the rest of physics and to engineering. This flow of ideas and methods across disciplinary boundaries speaks to the centrality of biological physics in the current scientific landscape.

A Brief History

For centuries, the phenomena of life were seen as fundamental challenges to human understanding of the physical world. In the 17th century, some of the first objects to be discovered with the light microscope were living cells. In the 18th century, modern ideas about charge and voltage emerged in part from explorations of "animal electricity." In the 19th century, the laws of thermodynamics were not fully established until one could balance the energy budget of animal movements, and the emerging understanding of light and sound was contiguous with the study of vision and hearing, including the nature of the inferences that the brain could draw from data collected by our eyes and ears. In the 20th century, both physics and biology were revolutionized, and there was an extraordinarily productive interaction between the disciplines. From the double helical structure of DNA to magnetic resonance images of our brain in action, the results of this collaboration between physics and biology are central to the modern view of life. But these great successes were codified as parts of biology, not physics. As the 20th century drew to a close, this began to change. Today, the phenomena of life are seen once again as challenges to physics itself.

What is it that changed, making it possible for the study of living systems to be part of physics, rather than being seen only as the application of physics to biology, or some interdisciplinary amalgam? The changes seem to have been gradual, with no obvious single "eureka" moment. Components of these changes were internal to separate communities of physicists and biologists.

From a biological perspective, the sequencing of DNA for whole organisms—studying genomes rather than individual genes—meant that the project of enumerating the molecular building blocks of life was approaching completion. At the same time, cell biologists and physiologists who had focused on how large assemblies of molecules come to life began to exploit new methods adapted from molecular biology, and classical tools of microscopy were revolutionized, in part with ideas from physics. Independently, neurobiologists appreciated that progress in understanding the brain would require making quantitative connections between the dynamics of neural circuits and the macroscopic phenomena of behavior, from perception and decision-making to learning, memory, and motor control. By focusing on microorganisms that reproduce quickly, evolution was being turned into a subject for laboratory experiments rather than being restricted to field observations. This list is illustrative but far from complete; by the start of the 21st century, almost every subfield of biology was being transformed, dramatically, when compared to the state of the field just one decade before.

From the physicist's perspective, the last decades of the 20th century brought new views of the interplay between simplicity and complexity. Some of the most beautiful and compelling macroscopic phenomena in real materials, such as critical behavior near the transitions between phases of matter, proved to be universal, quantitatively independent of microscopic detail, and a theoretical framework—the renormalization group—was developed within which these behaviors could be explained and predicted. Strikingly, these same theoretical ideas proved central to the development of the standard model of elementary particle interactions, uniting macroscopic and microscopic physics. In parallel, physicists explored more complex systems, including polymers and liquid crystals, that traditionally had been the domain of chemists and engineers. Where impurities and disorder had once been viewed as distractions from the idealized behavior of perfect crystals, it became clear that disordered materials had their own beautiful and profound regularities. The striking patterns that form in a variety of dynamical systems, from the layering of fluid flows to the branching of snowflakes, were brought into focus as physics problems. Even chaos itself could be tamed. These successes certainly emboldened the physics community, providing examples where it was possible to "find the physics" in ever more complex systems. Rekindling old dreams, many physicists began to wonder if the marvelous phenomena of life itself might be within reach. Looking back, it seems clear that the parallel revolutions in physics and biology encouraged physicists to start asking fundamental questions about some of the most complex systems on Earth, living organisms.

Doing Physics in a Biological Context

The previous paragraphs outline an optimistic view of history, with parallel developments in physics and biology leading toward the emergence of biological physics as a branch of physics. In fact, this was once a minority point of view. Some biologists saw the physicist's search for simplicity and universality as poorly matched to the complexity and diversity of life on Earth; theory, a crucial part of the physics culture, was openly derided in many, though not all, areas of biology. From the other side, some physicists worried that the phenomena of life might best be described as an accumulation of mechanistic details, with no hint of the unifying principles that are characteristic of physics more broadly. In this view, the complexity and diversity of life are evidence that biology and physics are separate subjects, and will remain separate forever. If this view were correct, then there are useful applications of physics to the problems of biology, but there is no hope for a physics of life. Establishing the physics of living systems as a field of physics has required a systematic response to this skepticism.

A first step is to turn qualitative impressions into quantitative measurements, taming the complexity and organizing the diversity of life. This proved to be a decades-long project. Many physicists began by searching for the microscopic building blocks of life. This approach is most famously connected to the emergence of molecular biology (see below), but it was repeated many times—the reaction center that allows photosynthetic organisms to capture the energy of sunlight, the ion channels that provide the basis for all electrical activity in the brain, and more. It is an extraordinary discovery that these molecular components are universal and often interchangeable, even across eons of evolutionary change. In many cases, it became possible to make precise measurements on the behavior of these basic molecules, one at a time, giving us a remarkably precise view of life's mechanisms. These single-molecule experiments have, unambiguously, the "look and feel" of physics experiments.

But much of what fascinates us about life is not visible in the behavior of its isolated parts. To capture these phenomena requires doing physics experiments on intact biological systems, in all their complexity, and this is where there has been dramatic progress in recent years. Thus, beyond observing single molecules of messenger RNA (mRNA) being transcribed, it now is possible to see and count essentially every one of these mRNA molecules, representing information being read out from thousands of genes, across many single cells. Beyond the electrical currents flowing through single ion channels, or the voltages generated by single cells, new methods make it possible to monitor, simultaneously, the electrical activity of hundreds or even thousands of individual neurons in the brain as an animal executes complex behaviors. Beyond the classic experiments of tracking the behavior of a single bacterium, experiments now monitor thousands of individual cells in a growing bacterial community, or thousands of individual birds in a flock, as they engage in collective behaviors. This list is illustrative rather than exhaustive. Importantly, this progress has emerged from a rich interplay between the intellectual traditions of physics and biology, and has touched phenomena across a wide range of spatial and temporal scales.

The last decades thus have seen enormous progress in our ability to "do physics" on intact, functioning biological systems in all of their richness. This involves uncovering precise and reproducible quantitative features of functional behavior in particular systems, and connecting these experimental observations to theoretical ideas grounded in more general principles. This approach has made inroads into the exploration of life on all scales, from single molecules to vast groups of diverse organisms. Ideas about information flow and collective behavior cut across these scales, holding out hope for a more unified understanding. Interaction between theory and experiment now happens, frequently, at the level of detail that we expect in physics, but which not so long ago seemed impossible in a biological context.

What emerges from all of this excitement is a new subfield of physics—biological physics, or the physics of living systems—which takes its place alongside established subfields such as astrophysics, atomic physics, condensed matter physics, elementary particle physics, nuclear physics, and plasma physics. Biological physics, as with the rest of physics, is both a theoretical and an experimental subject, and necessarily engages with the details of particular living systems; searching for simplicity does not mean ignoring complexity, but taming it. The physics of living systems is focused not only on what is interesting for physicists but on what matters in the lives of organisms.

Physics has made progress in part by going to extremes. Elementary particle physics strives to observe matter on the very smallest scales of length and time, while astrophysics and cosmology probe the very largest scales. One theme of modern atomic physics is the study of matter at extraordinarily low temperatures, while plasma physicists are interested in temperatures comparable to those inside the sun. Condensed matter physics studies the unexpected behaviors that emerge when very many particles interact with one another. In this catalogue of extremes, biological physics is concerned with matter that is extremely organized, and organized in ways that make possible the remarkable functions that are the everyday business of life. Biological physics, or the physics of living systems, aims to characterize this organization, to understand how it happens, or even how it is possible.

Reductionism, Emergence, and What Is Special About Life

Perhaps the defining problem of biological physics is to discern what distinguishes living systems from inanimate matter: What are the essential physical principles that enable the remarkable phenomena of life? As in physics more broadly, one can identify two very different approaches to the physics of living systems, reductionism and emergence. The reductionist approach searches for the fundamental building blocks of life and characterizes their interactions, in the spirit of elementary particle physics. In focusing on emergence, the goal is to classify and understand behaviors that arise when many of the building blocks interact, in the spirit of condensed matter physics. Each of these approaches to the physics of life has a substantial history, starting well before biological physics was accepted as a part of physics. Since this is the first decadal survey of the field, it seems appropriate to provide a coherent view of the current state and its historical context.

Reductionism

In the 1950s and 1960s, X-ray diffraction made it possible, for the first time, to visualize the events of life as being embodied in the structure and dynamics of particular molecules. While X-ray diffraction had been discovered and understood in the early years of the 20th century, new physics and chemistry would be needed to determine the structure of such large and complex molecules from their diffraction patterns. The fact that DNA is a helix was evident from the famous photograph 51 (see Figure I.1A), but only because the theory of diffraction from a helix had been developed not long before, in an effort to understand the structure of proteins. The double helix (see Figure I.1B) immediately suggested a theory for how genetic information is encoded in DNA and transmitted from one generation to the next. These experimental and theoretical developments, emerging in large part from physics departments, provided the foundations for modern molecular biology. There was a long path from these early structures to revealing the positions of many thousands of atoms in a protein, going far beyond anything that had been done in the established methods of physics and chemistry. The revolution in our view of biological molecules made possible by X-ray diffraction would be extended by new synchrotron light sources; by developments in nuclear magnetic resonance (NMR) which made it possible to infer the structure of molecules free in solution rather than confined in crystals; and most recently by the addition of new detectors to cryogenic electron microscopes, which have made cryo-EM a widely used tool for determining the structure of proteins at atomic resolution.

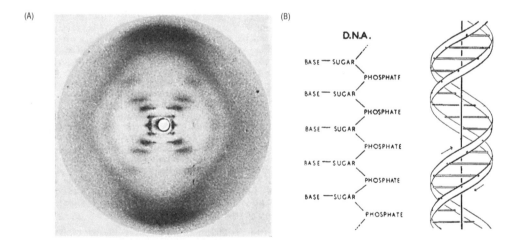

FIGURE I.1 Determining the structure of DNA was a seminal moment for nascent biological physics, as it captured the attention of the physics community. (A) Photograph 51, showing the "X" pattern of diffraction spots characteristic of a helical structure, whose dimensions can be read from the positions of the spots. (B) Schematic of the DNA double helix. Bases form the rungs of the ladder, and the phosphates and sugars are covalently bonded into the ribbons at the outside. SOURCES: (A) Reprinted by permission of Springer from R.E. Franklin and R.G. Gosling, 1953, Molecular configuration in sodium thymonucleate, *Nature* 171:740, copyright 1953. (B) Reprinted by permission of Springer from J.D. Watson and F.H.C. Crick, 1953, Genetical implications of the structure of deoxyribonucleic acid, *Nature* 171:964, copyright 1953.

Beyond visualizing these building blocks of life, a new generation of single-molecule experiments made it possible to observe and manipulate individual molecules as they carry out their functions—controlling the electrical currents flowing in neurons and muscles, generating forces, reading the information encoded in DNA, and more. This reductionist program continues to generate a stream of exciting results, characterizing the structures and dynamics of ever more complex molecular machines that carry out remarkable functions in living cells. These results are driven by new experimental methods, including a whole family of imaging methods that circumvent the diffraction limit to the spatial resolution of microscopes, allowing us literally to see the mechanisms of life in unprecedented detail (see Figure I.2). These methods required new mastery over the physics of light, connecting to core problems in atomic, molecular, and optical physics. In parallel, these experiments sharpen new theoretical questions about the physical principles that govern these nanoscale systems, connecting to a renaissance in non-equilibrium statistical physics and the thermodynamics of small systems.

Emergence

Life is more than the sum of its parts. This is evident on very large scales, as with the ordered yet fluid behavior of birds in a flock, but also on very small scales, as protein structures emerge from interactions among many amino acids. It is an old dream of the physics community that such emergent phenomena in living systems could be described in the language of statistical mechanics. In the 1980s, concrete statistical mechanics approaches to neural networks led to deeper understanding of memory, perception, and learning, and eventually to today's revolutionary developments in artificial intelligence. In the 1990s, flocks of birds, swarms of insects, and populations of bacteria inspired genuinely new statistical mechanics problems, which launched the field of active matter. Neural networks and active matter have been the source of new and profound problems in statistical physics, emphasizing how the effort to understand or even describe the phenomena of life leads to new physics.

Today, observations on real flocks and swarms in their natural environments are revealing surprising collective behaviors, beyond the predictions of existing theories of active matter (see Figure I.3), showing that we have not exhausted the new physics to be discovered in these systems. Theories of neural networks, grounded in statistical physics, are having greater impact on thinking about the brain itself as new experimental methods make it possible to monitor, simultaneously, the electrical activity in thousands of individual neurons. On a smaller scale, collective behaviors of molecules are manifest in the discovery that some of the organelles inside cells, and inside the nuclei of cells, are condensed droplets of proteins and nucleic acids (see Figure I.4), an idea that has swept through the cell biology and biological physics communities in less than a decade. There are new statistical phys-

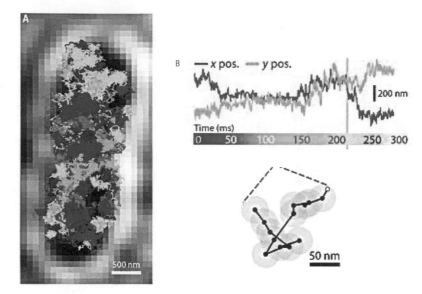

FIGURE I.2 A new family of imaging methods lets researchers track a single molecule inside the cell with near-nanometer precision using fluorescence microscopy. The position of the fluorescent molecule is computed from the spatially structured excitation light whose location is continually adjusted so that the molecule is only minimally excited. In this way, the fluorescent molecule can be precisely localized in space even while it is kept in the dark. (A) Trajectories of 77 individual fluorescently labeled ribosomes in a single bacterial cell, shown superposed on a transmission image of the whole bacterium. (B) A short (0.3 s) segment of one of the ribosome trajectories in (A), showing significant variations in mobility. At bottom, a 2 ms segment of the trajectory is expanded, with gray circles showing that ribosomes are localized to better than 50 nm even with time steps as small as 0.125 ms. SOURCE: F. Balzarotti, Y. Eilers, K.C. Gwosch, A.H. Gynnå, V. Westphal, F.D. Stefani, J. Elf, and S.W. Hell, 2017, Nanometer resolution imaging and tracking of fluorescent molecules with minimal photon fluxes, *Science* 355:606, reprinted with permission from AAAS.

ics approaches to the evolution of protein families and the persistence of ecological diversity, to the dynamics of chromosomes and the mechanics and movement of cells in tissues, and more. Abstract theoretical formulations are being connected to experiments on particular living systems, in unprecedented quantitative detail. Statistical physics provides a unifying language, connecting phenomena across the full range of scales, identifying new kinds of order, and locating living systems in the phase diagram of possible systems.

Searching for What Is Special About Living Systems

Biological physics encourages us to view living systems as examples drawn from a much larger class of possible systems. This view makes clear that what we see in

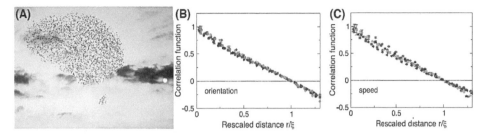

FIGURE I.3 Observations of bird flocks reveal large-scale collective, emergent behavior. Multiple cameras are used to reconstruct the three-dimensional positions of thousands of individual birds in a flock of starlings. On average the birds fly in the same directions, and variations around this average can be decomposed into fluctuations of flight direction (orientation) and speed. (A) A flock, silhouetted against the sky at moonrise. (B) The correlation C(r) between orientation fluctuations in birds separated by a distance r. In flocks of different sizes, correlations extend over different distances, but are the same once rescaled by this correlation length ξ. The correlation length itself is proportional to the size of the flock. Different colors correspond to different flocking events. (C) Same as in (B) for the correlation in speed fluctuations. Scaling behavior for orientational fluctuations is expected in a large class of statistical physics models for flocking, because the agreement to fly in a particular average direction is an example of spontaneous symmetry breaking, and the orientational fluctuations are the associated "massless" mode. The surprise is that exactly the same scaling is found for speed fluctuations, suggesting that the flock is not in a generic ordered state. SOURCES: (A) A. Cavagna and I. Giardina, 2014, Bird flocks as condensed matter, *Annual Review of Condensed Matter Physics* 5:183. (B and C): Reprinted from A. Cavagna, I. Giardina, and T.S. Grigera, 2018, The physics of flocking: Correlation as a compass from experiments to theory, *Physics Reports* 728:1, copyright 2018 with permission from Elsevier.

real organisms is not typical of random choices from this larger space of possibilities. Random sequences of amino acids do not fold into well-defined structures, unlike real proteins; networks of neurons with random connections are chaotic rather than functional. Specifying functions, and performance at these functions, points to limited regions in the space of possible systems. In this way, natural selection can stabilize behaviors that are inaccessible or unstable in the world of inanimate matter, revealing new physics.

A modest example from the 1950s is the idea that the size of lenses in insect eyes is chosen to maximize the quality of images subject to the constraints set by diffraction. This example connects otherwise arbitrary facts about the living world to basic physical principles, quantitatively, and is based on the idea that evolution can select for structures and mechanisms that come close to the physical limits on their performance at tasks crucial in the life of the organism. Ideas in this spirit now appear more widely, both in the abstract and in detailed partnership with experiment: Amino acid sequences could be selected to minimize the competing interactions that would frustrate protein folding; gene expression levels in bacteria could be tuned to maximize the conversion of nutrients into growth; the dynamics

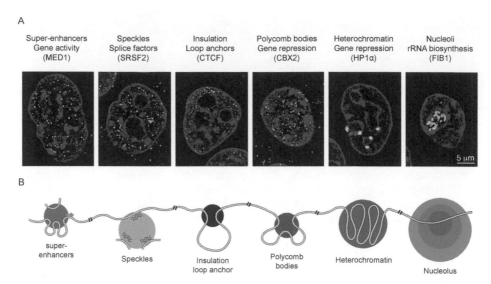

FIGURE I.4 Collective, emergent behaviors can manifest on a small scale within cells and, in this case, condensates in the nucleus. (A) Structured illumination microscopy images of immunofluorescence for the protein indicated in parentheses in murine embryonic stem cells. Immunofluorescence for the indicated protein is colored green, and the signal from Hoechst, a DNA stain, is colored dark blue. Condensates are denoted by their name (e.g., Super-enhancers), their function (e.g., gene activity), and the protein that provides the immunofluorescent signal (e.g., MED1). (B) Cartoon depiction of how various nuclear condensates organize and are organized by different chromatin substrates. The gray line represents the chromatin fiber, green arrow designates active transcription start site, and red squiggled lines represent RNA. Abbreviations: CBX2, chromobox protein homolog 2; CTCF, CCCTC-binding factor; FIB1, fibrillarin; HP1$_\alpha$, heterochromatin protein 1$_\alpha$; MED1, mediator of RNA polymerase II transcription subunit 1; SRSF2, serine/arginine-rich splicing factor 2. SOURCE: Reprinted from B.R. Sabari, A. Dall'Agnese, and R.A. Young, 2020, Biomolecular condensates in the nucleus, *Trends in Biochemical Sciences* 45:961, copyright 2020 with permission from Elsevier.

of immunological memory could be selected to optimize the response to the likely time course of antigenic challenges; neural codes could provide efficient representation of information in patterns of electrical activity, or efficient storage of memories in patterns of connections between neurons. Even if real organisms do not reach true optima, these theories provide useful idealizations and more precise ideas about what physics problems organisms need to solve in order to survive. Related work has explored the full functional landscape, constructing models for maximally diverse populations of organisms that reach some criterion level of performance, on average. There is a separate notion of optimizing the volume that this population occupies in parameter space, creating the largest possible target for evolution to find. Theories of evolutionary dynamics explain how organisms can find these targets, and when they cannot. These and other approaches to defining what is special about living systems generate controversy, but also exciting new experiments and theory.

Exploring Big Questions

To emphasize the uniqueness and coherence of biological physics, this report is organized not around particular biological systems but around larger conceptual questions, in the intellectual tradition of physics. These four questions echo the themes above, but are not quite the same. This choice of questions is not exhaustive, nor is it meant to be canonical.[1] Rather, the goal is to capture the spirit of the field at an exciting moment in its development, reviewing successes and pointing to exciting open problems. The committee hopes that the questions are broad enough to encompass the breadth of the field, but specific enough that we will see crisp answers over the next decade. Full exploration of these questions appears in Part I.

What are the physics problems that organisms need to solve? In order to survive in the world, organisms have to accomplish various tasks. They must move toward sources of food, sometimes over long distances, guided only by weak cues about the location of the source. They have to sense useful signals in the environment, and internal signals that guide the control of their own state. They often need to generate dynamics on time scales, which are not the natural scales given by the underlying mechanisms. All of these tasks consume energy, and hence require the organism to extract this energy from the environment. We refer to these various tasks of the organism as "functions," and this notion of function is an essential part of what sets living matter apart from non-living matter. One of the central problems in biological physics is to turn qualitative notions of biological function into new and precise physical concepts, as described in Chapter 1.

How do living systems represent and process information? A traditional introduction to physics emphasizes that the subject is about forces and energies. This might lead us to think that the physics of living systems is about the forces and energies relevant for life, and certainly this is an important part of the subject. But life depends not only on energy; it also depends on information. Organisms and even individual cells need information about what is happening in their environment, and they need information about their own internal states. Many crucial functions operate in a limit where information is scarce, creating pressure to represent and process this information efficiently; new physics emerges as mechanisms are selected to extract the maximum information from limited physical resources. Understanding the physics of living systems requires us to understand how information flows across many scales, from single molecules to groups of organisms, as described in Chapter 2.

---

[1] In particular, the committee chose to resist the language of "grand challenges." The committee hopes that its account of the many exciting things happening in the search for a physics of life conveys a sense of grandeur for the enterprise as a whole. The grouping of topics into conceptual questions is meant to aid, rather than constrain, the readers' imagination.

How do macroscopic functions of life emerge from interactions among many microscopic constituents? One of the great triumphs of science in the 20th century was the enumeration and characterization of the molecular components of life. But much of what strikes us as most interesting about living systems emerges from interactions among many of these molecular components. For us as humans, much of what we do happens on the scale of centimeters or even meters. For a single cell this behavioral scale is on the order of microns, something we can only see through a microscope but still a thousand times larger than the nanometer scale of individual molecules. Efforts to bridge these scales, from microscopic to macroscopic, have led to discovery of new physics in novel kinds of ordering, in phenomena ranging from protein folding to flocks and swarms. Many of the central questions in biological physics are aimed at understanding these emergent phenomena, as described in Chapter 3.

How do living systems navigate parameter space? Any attempt at a "realistic" description of biological systems leads immediately to a forest of details. If we want to make quantitative predictions about the behavior of a system, it seems we need to know many, many numerical facts: how many kinds of each relevant molecule we find inside a cell; how strongly these molecules interact with one another; how cells interact with one another, whether through synapses in the brain or mechanical contacts in a tissue; and more. The enormous number of these parameters that we encounter in describing living systems is quite unlike what happens in the rest of physics. It is not only that as scientists we find the enormous number of parameters frustrating, but the organism itself must "set" these numbers in order to function effectively. These many parameters are not fundamental constants; instead, they are themselves subject to change over time, and in different contexts these processes of parameter adjustment are called adaptation, learning, and evolution. Many different problems in the physics of living systems, from bacteria to brains, revolve around how organisms navigate parameter space, leading to physics problems that have no analog in the inanimate world, as described in Chapter 4.

Answers to these abstract questions will have concrete consequences. Physics provides not only compelling explanations for what we see in the world around us, but also precise predictions about what we will see when we look in new places, or engineer new devices and environments. As examples:

• Understanding the physics problems that organisms must solve will identify principles that can be emulated in technology, and constraints that must be obeyed as we try to harness life's mechanisms for applications.

• Understanding the representation and processing of information in living systems will continue to have impact on artificial intelligence, but also on our ability to control the decisions that cells make in determining health and disease.

- Understanding how macroscopic functions emerge from interactions among microscopic constituents will provide a framework for engineering on many scales, from designing new proteins to coordinating swarms of robots.
- Understanding how living systems navigate parameter space already is having an impact on our ability to predict the evolution of viruses, including those that cause the flu and the current COVID-19 pandemic, and will define the landscape within which medical treatments can be personalized.

More deeply, biological physics holds the promise of unifying ideas, seeing new and common physical principles at work in disparate biological systems. There is the hope of understanding not only particular mechanisms at work in living systems but the principles that stand behind the selection of these mechanisms. Along the way to realizing this promise, the biological physics community will develop new experimental methods that expand our ability to explore the living world, and new theories that expand the conceptual framework of physics. Success will lead to a redrawing of the intellectual landscape, likely in ways that will surprise us. Ultimately, a mature physics of life will change our view of ourselves as humans.

An Emerging Community

The emergence of biological physics is visible not only in its intellectual development but also in the changing sociology of the scientific community. As noted above, some of the first major steps in the modern reductionist approach to the physics of life emerged in the 1950s, especially in the United Kingdom. While housed in physics departments, many of these efforts were supported by the UK Medical Research Council; what grew out of this were institutions such as the Laboratory of Molecular Biology that hosted a new style of biological research, but outside of physics. In this same period, physics departments in the Netherlands had small biophysics groups with efforts on vision, hearing, and photosynthesis, and the descendants of these groups are active today, still in physics departments. The first major U.S. physics department to invest in biological physics was at the University of California, San Diego, which started building a group in the late 1960s. By the late 1970s, there was a substantial effort in biological physics at Bell Laboratories. Today, there is at least some representation of the field on the faculty of almost every one of the top 100 physics doctoral programs in the United States; parallel developments have occurred around the world. There is increasing representation of biological physics at annual conferences of the American Physical Society (see Figure I.5) and at other gatherings of physicists around the world.

A growing number of physics PhD students do their thesis work in biological physics. The National Center for Science and Engineering Statistics (NCSES) tracks

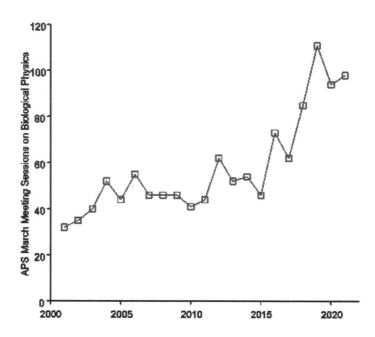

FIGURE I.5 Monitoring the growth of biological physics as a subfield of physics in the 21st century. Number of different scientific sessions on biological physics at the March Meeting of the American Physical Society; each session includes presentations from many different individual researchers and groups. SOURCE: Data from the American Physical Society's March Meeting programs, available at https://www.aps.org.

the awarding of PhDs in the United States by field and subfield; since 2004 NCSES has tracked biological physics as a subfield of physics, as discussed in Chapter 8. Although many people have the sense that biological physics is a nascent or minor activity in the physics community, in fact the number of students receiving physics PhDs with a specialization in biological physics has grown, in just 15 years, to a volume comparable to that of well-established subfields (see Figure 8.1 in Chapter 8). Biological physics today is producing the same number of new PhDs annually as did elementary particle physics in the years 2000–2005, and is continuing to grow.

Biology continues to be a much larger enterprise, producing, for example, nearly five times as many PhDs per year as in physics, spread across many more distinct subfields. Many of these subfields—molecular biology, structural biology, cell biology, systems biology, neurobiology, and more—have had, and continue to have, important input from the ideas and methods of physics. Many of these activities are categorized as "biophysics," and some of the practitioners identify themselves as biophysicists. There is thus a biophysics that is a field within the biological or biomedical sciences, and a biological physics that is a field within physics. Although it might be more accurate to view all of this activity as a continuum, the NCSES tracks the number of PhDs given in "biophysics (biological sciences)" as well as in "biophysics (physics)." Over the past decade the number of PhDs in biophysics (biological sciences) has declined slowly, while the number in biophysics (physics) has increased, with the total staying relatively constant (see Figure 8.2 in Chapter 8). Physics students who become fascinated by the phenomena of life now have more opportunities to pursue their interests either as physicists or as biologists.

CONNECTIONS

No healthy scientific field exists in isolation. This is especially true for biological physics, which by definition connects both to other parts of physics and to the enormously diverse enterprise of biology. Connections reach even further, to chemistry and engineering, and to medicine and technology, ultimately having deep implications for society. The field has been both a generous provider and an eager recipient of new concepts and principles, new instruments and tools. Part II of this report explores how biological physics has contributed to and benefited from its relationships with other scientific fields, and then describes the relationship of the field to human health, industry, and society more broadly.

Part of the beauty of physics is its interconnectedness, and biological physics is no exception (Chapter 5). Experimentalists studying the molecules of life use X-ray detectors that grew out of elementary particle physics, and particle physicists process their data with machine learning methods that grew out of theories for networks of neurons in the brain. Experiments from the biological physics community provide the most compelling measurements of the entropic elasticity of polymers, a problem that appears in almost all statistical physics textbooks. Polymers, membranes, and other materials inspired by biological molecules were at the origins of soft matter physics, while attempts to describe the collective behavior of flocks and swarms provided a foundation for the field of active matter; soft and active matter now are burgeoning fields of physics, independent of their connection to the phenomena of life. Neurons, the heart, and even slime molds have been a source of problems for the nonlinear dynamics community. Soft matter, nonlinear dynamics, and biological physics come together as these communities try to understand how organisms move, and control their movements, through granular media such as soils and sands. Efforts to observe living systems at ever-higher resolution were one of the primary drivers of a revolution in microscopy, making it possible to see beyond the diffraction limit, routinely (see Figure I.2).

Theories of neural networks are a rich source of problems in statistical physics, so much so that one of the categories for papers on the electronic archive of physics papers, arXiv.org, is "neural networks and disordered systems." Living systems provided crucial inspiration for early ideas in the thermodynamics of computation, and more recently for developments in non-equilibrium statistical mechanics. As with the elasticity of polymers, some of the most decisive tests of these ideas have come in single molecule experiments on biological molecules, adapting methods developed in the biological physics community.

All of these examples, and more, emphasize that the physics of living systems is not only a recipient of ideas and methods from other fields of physics, but also a source. The physics problems that arise in thinking about the phenomena of life not only contribute to shaping biological physics, but can take on a life of their own and contribute to the rest of physics.

Biological physics forms a nexus between physics and the many different subfields of biology and chemistry (Chapter 6). In some cases, the flux of ideas and results is from biology to physics: The biological physics community is able to ask new questions about the phenomena of life because of the foundations laid in the mainstream of biology. In return, biological physicists have given the broader biology community new tools for discovery (Chapter 6), and new ideas. The result is a continuum of activity, with different components acquiring different labels at different times (see Box I.1). As emphasized above, the committee takes a broad view of "biological physics" or the "physics of living systems," terms that we use interchangeably. We intend these terms to describe the exploration of problems that can be seen as part of physics more broadly, even if they can also be seen as parts of other disciplines.

In describing the relationship between biological physics and biology, Chapter 6 focuses on examples where ideas, methods, and results from the physics community have enabled new developments in a broader community of biologists. Increasingly this involves both experimental tools and theoretical structures (see Box I.2). The classic example is how X-ray diffraction, nuclear magnetic resonance, and cryogenic electron microscopy have made it possible to determine the structure of crucial biological molecules down to the positions of individual atoms; these methods have been exported to create structural biology, enabling exploration of a wide range of biology problems (Chapter 6). Thinking about the dynamics of these biomolecules has been driven by the theoretical ideas about energy landscapes which emerged from the biological physics community. The methods of single molecule manipulation and visualization, as well as microfluidics, are at the heart of modern, genome-wide surveys of gene expression in single cells, which are reshaping ideas about the identity of cells and cell types in complex organisms (Chapter 6). The realization that cellular organelles such as the nucleolus, known for a century, really are condensed droplets of proteins and nucleic acids has revolutionized cell biology (Chapter 6). The discovery by biological physicists that cells can sense and respond to the rigidity of their environment has had profound impacts both on developmental biology and on the biology of tumors.

The connections between biological physics and biology have led to numerous applications in public health, which have been particularly highlighted in the past

BOX I.1
Biological Physics, Biophysics, and Quantitative Biology

The 20th century saw extraordinary discoveries emerge from the interaction between physics and biology, and an important part of this effort came to be called "biophysics." Biophysics often is characterized as the application of methods from physics to answer questions in biology, but the history is richer than this description. Physicists confronting the phenomena of life often posed new and different questions, and many of the methods used in answering these questions went far beyond what had been done previously in exploring the physics and chemistry of the inanimate world. In many striking examples, success meant that the results of this interaction between physics and biology were absorbed into the mainstream of biology, leaving physics unchanged. In this way, biophysics came to be seen as a biological science.

In the 21st century, we have seen a rapid expansion of experimental tools that make much more of the living world accessible to quantitative experiments. This is driven by a complex interplay of biology, chemistry, engineering, and physics. As with biophysics, success often means that ideas and methods are absorbed into biology, independent of their parent disciplines. The result is a field described broadly as "quantitative biology," although parts of this effort can be seen in computational and systems neuroscience, systems biology, computational biology, and many other subfields.

Biophysics and quantitative biology have a continuing overlap with biological physics, both in methods and in goals. Many young scientists coming out of the biological physics community are making their careers as quantitative biologists, bringing the fields yet closer together.

The dramatic expansion of quantitative biology, not coincidentally, has occurred over the same period that biological physics emerged as part of physics. As more of the living world becomes accessible to quantitative exploration, searching for an understanding of life that parallels our understanding of inanimate matter becomes a more realistic research program. Far beyond applying known physics to problems outside the discipline, many of the developments in this field now are seen as part of physics itself. More deeply, there is growing appreciation that the physics of life is as inspiring as the physics of the early universe, as rich and varied as the physics of electrons in solids.

This report takes a broad view of the biological physics community, recognizing that many exciting developments also could be categorized as biophysics or quantitative biology. This inclusive view is used also in describing the historical development of the field, recognizing that early work by physicists who asked new questions and brought new tools to explore the phenomena of life should be seen as continuous with the 21st-century version of biological physics.

year by the community's contributions to the response to the COVID-19 pandemic. As described in Chapter 7, the biological physics community has contributed to epidemiological models that have helped guide public health decisions, and to our understanding of the physics of aerosol droplets and their role in disease transmission. Individual researchers have been central to many local university efforts to expand testing and establish safe campus reopening plans, and to the analysis of the evolutionary dynamics by which variants of concern continue to arise. Work in biological physics also has been critical to our understanding of the SARS-CoV-2 spike protein structure, its interactions with host receptors, with neutralizing anti-

---

BOX I.2
Evolving Views of Theory

For decades, one obvious difference between physics and biology concerned the role of theory. While there has always been good-natured rivalry between theorists and experimentalists in physics, there has never been any doubt about the essential features of the partnership. In contrast, the notion of theory as a partner to experiment in the exploration of life has been widely ridiculed. Many areas of biology have been described as too complex and too messy to be the subject of theories in the way that is familiar from physics. The emergence of new experimental tools for more quantitative and larger scale measurements is driving a change in these attitudes. At a practical level, the sheer volume of data demands new data analysis methods, but theory is more than data analysis. As expressed by Sydney Brenner,[a] one of the founding figures of molecular biology,[b]

Biological research is in crisis ... we are drowning in a sea of data and thirsting for some theoretical framework with which to understand it. Although many believe that "more is better," history tells us that "least is best." We need theory and a firm grasp on the nature of the objects we study to predict the rest.

---

[a] See The Nobel Prize, "Sydney Brenner Biographical," https://www.nobelprize.org/prizes/medicine/2002/brenner/biographical.
[b] S. Brenner, 2012, Life's code script, *Nature* 482:461.

---

bodies, and with potential drugs. It was also fortunate that fundamental research in structural biology, supported by decades of developments in X-ray crystallography and cryo-EM methods, provided a running start to the development of vaccines for COVID-19 (see Figure I.6). Many of the novel vaccine strategies that were used to deliver vaccines in record time were informed by very detailed knowledge of the structure and dynamics of the spike protein.

Beyond the scale of molecules and cells, physicists have been fascinated by the brain, and in trying to answer their own questions have created methods that have swept through neuroscience more generally, even reaching to psychology (Chapter 6). This has involved a mix of theory and experiment, observing the dynamics of single channels but also building mathematical models that explain how these dynamics shape the computations done by neurons; making it possible to record, simultaneously, the electrical activity of thousands of neurons but also providing theoretical frameworks within which to search for meaningful collective dynamics in these large data sets. Functional magnetic resonance imaging (fMRI) provides a unique ability to observe the activity of the human brain during sensory perception, diverse modes of cognition and decision-making, language processing, social interactions, and motor control tasks. These developments completely reshaped modern discussions of the brain and mind.

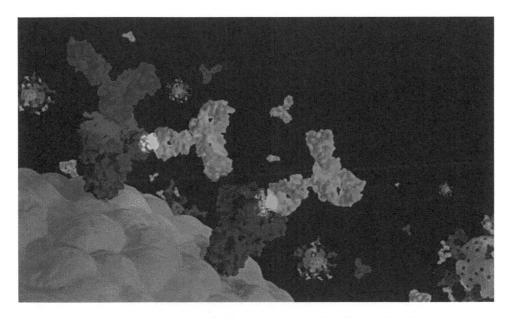

FIGURE I.6 Biological physics research is deeply connected with public health matters. For example, antibodies (multiple colors) are shown interacting with the SARS-CoV-2 spike protein (red). These structures, determined using cryogenic electron microscopy, provided a running start to the development of COVID-19 vaccines. SOURCE: Courtesy of Micah Rapp, Simons Electron Microscopy Center, New York Structural Biology Center.

Ideas and results from the biological physics community have had extensive impact on health, medicine, and technology more generally (Chapter 7). Walking into a doctor's office or a hospital, one encounters a myriad of instruments that have grown out of the biological physics community. From Doppler sonography to detect the heartbeat of an infant in utero to surgeries that are guided by sophisticated optical and X-ray imaging, medicine has been revolutionized by the physics-based ability to see inside the human body. The same imaging methods that make it possible to observe live cells at unprecedented resolution have applications in pathology and other diagnosis methods. Prosthetic devices, from cochlear implants for the deaf to brain-computer interfaces for quadriplegics, depend on recording/stimulation methods that grow out of techniques developed in the biological physics community, and theoretical ideas about how information is represented in the brain.

Tools and ideas developed in studying the physics of living systems provide a foundation for the design of new molecules with useful functions, and there is a particularly close connection between theoretical ideas about protein folding and the design of new proteins. Experimental methods for structure determination lead to structure-based drug design, which has become central to the pharmaceutical industry. Understanding of dynamics and information flow in genetic networks provides tools for synthetic biology, with applications ranging from biofuels to personalized medicine. As noted above, statistical physics approaches to evolution have advanced to the point of predicting the evolution of viruses, feeding into vaccine design.

Ideas and results from biological physics reach beyond health and medicine to technology more broadly. The ease with which we walk or run through complex environments belies the enormously challenging physics problems that complicate efforts to build robots. There are clear paths from ideas in the biological physics community about the mechanics and neural control of movement to robots that implement a broad range of locomotion strategies, from insect-like hexapods to snake-like limbless robots. Similarly, there is a path from physics-based models of neural networks in the brain to the artificial networks that are driving the deep learning revolution, changing forever how humans interact with machines. The circle is closing, as these deep networks become tools in the biological physics laboratory.

Every piece of technology in the modern world has at its foundation remarkable developments in basic science. The path from science to useful technology can be long, and requires its own unique innovations, but without the scientific foundation none of this is possible. While it is not so difficult to trace back in time from useful technology to foundational scientific discoveries, it is much harder to predict which discoveries or even which areas of research will lead to useful technology. More subtly, the more sophisticated the technology the more different threads that need to be woven together. One should thus be careful of the claim that the results of any single scientific field are uniquely responsible for a technological advance. Nonetheless, biological physics has been one essential component of many revolutionary developments.

CHALLENGES

Biological physics is healthy, growing, and exciting. Realizing the promise of the field, however, requires addressing fundamental challenges in how the community is organized, how it is funded, how students are taught, and more generally how aspiring scientists and welcomed and nurtured. These challenges are addressed in Part III of this report.

Building a new scientific field is a multigenerational project. Success in communicating the enticing intellectual opportunities of biological physics, and thus attracting talented young scientists to the field, depends on effective integration of biological physics into physics education, and into education more generally. The importance of this challenge is reflected in the fact that the majority of input that the committee received from the community was about education. This input came from colleagues at all career stages—from senior faculty to beginning students—and from a wide range of institutions, including community colleges, primarily undergraduate institutions, and major research universities.

There is no unique model for biological physics education that would fit the enormous variety of educational institutions, but there is a foundation on which to build these efforts. Although teaching physics of course involves teaching particular things, there is a unique physics culture at the core of our teaching. This culture emphasizes general principles, and the use of these principles, to predict the behavior of specific systems; the importance of numerical facts about the world, and how these facts are related to one another through the general principles; the value of idealization and simplification, sometimes even to the point of over-simplification; and the deep connections between distant subfields of physics. It is vital that this unifying culture is transmitted to students in biological physics.

Chapter 8 emphasizes that what is needed is not added specialization, but integration: Biological physics needs to be integrated into the core physics curriculum, at all levels. Ideas and results from the physics of living systems convey central ideas in physics: Flying, swimming, and walking provide an engaging universe of examples in classical mechanics; the dynamics of neurons provide examples of electric circuits and current flow; optical trapping and super-resolution microscopy illustrate deep principles of electromagnetism and optics; and the broad optical absorption bands of biological molecules, which literally provide color to much of our world, provide opportunities to build quantum mechanical intuition beyond the energy levels of isolated atoms. The concepts and methods of statistical physics, in particular, are illustrated by numerous phenomena from the living world, on all scales from protein folding to flocking and swarming. In an important counter to the impression that experiments on biological systems are messy, some of the most quantitative tests of simple polymer physics models, and the notion of entropic elasticity, have been done with DNA.

The emergence of a new field also provides the opportunity to rethink the physics curriculum more broadly. In statistical mechanics and optics—two topics that are central to biological physics—there have been revolutionary developments that are hardly reflected in current undergraduate teaching. Modernizing the teaching of these subjects would be good not only for the progress of biological physics but for physics more generally. It is important to emphasize that all of this curricular innovation will require institutional support.

While many of the educational challenges and opportunities created by the emergence of biological physics are internal to physics departments, there also are clear connections to the larger project of nurturing a more quantitative biology. Building on ideas articulated nearly 20 years ago, there continue to be important opportunities for physicists and biologists to collaborate, especially on introductory courses, and this collaboration needs to be supported by college and university administrators, above the level of departments. These opportunities cannot be realized by making longer lists of courses from multiple departments, but again require an integrated approach, weaving biological physics into the fabric of science education in ways that truly add value for students from all backgrounds.

Realizing the promise of biological physics obviously depends on having sufficient financial support. As explained in Chapter 9, research on the physics of living systems is supported by a surprisingly wide array of federal agencies and private foundations. While this diversity of funding sources has advantages, and speaks to the impact of the field on many different agency missions, it also creates problems. Only one program—the Physics of Living Systems program within the Physics Division of the National Science Foundation (NSF)—sees the field in the broad and coherent form outlined above and described more fully in Part I. But this program represents only a small fraction of the total funding for the field. Other existing funding structures fragment the field in various ways, obscuring its coherence and perhaps even slowing the emergence of unifying ideas and methods. Chapter 9 addresses various ways in which this problem can be overcome, within the overall structures provided by the federal agencies, and touches on important roles of funding agencies like the National Institutes of Health (NIH) and the Department of Energy (DOE) for specific sections of biological physics, even if they do not have a comprehensive and broad-reaching program like NSF.

There also are issues about how support is distributed across the different dimensions of the scientific enterprise. While there is widespread agreement that engagement of undergraduates in research is good for the students and good for the scientific enterprise, there is less appreciation that this engagement builds on high-quality coursework, especially at the introductory level. The committee finds that there remains too sharp a boundary between support for "scientific workforce development" and support for education. At the graduate level, different agencies have different approaches, and there are opportunities to combine best practices. As in many fields, there is a challenge in maintaining a portfolio of mechanisms to fund the spontaneity of individual investigators, the supportive mentoring environments of research centers, and the ambitious projects requiring larger collaborations. As biological physics matures, theory plays a more central role, not just as a tool for data analysis but as an independent activity, and there is an opportunity to develop programs that support this independence, as in other subfields of physics.

Much of the justification for the support of science rests on its impact in technology, medicine, and the economy more broadly, and biological physics has been a major contributor in these areas. As noted above and described more fully in Chapter 7, these contributions range from ultrasound imaging to robotics, from the design of vaccines to artificial intelligence, and more. To maintain the flow of ideas, methods, and results into these more practical domains may require new structures, but certainly will require maintaining robust support for the basic science.

Finally, scientific fields are defined as much by their community as by the list of questions they address. It is important to build a community that is not only productive, but welcoming—welcoming of aspiring scientists from all over the world, and from all the different segments of global society, including those who have been the victims of historical and ongoing injustices. These human dimensions of the scientific enterprise are explored in Chapter 10.

A DECADAL SURVEY IN CONTEXT

This volume stands in a long history of efforts to grapple with the astonishing breadth and depth of physics. The first such survey was released by the National Research Council in 1966.[2] Twenty years later,[3] *Physics Through the 1990s* divided physics into six subdisciplines, each described in a separate volume: elementary particle physics; nuclear physics; condensed matter physics; atomic, molecular, and optical physics; plasmas and fluids; and gravitation, cosmology, and cosmic-ray physics. There was, in addition, a volume devoted to "scientific interfaces and technological applications," and that volume contained a section about biophysics. This organization enshrined the view that physicists who became engaged with the phenomena of life became applied physicists, using the tools of the discipline to answer questions outside its boundaries. Such applications could be profound, and they could have enormous impact on society, but they were not seen as being in the intellectual core of physics.

The next cycle of decadal surveys culminated in the report *Physics in a New Era*, released in 2001.[4] Strikingly, the physics of biological systems had moved from the chapters on physics and society into the lead section on physics frontiers. By the next cycle, the survey volume on condensed matter and materials physics identified the physics of life as one of six scientific challenges to the field. That discussion concluded:[5]

> We have passed the point at which the interaction between physics and biology can be viewed as "merely" the application of known physics. Rather, the conceptual challenges of the phenomena of life are driving the emergence of a biological physics that is genuinely a subfield of physics.

---

[2] National Research Council (NRC), 1966, *Physics: Survey and Outlook—A Report on the Present State of U.S. Physics and Its Requirements for Future Growth,* National Academy Press, Washington, DC.

[3] NRC, 1986, *An Overview: Physics Through the 1990s,* National Academy Press, Washington, DC.

[4] NRC, 2001, *Physics in a New Era: An Overview,* National Academy Press, Washington, DC.

[5] NRC, 2007, *Condensed-Matter and Materials Physics: The Science of the World Around Us,* The National Academies Press, Washington, DC.

Finally, the present volume marks the first time that the National Academies of Sciences, Engineering, and Medicine are surveying biological physics, or the physics of living systems, as a distinct branch of physics, standing alongside other subfields in the decadal survey of physics as a whole. This completes a process that has taken a generation.

As with all decadal surveys, this report is responding to specific questions, and the outline follows this statement of task (Appendix A). As part of the process, the committee gathered community input through written submissions and at two town hall meetings, one at the Biophysical Society Meeting (February 16, 2020) and one held online through the Division of Biological Physics of the American Physical Society after the pandemic resulted in cancellation of the American Physical Society March Meeting (April 16, 2020). The clarity and consistency of community input, on several themes, was striking.

The community admonished the committee to view the field in the broadest possible terms, to articulate the rich connections to other fields while clarifying what makes this field distinct, to emphasize the special role of theory, and to point forward to exciting opportunities. There were clear hopes that the report would highlight the extraordinary contribution that the tools and methods of physics have made to the exploration of life, but not characterize the field merely as the application of physics to biology. Instead, the committee was advised to emphasize the many places where physicists have asked new questions about the living world, introducing new concepts and searching for more general principles that connect myriad particular systems.

The largest component of community input drew attention to the challenges connected to education, with clearly articulated concerns and suggestions coming from young students and from senior faculty, from colleagues at research universities, at primarily undergraduate institutions, and at community colleges. Thoughts about federal support for research rose above the usual concerns about the amount of funding, pinpointing the mismatches between narrowly defined funding structures and the broad scope of the field. Finally, many community members spoke to the unique appeal of the field. Where the physicist's intellectual style can seem demanding and inaccessible, taking inspiration from the phenomena of life connects these scientific ambitions to the imagination of a wider audience. Many members of the biological physics community see this as a special path to exciting the public about science more generally, to recruiting and retaining a more diverse community of students, and ultimately to shaping how we think of ourselves as humans.

All of these ideas found resonance with the experiences of the committee. We hope that we have done justice to the breadth and depth of the community's views.

FINDINGS, CONCLUSIONS, AND RECOMMENDATIONS

This introduction and overview concludes with a collection of the committee's findings, conclusions, and recommendations. Complete discussions of these issues can be found at appropriate places in the main text, as indicated. A compact summary of the recommendations alone can be found in Appendix B.

Transparency

Summarizing the results of the committee's deliberations provides an opportunity to reflect on the process that led to this report. The National Academies have established a framework for these efforts that is designed to minimize the risks of conflict and bias, and this process has been honed over the long history of the National Academies' role in advising the nation.[6] Nominations to the committee are solicited from a broad cross-section of the community, the membership is vetted by the National Academies' staff and ultimately approved by the President of the National Academy of Sciences in her role as Chair of the National Research Council. The goal is to assemble a group that is balanced along relevant axes, and an important part of the committee's initial meeting is a full disclosure to one another about individual affiliations and commitments that may bias its views. These disclosures are updated over the course of the committee's work, and the final editing of the report provided a chance to revisit these issues, which can be subtle. As an example, not only have all committee members been supported by one or more of the federal agencies and private foundations whose funding programs are described in Chapter 9, several committee members have provided advice to these agencies and foundations. In accord with National Academies' policies, all of these potential problems have been disclosed, and the character of the committee's discussion was such that no single member's views were taken as authoritative on any issue. This is a consensus report, and as such, all findings, conclusions, and recommendations have been agreed to by all members.

Emergence of a New Field

The enormous range of phenomena encountered in living systems—phenomena that often have no analog or precedent in the inanimate world—means that the intellectual agenda of biological physics is exceptionally broad, even by the

---

[6] For a brief summary, see National Academies of Sciences, Engineering, and Medicine, "Our Study Process," https://www.nationalacademies.org/about/our-study-process.

ambitious standards of physics. Part I of this report surveys these exciting developments, organized around the four conceptual questions outlined above, which serve to define biological physics as a field of physics. The seemingly disparate examples encountered in Part I—from the first femtoseconds of photosynthesis to evolution over thousands of generations—are united by their intellectual style. Rekindling century-old dreams, the search for a broad and unifying physics of life now is a realistic agenda. This leads to the committee's first, overarching conclusion:

> **Conclusion:** Biological physics, or the physics of living systems, now has emerged fully as a field of physics, alongside more traditional fields of astrophysics and cosmology; atomic, molecular, and optical physics; condensed matter physics; nuclear physics; particle physics; and plasma physics.

At the same time that this report marks the emergence of biological physics as a distinct enterprise, it is essential to remember that all fields of physics have extensive connections to one another, and to other disciplines:

> **Conclusion:** Explorations in the physics of living systems have produced results, ideas, and methods that have had enormous impact on neighboring fields within physics, many fields of biology, on the sciences more generally, and on society through medicine and industry.

These observations about the field lead to the first general recommendations:

> **General Recommendation: Realizing the promise of biological physics requires recognition that is distinct from, but synergistic with, related fields, both in physics and in biology. In colleges and universities it should have a home in physics departments, even as its intellectual agenda connects profoundly to efforts in many other departments across schools of science, engineering, and medicine.**

> **General Recommendation: Physics departments at research universities should have identifiable efforts in the physics of living systems, alongside groups in more traditional subfields of physics.**

While biological physicists find homes in a wide range of academic departments, research institutes, and laboratories, representation in physics departments is important for the development of the field. This representation will take different forms in different institutions, but positioning biological physics as a core component of the physics community reinforces an approach to the beautiful and complex phenomena of the living world through the "physics mindset" that prizes not just simplification but unification—the search for analogies and deeper commonalities among diverse systems. At the same time, this is not enough to ensure the health of the biological physics community.

> **Specific Recommendation: The biological physics community should support exploration of the full range of questions being addressed in the field, and assert its identity as a distinct and coherent subfield embedded in the larger physics community.**

Educating the Next Generation

Establishing a new field and stretching the boundaries of well-established disciplines are multigenerational projects. As emphasized above, the largest components of community input to the committee focused on these educational issues, which are the topic of Chapter 8. The survey of the current educational landscape reveals both striking progress and startling gaps. Some issues are specific to realizing the promise of the field, and some are more general. The analysis begins with issues that are internal to physics departments, and then turns to challenges that can be addressed only by collaboration among faculty across multiple departments. There are special concerns and opportunities in the integration of education and research, and about the trajectories of young scientists after earning their PhD.

> **Finding:** There has been considerable growth in the number of PhD students working in biological physics, so that the field now is comparable in size to well-established subfields of physics. This growth has occurred in less than a generation, and is continuing.

> **Finding:** Biological physics remains poorly represented in the core undergraduate physics curriculum, and few students have opportunities for specialized courses that convey the full breadth and depth of the field.

> **Conclusion:** The current physics curriculum misses opportunities to convey both the coherence of biological physics as a part of physics and its impact on biology.

The lack of coherence in the presentation of biological physics to students is impeding the progress of the field. It is possible for students to receive an undergraduate degree in physics and not even realize that there *is* a physics of living systems. To address these issues requires rethinking of the core physics curriculum.

General Recommendation: All universities and colleges should integrate biological physics into the mainstream physics curriculum, at all levels.

This integration necessarily will take different forms at different institutions, although there are guiding principles. Chapter 8 explores several places in the physics curriculum where the phenomena of life, and the progress of biological physics, can be used to convey core physics principles, not just in the introductory courses but continuing into more advanced undergraduate material on classical mechanics, electricity and magnetism, quantum mechanics, and statistical mechanics.

Conclusion: There is a need to develop, collect, and disseminate resources showing how examples from biological physics can be used to teach core physics principles.

Specific Recommendation: Physics courses and textbooks should illustrate major principles with examples from biological physics, in all courses from introductory to advanced levels.

Finding: Current undergraduate courses in statistical mechanics often do not reflect our modern understanding of the subject, or even its full historical role in the development of physics. Among other neglected topics, Brownian motion, Monte Carlo simulation, and the renormalization group all belong in the undergraduate curriculum.

Finding: Statistical mechanics courses typically come late in the undergraduate curriculum, limiting the window in which students can explore biological physics with an adequate foundation.

Specific Recommendation: Physics faculty should modernize the presentation of statistical physics to undergraduates, find ways of moving at least parts of the subject earlier in the curriculum, and highlight connections to biological physics.

Finding: Current treatment of optics in the undergraduate physics curriculum does not reflect modern developments, many of which have strong connections to biological physics. Among other neglected topics, optical traps and tweezers, laser scanning, nonlinear optical imaging modalities, and imaging beyond the diffraction limit all belong in the undergraduate curriculum.

Specific Recommendation: Physics faculty should modernize undergraduate laboratory courses to include modules on light microscopy that emphasize recent developments, and highlight connections to biological physics.

Physics departments offer advanced courses that introduce both undergraduate and graduate students to the distinct fields of physics, bridging some of the gap between the core curriculum and the frontiers of modern research. Biological physics courses now stand alongside more traditional courses on astrophysics and cosmology, condensed matter, elementary particles, and so forth.

> **Conclusion:** The great breadth of the field poses a challenge in teaching an introduction to biological physics for advanced undergraduates or beginning graduate students.

> **General Recommendation: Physics faculty should organize biological physics coursework around general principles, and ensure that students specializing in biological physics receive a broad and deep general physics education.**

Almost all fields of science are being revolutionized by the opportunity to gather "big data." While this often is presented as a recent development, experimental high energy physics and cosmological surveys entered the big data era before it had a name. Today, biological physics is following a similar path, with even modest experiments generating terabytes of data in an afternoon and many experiments reaching the petabyte scale.

> **Conclusion:** Biological physics, and physics more generally, faces a challenge in embracing the excitement that surrounds big data, while maintaining the unique physics culture of interaction between experiment and theory.

Parallel with the emergence of biological physics, biology itself has experienced dramatic changes, with new experimental methods making it possible to explore the living world on an unprecedented scale. Twenty years ago, the *BIO 2010* report brought attention to the educational challenges that follow from these developments, emphasizing that quantitative measurements and mathematical analyses would play a central role in the future of the biomedical sciences.[7] The intervening decades have seen even more rapid progress, in directions that have strong overlap with the interests of the biological physics community. These developments underscore the continued relevance of the message in *BIO 2010*. As detailed in Chapter 8, there is an opportunity for biology and physics faculty to work together, especially in the design of introductory courses.

> **Conclusion:** There still is room to improve the integration of quantitative methods and theoretical ideas into the core biology curriculum, continuing the spirit of *BIO 2010*. This remains crucial in preparing students for the bio-

---

[7] NRC, 2003, *BIO 2010: Transforming Undergraduate Education for Future Research Biologists,* The National Academies Press, Washington, DC.

medical sciences as they are practiced today, and as they are likely to evolve over the coming generation.

In reaching outside the physics department, it is crucial to present not just the application of physicist's tools to biological problems, but the physicist's approach to asking questions. While mathematical models and computational analyses have become more widespread in biology and in biology education, there remains a significant challenge in communicating the physics community's view of the interactions among quantitative experiment, data analysis, and theory. All of these initiatives will need support.

Conclusion: The biological physics community has a central role to play in initiatives for multidisciplinary education in quantitative biology, bioengineering, and related directions.

General Recommendation: University and college administrators should allocate resources to physics departments as part of their growing educational and research initiatives in quantitative biology and biological engineering, acknowledging the central role of biological physics in these fields.

One of the most important products of the research enterprise is educated people. Research and education are intertwined, and this connection has deep implications for our society.

Finding: Meaningful engagement with research plays a crucial role in awakening and maintaining undergraduate student interest in the sciences.

Conclusion: Biological physics presents unique opportunities for the involvement of undergraduates in research at the frontier of our understanding, offering more intimate communities through smaller research groups and providing opportunities for students to enter with varying levels of background knowledge and from a range of undergraduate majors.

Conclusion: Equality of opportunity for students to engage with physics, including biological physics, depends on high-quality introductory courses, emphasizing the interconnectedness of education and research.

Finding: Current models for support of undergraduate research perpetuate a sharp distinction between the core curriculum (education) and the development of the scientific workforce (research). This extends to the fact that science and education are overseen by different standing committees in Congress.

Conclusion: Support for the development of the scientific workforce will require direct federal investment in the core of undergraduate education, especially at an introductory level.

Specific Recommendation: Universities should provide and fund opportunities for undergraduate students to engage in biological physics research, as an integral part of their education, starting as soon as their first year.

Specific Recommendation: Funding agencies, such as the National Institutes of Health, the National Science Foundation, the Department of Energy, the Department of Defense, as well as private foundations, should develop and expand programs to support integrated efforts in education and research at all levels, from beginning undergraduates to more senior scientists migrating across disciplinary boundaries.

Supporting the Field

The health of a scientific field depends on financial support, which needs to match not just the scale of the opportunities but also their character. With the relatively recent emergence of biological physics as an identifiable field, it perhaps is not surprising that existing funding structures are not ideal. This report's survey of funding, collaboration, and coordination in Chapter 9 begins, however, with some of the many positive features of the current funding environment. In the first instance, this survey is organized agency by agency; note that larger facilities are the subject of a separate discussion:

Finding: The Physics of Living Systems program in the Physics Division of the National Science Foundation is the only federal program that aims to match the breadth of the field as a subfield of physics.

Finding: The United States has had a long-standing role as a leader in the area of biological physics at the molecular scale. Crucial support for this effort comes from Department of Energy investment in programs and user facilities.

Finding: The National Institutes of Health provide strong support for many individual investigators in biological physics, through multiple institutes and funding mechanisms.

Finding: Department of Defense agencies have highlighted multiple areas where the interests of the biological physics community intersect their missions.

Finding: Private foundations have supported programs that engage the biological physics community, often before such programs become mainstream in federal agencies, and have explored different funding models.

Finding: Biological physics has benefited from funding programs that are shared across divisions within individual federal funding agencies, between agencies, and between federal agencies and private foundations.

At the same time, there are features of the funding environment that work against realizing the full promise of the field:

Finding: National Science Foundation award sizes for individual investigators in biological physics have reached dangerously low levels, both in contrast to the National Institutes of Health and in absolute terms.

Finding: Support for the physics of living systems is scattered widely across the National Institutes of Health, making it difficult for investigators to find their way and obscuring the coherence of the field.

Finding: The Department of Energy (DOE) has become a major sponsor of research in biological physics, especially through facilities, without acknowledging the field's supporting contribution to the DOE mission.

As an alternative to looking at individual agencies, it also is useful to look at how funding is distributed across other dimensions of the scientific enterprise:

Conclusion: As in many areas of science, there is a challenge in maintaining a portfolio of mechanisms to fund the spontaneity of individual investigators, the supportive mentoring environments of research centers, and the ambitious projects requiring larger collaborations.

Finding: Physics programs do not have the stable, programmatic support for PhD students that is the norm in the biomedical sciences.

Conclusion: Accelerating young researchers to independence is critical to empowering the next generation of biological physicists. As in other fields of physics, independent, individual fellowships are an effective mechanism.

Finding: Physics has a unique view of the relationship between theory and experiment, and in many fields of physics this is supported by separate programs funding theorists and experimentalists. This structure does not exist in biological physics.

Finding: Large-scale physical tools, particularly those for imaging and advanced computing and data, are an important part of the infrastructure supporting thousands of researchers exploring the living world.

Conclusion: There is an opportunity for Department of Defense agencies to use the Multidisciplinary University Research Initiatives Program to support biological physics, and for the National Science Foundation and the National Institutes of Health to expand their support of these mid-sized collaborations.

Stepping back once more to survey the funding landscape as whole:

Finding: Total support for biological physics is barely consistent with the minimum needed to maintain a steady flow of young people into the field. This approximate balance of needs and support leaves significant gaps, and provides little room for new initiatives.

Conclusion: Biological physics is supported by multiple agencies and foundations, but this support is fragmented, obscuring the breadth and coherence of the field. It is dangerously close to the minimum needed for the health of the field.

These observations lead to the committee's recommendations about financial support for the field (Chapter 9), starting with an overarching response to the concerns outlined above:

General Recommendation: Funding agencies, including the National Institutes of Health, the National Science Foundation, the Department of Energy, and the Department of Defense, as well as private foundations, should develop and expand programs that match the breadth of biological physics as a coherent field.

This recommendation is embodied differently in relation to different agencies:

Specific Recommendation: The federal government should provide the National Science Foundation with substantially more resources to fulfill its mission, allowing a much needed increase in the size of individual grant awards without compromising the breadth of its activities.

Specific Recommendation: The National Institutes of Health should form study sections devoted to biological physics, in its full breadth.

Specific Recommendation: Congress should expand the Department of Energy mission to partner with the National Institutes of Health and the National Science Foundation to construct and manage user facilities and infrastructure in order to advance the field of biological physics more broadly.

Specific Recommendation: The Department of Defense should support research in biological physics that aims to discover broad principles that can be emulated in engineered systems of relevance to its mission.

Specific Recommendation: Industrial research laboratories should reinvest in biological physics, embracing their historic role in nurturing the field.

Supporting the full range of activities in biological physics also involves issues that potentially cut across the agencies:

Specific Recommendation: Federal funding agencies should establish grant program(s) for the direct, institutional support of graduate education in biological physics.

Specific Recommendation: Federal agencies and private foundations should establish programs for the support of international students in U.S. PhD programs, in biological physics and more generally.

Specific Recommendation: Federal agencies and private foundations should develop funding programs that recognize and support theory as an independent activity in biological physics, as in other fields of physics.

Finally, an essential part of the justification for federal support of science is that it generates useful products. The modern vision of the connections among science, technology, and society was articulated 75 years ago, in what can be seen as the founding document for our current system of federal science funding, *Science— The Endless Frontier*.[8] Today, a large fraction of the nation's economy is driven by and depends on technology, and with the benefit of hindsight each of these many technological advances can be traced back to foundational advances in the basic

---

[8] V. Bush, 1945, *Science—The Endless Frontier. A Report to the President on a Program for Postwar Scientific Research*, U.S. Government Printing Office, Washington, DC.

sciences. But it would have been difficult if not impossible to plan these trajectories from science to technology, to health care, and to economic growth.

General Recommendation: To maintain the flow of concepts and methods from biological physics into medicine and technology, the federal government should recommit to the vigorous support of basic science, including theory and the development of new technologies for experiments.

Human Dimensions of Science

Science is a human activity. Progress depends on recruiting, welcoming, and nurturing a continuous flux of new talent. At the same time, the scientific community has stewardship of precious resources—access to high-quality science education and the opportunity for individuals to pursue their intellectual passions as professional scientists. It is crucial both to maximize the progress of science and to exercise stewardship with justice. These goals are not in conflict. Chapter 10 explores the human dimensions of science, focusing on international engagement and equality of opportunity. Many of the issues are immediately relevant to biological physics, but also much more general, and need to be addressed across science as a whole.

Policies regarding the nation's engagement with the international scientific community should be grounded in historical facts:

Finding: Science in the United States has long benefited from the influx of talented students and scientists from elsewhere in the world.

Finding: International students have made substantial contributions to the economy of the United States.

Indeed, across the second half of the 20th century in particular, the United States held a privileged position on the world's scientific stage. This position is at risk:

Finding: Applications to U.S. physics graduate programs from international students have decreased since 2016.

Finding: Many international students find the United States unwelcoming and feel that they have better opportunities outside the United States.

These changes are not coincidental. They have occurred against a backdrop of dramatic changes in U.S. immigration policy, even more dramatic changes in rhetoric, and the prosecution of scientists under the Department of Justice "China Initiative." While there are specific incidents that need to be addressed, there is a danger that normal components of academic interaction and scientific collaboration are being criminalized.

Finding: Discussions of U.S. policy toward international students and scientists are being driven by concerns about national and economic security.

Conclusion: The open exchange of people and ideas is critical to the health of biological physics, physics, and the scientific enterprise generally. This exchange has enormous economic and security benefits.

General Recommendation: All branches of the U.S. government should support the open exchange of people and ideas. The scientific community should support this openness by maintaining the highest ethical standards.

Concrete steps to implement this recommendation are discussed in Chapter 10. The committee's suggestions echo and extend those articulated in the recent decadal survey of atomic, molecular, and optical physics.[9]

In principle, issues surrounding international engagement are quite general. In practice, as noted above, current attention is focused on relations with China. As this report is being written, the United States is experiencing a dramatic increase in anti-Asian violence on American streets, even in cities that are home to well-established Asian American communities. This suggests that movement toward more productive policies concerning academic exchange and international collaboration will require reckoning with larger issues about race in our society.

Discrimination based on race has a long history, and this history will not be overcome by actions of the scientific community alone. The challenge for our community is to do everything possible to welcome, support, and nurture talented young people from around the world and from U.S. citizens of all ethnic groups. The structure of physics education creates special circumstances:

Finding: Physics education is layered, with one layer building strongly on the one below. Inequality of access or resources is compounded.

The compounding effect of inequality creates burdens that fall with greater weight on those already subject to systemic discrimination.

---

[9] National Academies of Sciences, Engineering, and Medicine, 2020, *Manipulating Quantum Systems: An Assessment of Atomic, Molecular, and Optical Physics in the United States,* The National Academies Press, Washington, DC.

Finding: Recent data indicate that while the number of Black students earning physics bachelor's degrees is growing, the percentage has not increased.

Finding: Historically Black Colleges and Universities have played a crucial role in the scientific and professional education of Black Americans.

Finding: The total number of physics bachelor's degrees awarded by Historically Black Colleges and Universities has shrunk.

Conclusion: Inequalities of educational opportunity continue to limit the accessibility of physics education for Black students.

Although the experience of each group is unique, one can find related problems for all of the underrepresented groups in the biological physics community. Parallel to the role of Historically Black Colleges and Universities (HBCUs) for Black students are the broader collection of Minority Serving Institutions (MSIs) and Tribal Colleges and Universities (TCUs). There is also a strong connection between the committee's specific concerns about the education of underrepresented groups and its general concerns about the lack of proper support for core undergraduate education as part of scientific workforce development, as described above.

General Recommendation: Federal agencies should make new resources available to support core undergraduate physics education for underrepresented and historically excluded groups, and the integration of research into their education.

Specific Recommendation: Recognizing the historical impact of Historically Black Colleges and Universities, Minority Serving Institutions, and Tribal Colleges and Universities, faculty from these institutions should play a central role in shaping and implementing new federal programs aimed at recruiting and retaining students from underrepresented and historically excluded groups.

In addition to underrepresentation of ethnic minority groups, it is well known that women continue to be underrepresented in the sciences, and that this gap is particularly large in physics.

Finding: The fraction of women who take a high school physics course is almost equal to the fraction of men, but women comprise only ∼25 percent of students in the most advanced high school courses.

Finding: After steady growth for a generation, the fraction of bachelor's degrees in physics earned by women plateaued in 2007 at ~20 percent. The fraction of PhDs in physics earned by women has continued to grow, now matching the fraction of bachelor's degrees.

Specific Recommendation: In implementing this report's recommendations on introductory undergraduate education and its integration with research, special attention should be paid to the experience of women students.

Finally, the committee notes that these findings, conclusions, and recommendations regarding the human dimensions of science apply in large part to all areas of physics, and in many cases to the scientific community more generally. There is a sense, however, that biological physics has a special role to play in welcoming a broader community.

Conclusion: The biological physics community has a special opportunity to reach broader audiences, leveraging human fascination with the living world to create entrance points to physics for a more diverse population of students and for the general public.

PART I

EXPLORING BIG QUESTIONS

1

What Physics Problems Do Organisms Need to Solve?

In order to survive in the world, organisms have to accomplish various tasks. They have to move toward sources of food, sometimes over long distances, guided only by weak cues about the location of the source. They have to sense useful signals in the environment, and internal signals that guide the control of their own state. They often need to generate dynamics on time scales, which are not the natural scales given by the underlying mechanisms. All of these tasks consume energy, and hence require the organism to extract this energy from the environment. These various tasks of the organism often are described as "functions," and this notion of function is an essential part of what sets living matter apart from non-living matter. To a remarkable extent, carrying out these functions requires the organism to solve physics problems (see Box 1.1), although it is more precise to say that evolution has selected organisms that achieve effective solutions to these problems. One of the central problems in biological physics is to turn qualitative notions of function into precise physical concepts. Along the way, we will see that these physical concepts often give us absolute notions of performance, such as the efficiency of energy conversion or the precision of chemical sensing in relation to the limits set by random arrival of molecules at their targets. It is a remarkable fact about living systems that evolution can in some cases select for mechanisms that approach the bounds of what is allowed by the laws of physics. In many cases what is understood are just the first steps in how these functions are achieved, and new and open physics problems emerge as one pushes beyond these. The list of physics problems that organisms must solve is far from exhausted by the examples in this chapter, and this emphasis on function will carry through all of the subsequent discussion. A sampling of the issues that we encounter is provided in Table 1.1.

BOX 1.1
Physical Principles and Biological Functions

Confronted with the beautiful phenomena of life, it is natural to ask how they work. But living systems are not random collections of mechanisms. They have been selected by evolution to function in their natural environment. As with engineered systems, understanding how they work is easier if one understands this functional context. One should ask not just how they work, but what they do.

Horace Barlow, known for many contributions to our understanding of vision and the brain, gave these ideas about the relation of function, mechanism, and physics an eloquent formulation:

A wing would be a most mystifying structure if one did not know that birds flew…. Without understanding something of the principles of flight, a more detailed examination of the wing itself would probably be unrewarding.[a]

_____

[a] H.B. Barlow, 1961, "Possible Principles Underlying the Transformation of Sensory Messages," p. 217 in _Sensory Communication_ (W. Rosenblith, ed.), MIT Press, Cambridge, MA.

TABLE 1.1 Physics Problems in the Life of the Organism

Physics Problems	Page Number	Broad Description of Problems	Frontier of New Physics in the "Physics of Life"	Potential Application Areas
Energy conversion	49	Living systems harness classical and quantum dynamics to achieve efficient energy conversion; nanoscale linear and rotary motors.	New methods for single molecule measurement and manipulation; new theoretical ideas on thermodynamics and information on small scales and away from equilibrium.	Photosynthesis, batteries, solar panels.
Mechanics, movement, and the physics of behavior	61	Finding simplicity in movements through complex, natural environments; mechanics and hydrodynamics in novel regimes.	Long time scales and hidden symmetries; optimal flow networks.	High throughput analysis of behavior; robotics.
Sensing the environment	70	Counting photons and molecules; sensing small forces and displacements; signal processing.	Physical limits to sensing and signaling; molecular mechanisms of amplification; separating signal from noise.	Bio-inspired sensors and signal processing strategies.
Structures in space and time	80	Emergence and control of structure on scales far removed from microscopic mechanisms.	Self-assembly; scaling; molecular clocks.	New materials and synthesis methods.

ENERGY CONVERSION

In the physicists' view of life, equilibrium is death. To maintain life, organisms capture energy from the environment and use this energy to keep themselves away from equilibrium, creating locally ordered states of matter and carrying out all the other functions crucial for survival. Starting with 19th-century concerns about the laws of thermodynamics, energy conversion in biological systems has been a continuous source of fascination for physicists. These problems range from quantum dynamics in the first steps of photosynthesis to the classical mechanics of swimming and flying.

Photosynthesis

Much of the energy that supports life on Earth comes from the sun, and photosynthetic organisms capture the energy of sunlight directly. This initial energy capture ultimately drives a chain of chemical reactions that convert carbon dioxide into the stuff of life, with spectacular efficiency. Along the way, many photosynthetic organisms emit oxygen as a waste product, and this is the source of almost all the oxygen in our atmosphere, making it possible for us to breathe. It is difficult to overstate the importance of photosynthesis to our lives, and to the health of the planet as a whole. In addition, photosynthesis provides inspiration for the design of artificial systems that capture solar energy.

Physicists have been fascinated by photosynthesis for nearly a century. These explorations have generated the remarkable result that photosynthetic organisms harness a subtle interplay of classical and quantum physics to achieve extraordinary efficiency. Parts of this understanding now are well established, providing a solid foundation for exploration of quantum effects in other biological processes. From a historical perspective, the emergence of this understanding straddles the emergence of biological physics as a part of physics, and thus some of the crucial insights are seen now as part of mainstream biology, or perhaps part of biophysics as a biological science. Many of the conceptual problems arise also in the behavior of large molecules more generally, and thus have strong connections to chemistry. The problem of categorizing these developments is highlighted by the fact that there are subjects (with specialized journals) called physical chemistry and chemical physics. In surveying the physics of living systems, what seems important here is that many crucial questions about photosynthesis came out of the physics community, along with methods—both theoretical and experimental—to address these questions. In this section, as in the rest of the report, the committee takes this broad view of biological physics as the engagement of physicists with the phenomena of life, even at moments when the field did not have a name.

The basic processes of photosynthesis are associated with chlorophyll molecules, but photons can be absorbed by other pigment molecules and drive these processes with nearly equal efficiency. Hints of this possibility date back into the 1800s, but conclusive evidence came only in the 1930s, and this triggered, in the theoretical physics community, the first discussions of what now is called fluorescence energy transfer. More generally, photosynthetic organisms contain many more chlorophyll molecules than those involved directly in the chemical reactions driven by light, leading to the picture of a large "antenna" composed of many chlorophylls, absorbing light and funneling energy to a "reaction center" that contains only a handful of these molecules.

The problem of energy transfer in the photosynthetic antenna would recapture the attention of the physics community in the 21st century, with the first direct evidence that the process involves quantum mechanical coherence. But effort first would be focused on the isolation of the reaction center. A crucial observation was that the initial events following photon absorption involved the transfer of an electron, and that this could happen even at very low temperatures. The fact that electron transfer continues at low temperature implied that it was happening not between two separate molecules that had to find one another in solution, but inside a single large molecule or molecular complex, the reaction center (see Figure 1.1).

Electron transfer was detected first by the presence of an electron paramagnetic resonance (EPR) signal from the resulting unpaired electron(s). This pointed the way to the use of sophisticated spectroscopic methods, such as electron-nuclear double resonance, to characterize the dynamics of the reaction center, and this was parallel to the development of these methods to characterize impurities in semiconductors. Furthermore, movement of electrons in large organic molecules is associated with changes in their optical properties. With the continuing development of faster pulsed lasers, it became possible to resolve photon-driven electron transfer in the reaction center on the microsecond time scale, then nanoseconds, then picoseconds. All of these spectroscopic experiments resulted in a picture of the reaction center as a large protein complex that holds several organic molecules, including chlorophylls, which act as electron donors and acceptors, summarized in Figure 1.1A and 1.1B. The reaction center was known to be embedded in a membrane, so that the net result of photon absorption is to separate charge across the membrane, and this provides the "battery" that drives all subsequent chemical reactions. This picture was confirmed, beautifully, when it became possible to crystallize the reactions centers and solve their structures by X-ray diffraction, shown in Figure 1.1B.

Careful investigation of the photon-driven electron transfer events in the photosynthetic reaction center revealed that these reactions not only happen at low temperatures, but the rates of these reactions are in many cases nearly independent of temperature. In the first example, a microsecond reaction time slowed to mil-

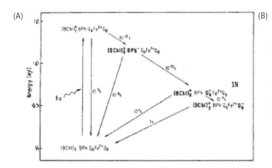

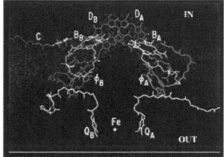

FIGURE 1.1 Photon-driven electron transfer in the photosynthetic reaction center. (A) Schematic of photon absorption by a dimer of bacteriochlorophyll [$(BChl)_2$] followed by electron transfer to a bacteriopheophytin (BPh) and then quinones ($Q^{A,B}$), with reaction time scales indicated; when the reaction center is isolated the electron eventually returns to $(BChl)_2$, but slowly. (B) The electron donor and acceptors in the three-dimensional structure of the reaction center. D^A and D^B form $(BChl)_2$; B is bacteriochlorophyll that acts as a virtual intermediate in the transfer to BPh (φ). The symmetry in arrangement of these components is broken by the dielectric properties of the protein, channeling electron transfer to one side. SOURCE: Reprinted by permission from Springer: G. Feher, J.P. Allen, M.Y. Okamura, and D.C. Rees, 1989, Structure and function of bacterial photosynthetic reaction centres, *Nature* 339:111, copyright 1989.

liseconds as the system was cooled, but then the time became temperature independent below an absolute temperature of around 100 K. To begin, it is astonishing that the mechanisms of life "work" at these low temperatures. As the faster reactions were resolved, it was found that these are nearly temperature independent at room temperature, in some cases even becoming slightly faster as the system is cooled. These results are in marked contrast to typical chemical reactions, where rates are exponentially sensitive to temperature changes, following the Arrhenius law. This is a sign that quantum mechanical effects are important, and through the 1970s and 1980s, the biological physics community reached a relatively complete understanding of this.

In the classical picture of chemical reactions, molecules have two possible structures, each of which is locally stable, being at a minimum of the energy, and there is an energy barrier between them. Thermal fluctuations cause random motions around these minima, and with some small probability these fluctuations have a large enough amplitude to allow escape over the barrier. The typical energy of the thermal fluctuations is $k_B T$, and the height of the barrier is called the activation energy E_{act}; the rate of the chemical reaction is $k \sim A e^{-E_{act}/k_B T}$, where the factor A is related to the time scales of random vibration around the local minima. But quantum mechanics has the possibility of systems visiting states which would be forbidden by classical physics, such as positions under the barrier when the molecule does not have enough energy to go over the barrier. This is called tunneling through the barrier, and will be the dominant reaction mechanism at sufficiently low temperatures. In another view, as the temperature is lowered the random thermal motion is frozen out, and all that remains is the quantum zero-point motion that is required by the uncertainty principle.

Electron transfer reactions, as in photosynthesis, have an additional feature, because the two possible structures of the relevant molecules are associated with distinct states of the electrons. Because of the large distances over which electrons are transferred, these states are very different and the "mixing" of the states is weak. The transferred electron itself always passes through a region of the molecule that would be forbidden by classical physics, and across the relevant range of temperatures there is never enough energy for the electron to go over these barriers. Thus, electron transfer always proceeds by electron tunneling. At high temperatures the prediction is that that the changes in molecular structure occur by thermal activation, but at lower temperatures there will be tunneling from one structure to the other. The crossover temperature is such that the thermal energy is comparable to the energy for one quantum of vibrational motion, and this is consistent with what is seen in the photosynthetic examples. The chemical reaction can be seen as converting the energy released by electron transfer into multiple quanta of vibrational energy, or phonons, which then relax into the surrounding medium. In large molecules such as the photosynthetic reaction center, there is a mix of high frequency and low frequency vibrations, and this can lead to the anomalous patterns of temperature dependence seen in this system. Thus, electron transfer in photosynthesis depends on an interplay of classical and quantum dynamics, at biologically relevant temperatures. This understanding provides a foundation for thinking about quantum effects in biological molecules more generally (see Box 1.2).

Reactions can release energy, and it takes time for the molecule to dissipate this energy, coming to internal equilibrium and to equilibrium with its surroundings. Our usual ideas about chemical reactions are based on a separation of time scales, in which this equilibration is understood to happen fast, much faster than the rate of the reaction, so that it makes sense, for example, to say that the molecule is at the temperature of its environment. However, the very first photon-driven electron transfer in the reaction center happens so quickly that one might worry whether this approximation is valid. More subtly, in a quantum mechanical description, the states where the electron is localized on the donor or acceptor mix coherently, and it takes time for this coherence to be destroyed; again, conventional ideas about chemical reactions assume that the time for loss of quantum coherence is much shorter than the reaction time.

Very careful fast pulse laser spectroscopy on the reaction center reveals oscillations reflecting the persistence of coherence in molecular vibrations on time

---

BOX 1.2
Proton Tunneling in Enzymes

Understanding the role of tunneling in the photosynthetic reaction center raises the question of whether such quantum effects could be important in other biochemical events. The probability of tunneling declines exponentially with the square-root of the mass of the tunneling particle. If electrons always tunnel, perhaps the next lightest of the relevant particles—protons, or hydrogen atoms—could also tunnel?

Hydrogen transfer reactions are central to many processes in living cells. There is a long history of probing these reactions by substituting deuterium or tritium for hydrogen, and measuring the change in reaction rate. These "kinetic isotope effects" are signatures of the underlying dynamics. In a reaction that proceeds by classical activation over a barrier, kinetic isotope effects typically are small. For tunneling, kinetic isotope effects are expected to be large, but independent of temperature.

What happens in the biological context, where hydrogen transfer is catalyzed by enzymes, or specialized proteins? It came as a surprise in the late 1980s and early 1990s when it was discovered that some enzyme-catalyzed hydrogen transfer reactions have large but temperature dependent isotope effects. After much back and forth between theory and experiment, this now is understood: Protons tunnel, but the fluctuations of the protein into an optimal configuration for tunneling are thermally activated. Efficient hydrogen transfer in biological molecules thus depends on an interplay between classical and quantum dynamics, in many ways parallel to the case of photosynthetic electron transfer.

---

scales comparable to the initial electron transfer rate. It is important that truly irreversible reactions cannot happen faster than the loss of coherence. The reaction center seems to be in a regime such that the mixing of electronic states, the loss of coherence, and the reaction rate all are on similar time scales. This generates the fastest possible rate given the structure of the relevant electronic states, and pushes us to think about the physics of quantum transitions in a new regime.

The question of quantum coherence was revitalized by observations on components of the photosynthetic antenna. As with chemical reactions, the usual regime for energy transfer between molecules is one in which the transfer rate is much slower than internal rates for the dissipation of energy and the destruction of coherence. Sophisticated spectroscopic experiments in the late 2000s showed that coherence in energy transfer through photosynthetic antenna complexes persists for surprisingly long time scales. The result is that transfer rates are comparable to the rate at which coherence is lost. Exploring these dynamics, and understanding their implications for the efficiency of energy harvesting, are topics of current research.

In the physics laboratory, the search for direct manifestations of quantum coherence often drives us to very low temperatures, and to settings in which the system of interest can be isolated from its surroundings. But life operates (primarily) in a narrow range of temperatures near room temperature, and biological molecules interact strongly with a surrounding bath of water molecules. These warm and wet conditions are not those under which quantum coherence is expected to survive, so the results of these experiments and the subsequent theoretical analysis were unexpected. It was discovered that spectrally tuned "noise" from the real world could in fact enhance both quantum coherence and energy transfer; they can be turned to advantage but are required together, and it seems that living systems have discovered this path and evolved molecular structures that execute this effectively. These analyses connect to ideas about coherence and dissipation in quantum measurement and quantum computing, part of a broader rethinking of how quantum systems couple to the macroscopic world.

The photosynthetic reaction center traps the energy of light by separating electronic charge across a membrane. In the intact system, this electronic charge is compensated by the movement of protons, so that energy is stored in a concentration difference, or chemical potential difference for protons. This chemical potential difference turns out to be a universal intermediary in how cells handle their energy supply. There are even bacteria that have direct light-driven proton pumps as an alternative to the more complex photosynthetic reaction center. In our cells, as in those of other eukaryotes, specialized structures called mitochondria extract energy from many different chemical sources, and then use this energy to pump protons across their membranes. A different protein in the membrane provides a channel for these protons to flow back along the gradient in their chemical potential and harnesses the energy that is released to synthesize the adenosine triphosphate (ATP) molecule; see Figure 1.5. ATP is the classical "energy currency" of biochemistry, and provides the direct fuel for processes ranging from the contraction of our muscles to the correction of errors in reading and copying genetically encoded information. In contrast, bacteria swim by rotating their flagella, and this rotation is powered directly by protons rather than ATP (see Figure 1.4). Even beyond its intrinsic importance, the physics of how the photosynthetic reaction center captures the energy of sunlight thus provides an entrance point for studying mechanisms of energy conversion that are shared across all forms of life on Earth.

Motors

Photosynthesis converts the energy of sunlight into chemical form. When organisms move, they (we!) convert energy from chemical form into mechanical form—forces and displacements. Careful measurements on mechanics and energy dissipation in muscle have their roots in 19th-century experiments that were instrumental in establishing the laws of thermodynamics and banishing ideas of vitalism. In the mid-20th century, X-ray diffraction and optical microscopy methods were developed to visualize the relative sliding of actin and myosin protein filaments, which provides the microscopic basis for muscle contraction. Today, these X-ray measurements can even be done in an insect flight muscle while the insect is flying, directly connecting nanometer scale filament movements to the macroscopic dynamics of force production and movement (see Figure 1.2).

Biochemists and cell biologists discovered that the same proteins, actin and myosin, are present not just in muscle but in all eukaryotic cells. Actin is a key component of the cellular cytoskeleton, which provides a cell with structural support, and myosin is a "motor protein," or a mechanoenzyme, which converts chemical energy into mechanical work. In muscle fibers, myosin motor proteins are assembled into filaments, responsible for force generation and sliding along

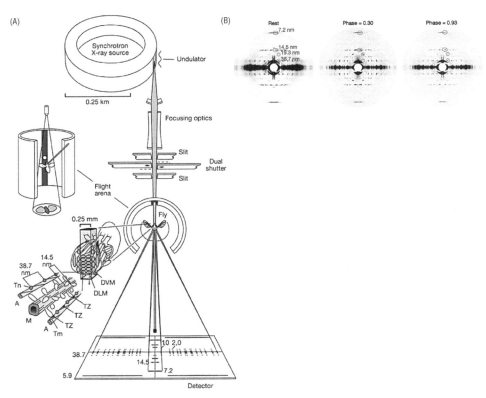

FIGURE 1.2 X-ray diffraction can be used to visualize microscopic-level muscle contraction, or the relative sliding of actin and myosin protein filaments. Here, synchrotron X-ray diffraction is used to image molecular-scale changes in muscle during flapping flight in a fly. (A) Experimental apparatus showing flight musculature in the tethered fly *Drosophila* scattering-ray radiation from the small angle instrument on the BioCAT undulator-based beamline 18-ID at the Advanced Photon Source, Argonne National Laboratory. (B) Sample diffraction patterns from live flies at rest (*left*) and at two phases in a wingbeat cycle. SOURCE: Reprinted by permission from Springer: M. Dickinson, G. Farman, M. Frye, T. Bekyarova, D. Gore, D. Maughn, and T. Irving, 2005, Molecular dynamics of cyclically contracting insect flight muscle in vivo, *Nature* 433:330, copyright 2005.

actin filaments. In addition, cells have other motor proteins, such as the kinesin molecules that transport intracellular cargo along microtubules, another structural element of cells. For example, the axons of nerve cells can reach one meter in length, from our spinal cord to our toes, and kinesin motor proteins haul cargo vesicles from the neuronal cell bodies in the spinal column to the tip of the big toe. As described in Chapter 3, the collective behavior of these motor molecules and filaments in living cells provides a prototypical example of active matter.

Analysis of macroscopic measurements on forces and displacements of single muscle fibers suggested that individual protein molecules produce forces on the scale of picoNewtons, and that the elementary molecular events involve displacements on the scale of nanometers. This means that the energy used in the elementary steps of movement is on the order of the thermal kinetic energy of single molecules, far from our usual intuition about macroscopic motors and engines. These basic facts about biological motors provided crucial motivation for new theoretical ideas about non-equilibrium statistical physics and thermodynamics in the stochastic regime. From the experimental side, the exploration of movements in cells and organisms was revolutionized by the realization that controlled forces on this scale could be applied by the interaction of light with matter, in optical traps or "tweezers" (see Box 1.3). The biological physics community made a major effort to develop these single molecule manipulation experiments, which have now been exported to the broader community of biologists. Figure 1.3 shows early measurements on single molecules of the proteins actin and myosin, which generate the force in our muscles. By holding a single actin filament between a pair of optical traps and bringing it into contact with myosin molecules attached to a plastic bead (Figure 1.3A), it was possible to see hints of stepwise motion as the myosin molecule "walks" from one actin monomer to the next along the filament. At low concentration of the fuel for these reactions (ATP), force generating

---

BOX 1.3
Optical Tweezers

It was understood in the 19th century that when light reflects from a mirror, it applies a mechanical force. But other mechanical aspects of light remained confusing. It was not until late in the 20th century that it was realized that a focused spot of light could *attract* a small object, such as a plastic bead. By trapping a particle with a focused laser beam and then moving the focal spot, one can manipulate micron-sized objects as if picking them up with "optical tweezers." This realizes, on the micro scale, the tractor beams of science fiction. Importantly, the scale of forces is comparable to the forces generated by individual motor proteins in cells, and the idea that optical traps would be a powerful tool for the exploration of life and force generation in cells was clear from their inception. In 2018, Arthur Ashkin shared the Nobel Prize in Physics "for his invention of the optical tweezers and their application to biological systems."

interactions are rare and there are signs of quantization (Figure 1.3B), as expected if these result from discrete molecular interactions. In subsequent data, all of these results have become sharper, and similar results have been found for other motors such as kinesin and dynein.

Myosin, kinesin, and dynein are linear motors. These linear motors power muscles, cell movements, and intracellular transport. They also power the whip-like motions of the cilia that, for example, move fluids and debris along the airways leading to our lungs, and the flagella that allow many single celled organisms to swim. It thus came as a huge surprise that the flagella which power the motion of bacteria are not waving, as they seem to be, but rather rotating. They are driven by the world's smallest rotary engine, as schematized in Figure 1.4. As noted above, this motor is powered directly by the difference in chemical potential for protons

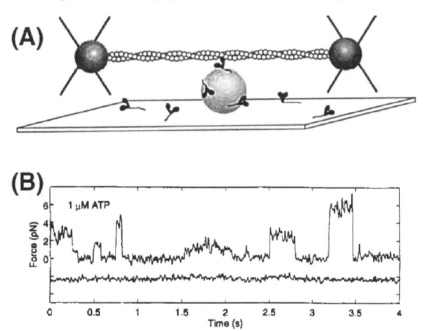

FIGURE 1.3 Single molecule experiments help us to understand molecular motors, including force generation in muscle. Optical trapping reveals molecular-scale forces and displacements generated by muscle proteins. (A) A single actin filament is stretched between two beads, each of which is held in an optical trap. The filament is brought into contact with a third bead, which is coated with myosin molecular motors. (B) Force exerted by myosin motors over time under (isometric) conditions of zero displacement; upper trace is force along the filament, lower trace perpendicular to the filament. At these low concentrations of ATP, force generating interactions between myosin and actin are rare, and discrete, suggestive of individual molecular events. SOURCE: Reprinted by permission from Springer: J.T. Finer, R.M. Simmons, and J.A. Spudich, 1994, Single myosin molecule mechanics: Piconewton forces and nanometre steps, *Nature* 368:113, copyright 1994.

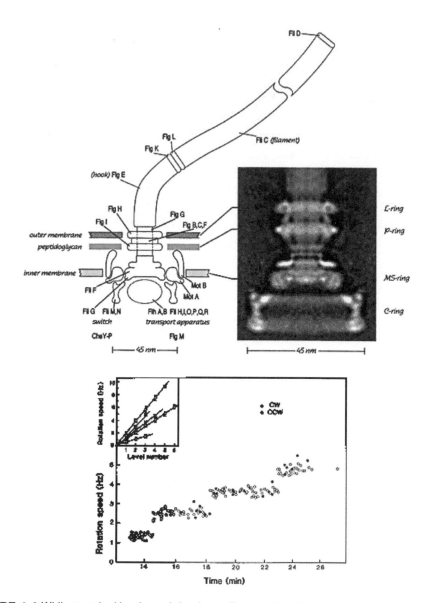

FIGURE 1.4 While myosin, kinesin, and dynein are linear motors that power muscles, cell movement, and intracellular transport, the flagella, which power the motion of bacteria, are powered by rotary motors. Schematic (*top left*) showing the protein components of the motor and its anchoring in the cell membranes. Structure of core components (*top right*) reconstructed from electron microscopes images. Cells that do not make the MotA protein cannot rotate their flagella. As they make more of the protein, the rotation rate (*bottom*, for a cell tethered to a glass slide) increases in steps as individual proteins are inserted into the structure, each contributing a discrete unit of torque. SOURCE: H.C. Berg, 2003, The rotary motor of bacterial flagella, *Annual Review of Biochemistry* 72:19.

between the inside and outside of the cell. A steady stream of mechanical measurements on single motors has led to very well developed theoretical ideas, which have been invigorated in the last 2 years by two major developments. First, there is a previously undetected feedback from the mechanical load on these motors to their internal dynamics, which has implications for how bacteria control their movements. Second, developments in cryogenic electron microscopy have delivered a structure of almost the entire motor in full atomic detail (see Figure 6.2), which will provide a literal scaffolding for understanding how the flow of protons is coupled to molecular rotation.

An important result of measurements on single motor molecules is the clear demonstration that these systems are engines in which a single working cycle delivers an energy that is larger than the thermal energy, but not by a very large factor. They thus operate in a regime where randomness is not negligible, and consequently the engine cycle has a stochastic duration. Such engines are bound by the same laws of thermodynamics as are the more familiar engines in our engineered, industrial world. But this regime of "Brownian motors" is very different, and this raises new and fundamental questions in statistical physics. On this small scale, for example, is the second law of thermodynamics only true on average? These questions have given rise to a new field of stochastic thermodynamics, and connects to a renaissance in non-equilibrium statistical mechanics. In return, optical trapping and the manipulation of single molecules have provided some of the most important experimental tests of emerging theoretical ideas. These connections are described in more detail in Chapter 5.

Movement Beyond Motors

Many molecules that are not functioning primarily as motors in fact generate forces and movements as an essential part of their function, and can thus be studied using the same methods developed for studying single motor molecules. Indeed, this is true for some of the most crucial molecules of life. As noted above, much of the ATP in all eukaryotic cells is synthesized by a membrane protein, called the F_0F_1-ATP synthase, that uses the chemical potential difference of protons across the membrane as an energy source. As with the proton-driven motor of bacterial flagella, this molecule rotates as it carries out its chemical function. This rotation is visible in single molecules that are fixed to a glass slide and running "backward," degrading ATP molecules to pump protons, as in Figure 1.5. Strikingly, unlike the linear motors myosin and kinesin, which have irreversible cycles and convert only a fraction of the energy from ATP into mechanical work, the ATP synthases are nearly 100 percent efficient; they are thus reversible and can be run in either direction (i.e., to consume or to produce ATP). In eukaryotic cells there are additional V-ATPases, rotary motor enzymes that create proton gradients using the energy stored in ATP. These molecules provide important motivating examples for stochastic thermodynamics.

Another crucial class of molecules that move as they carry out their function are those involved in copying, reading, and translating the information encoded in DNA and RNA. These include polymerases, ribosomes, helicases, gyrases, and topoisomerases. Single molecule measurements have provided an extraordinarily direct and precise view of this information processing, as discussed in Chapter 2. As explained there, an important function for many of these molecules is "proofreading," whereby the cell expends energy to achieve fidelity of information transmission beyond what would be possible from the equilibrium thermodynamic specificity of molecular interactions alone. In the same way the operation of biological motors with cycles that generate near-thermal energies provides motivation for the more general problems of stochastic thermodynamics, the interplay of energy

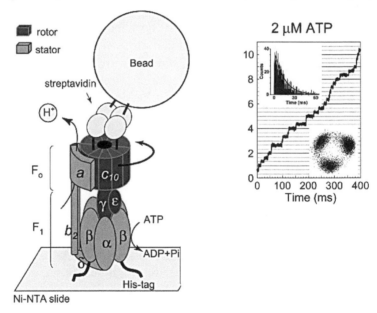

FIGURE 1.5 Many molecules that are not functioning primarily as motors generate forces and movements that allow them to be studied using the same methods developed for studying single motor molecules. The F0F1-ATP synthase, a membrane protein that synthesizes much of the ATP in all eukaryotic cells, is an example of a non-motor molecule that rotates to carry out its chemical function. At left, a schematic of the experiment shows the many protein subunits of the molecule assembled and bound to a glass slide at one side and a plastic bead (not to scale) at the other. At right, the angular position of the bead (measured by the number of rotations) versus time is shown. It is clear that movement pauses three times per full rotation. Insets show the distribution of pause lengths and the distribution of bead positions. SOURCE: H. Ueno, T. Suzuki, K. Kinosita, Jr., and M. Yoshida, 2005, ATP-driven stepwise rotation of F0F1-ATP synthase, *Proceedings of the National Academy of Sciences U.S.A.* 102:1333, Creative Commons License CC BY-NC-ND 4.0.

dissipation and fidelity in proofreading provided important motivation for deeper understanding of the connections between thermodynamics and information. These ideas continue to develop, and it is reasonable to expect that we will see the emergence of some deeper principles about how life harnesses non-equilibrium statistical physics to generate precise functions at minimal energy cost.

Perspective

The phenomena of energy conversion in living systems have inspired the development of new physics for more than a century, dating back to the origins of thermodynamics, and this continues to the present day. At one extreme, living systems have harnessed a subtle combination of classical and quantum dynamics to achieve efficient energy conversion in photosynthesis. This runs counter to the common intuition that quantum mechanics connects to the phenomena of life only by determining the rules of chemical bonding, highlights new physics in regimes not commonly encountered in the inanimate world, and points to new opportunities for engineered devices and molecular design (Chapter 7). In a different regime, linear and rotary motors provide examples of efficient energy conversion in non-equilibrium systems at fixed temperature. The effort to explore these nanomachines has led to the development of new methods for single molecule measurement and manipulation, and to sharp new theoretical ideas about the relations between thermodynamics and information on small scales and away from equilibrium. These developments in biological physics are continuous with a broader renaissance in non-equilibrium statistical mechanics (Chapter 5). The scarcity of resources places enormous pressure on living systems to be energy efficient, and clear articulation of the physical principles that underlie this efficiency will provide paths for engineering and synthesis (Chapter 7) that will surely help us rise to our global challenges in sustainability and climate change. While we understand much about individual processes, how these processes fit together in whole organisms, and in communities of organisms, is as yet only faintly sketched.

MECHANICS, MOVEMENT, AND THE PHYSICS OF BEHAVIOR

Humans have long been inspired by the soaring flight of birds and the elegant swimming of fish. Children are charmed by the dispersing seeds of a dandelion, and fascinated by columns of ants. The movements of fluid through the stems and leaves of plants are hidden from us, but no less vital. To be alive is to be in motion.

Organisms do not move in isolation. Swimming and flying depend in an essential way on interactions with the surrounding water and air, and have been the source of important problems in the development of fluid mechanics. Many organisms propel themselves through sand or soil, and our growing understanding of these movements is linked to current physics problems in the description of granular materials (Chapter 5); there is growing appreciation that similar problems arise for cells moving through tissues, including tumors. Our own interactions with the hard ground while walking seem simpler, but the persistent challenge of building robots that can walk on rough terrain suggests that the underlying dynamics are subtler than they first appear (Chapter 7).

Life at Low Reynolds Number

Most of the organisms on Earth experience moving through the world in a regime very different from what humans experience moving through air or water. A human swimmer, for example, can push off the wall of a swimming pool and move forward for a significant distance with no additional effort. Similarly, if swimmers stop moving their arms and legs, their whole bodies continue to move forward. These are manifestations of inertia. Eventually the drag or viscosity of the surrounding fluid wins out, but not before the swimmer has "coasted" a distance comparable to their own body length, or more. Life is very different for bacteria and other microorganisms.

As always in physics, when writing the equations of fluid flow—in air or water—there is freedom to choose the units of measurements. There are "natural" choices, for example using the typical swimming speed as a unit of velocity, and the size of an organism as the unit of length. In these natural units, the equations describing different organisms in different environments depend only on some unit-less combinations, or ratios. Perhaps the most important of these is the Reynolds number, which is the ratio of inertial to viscous forces (see Figure 1.6). At very large Reynolds number fluid flows become turbulent; at very small Reynolds number inertia is negligible and organisms need to work constantly against the viscosity of the surrounding fluid in order to keep moving.

For a human swimming in water, a typical Reynolds number is 10,000. For a bacterium, a typical Reynolds number is 100 million times smaller, roughly 0.0001. When humans stop swimming they can coast for several feet. If a bacterium stops rotating its flagella, it will coast only for a distance comparable to the diameter of an atom.

Life at low Reynolds number has many consequences. As an example, nutrient molecules arrive at the surface of a bacterium as a result of their random motion, or diffusion, and no reasonable expenditure of energy by the cell could stir the fluid enough to increase this flux. More profoundly, at very low Reynolds number, the viscous forces from the surrounding fluid balance the active forces that an organism generates in order to move, and this balance is enforced moment by moment. As a consequence, as the organism goes through one cycle of movement—one rotation of a bacterial flagellum, one beat of a eukaryotic cilium, one full squirm or writhe

FIGURE 1.6 One can make a rough estimate of the inertial and viscous forces involved when an object of size α moves at speed ν through a fluid with density ρ and viscosity η. The ratio of inertial to viscous forces is the Reynolds number $R = \alpha \nu \rho / \eta$; it is convenient to define the kinematic viscosity $\nu = \eta / \rho$. The sketches at the right indicate typical Reynolds numbers for swimming humans, fish, and bacteria. SOURCE: Reproduced from E.M. Purcell, 1977, Life at low Reynolds number, *American Journal of Physics* 45:3, https://doi.org/10.1119/1.10903, with the permission of the American Association of Physics Teachers.

of a more complex but still microscopic creature—the amount by which it moves forward depends on the sequence of movements and not on the speed with which these movements are executed. In particular, if the sequence looks the same when played forward and backward in time, the net displacement will be zero.

A fluid without viscosity obeys time-reversal invariance, so that a movie of the flow can be run in reverse and still be a solution of the underlying equations. With a little bit of viscosity, this time-reversal invariance is broken, and movies running backward are easily recognized as impossible. But in the limit that viscosity is large—the limit of low Reynolds number—time-reversal invariance is restored. To move through the world, microorganisms must wriggle and writhe in ways that actively break this symmetry on the macroscopic scale.

Bacteria swim by rotating their flagella (see Figure 1.4). The flagella themselves are helical, and thus have a handedness, twisting clockwise or counterclockwise when viewed from the cell body. The combination of rotational direction and helical handedness serves to break time-reversal invariance, and this is what allows the bacterium to propel itself. Similar symmetry breaking can be found in cilia (eukaryotic flagella). Thus, it is not enough to say that these organisms swim because they have motor proteins that generate forces. These forces need to be organized in ways to solve the underlying physics problem. It remains a challenging problem to link what is being learned about molecular motors (e.g., through single molecule experiments as in Figure 1.3) to macroscopic, functional movements.

Abstracting away from the details of particular movements is useful, and generates surprises. Self-propulsion is a process in which organisms change their shape periodically, as with the rhythmic strokes of a swimmer or the rotation of the bacterial flagellum. At low Reynolds number all that matters is the trajectory through this "space of shapes." But in describing the space of shapes, there is an arbitrariness in the choice of coordinates; any physical quantity, such as the actual movement of the organism as it swims, must be invariant to these choices. This kind of "gauge invariance" has a long history in physics, starting with the theory of the electromagnetic field, and continuing into the theories of the strong and weak forces among elementary particles. In the 1980s, it was realized that much of the beautiful mathematical structure of gauge theories can be translated into this new context, and turned into practical tools for calculation.

The first applications of the gauge theory approach focused on swimming by small deformations of the organism's surface. This is relevant for single celled organisms such as paramecia that are covered by a dense array of cilia that beat in co-ordinated patterns, approximating an undulating surface. In this regime it becomes possible to calculate the maximally efficient deformations, and the important lesson is that even the optimal energetic efficiency is very low—a physical rather than biological limitation. More recently, largely through the efforts of control theorists, the gauge theory formulation has been applied to larger amplitude motions, relevant for a much wider range of organisms. In parallel, it has come to be appreciated that organisms that slither and crawl over and through granular materials, such as snakes on the sand, also live at effectively low Reynolds number. Analyses of the movements of these organisms, coupled with gauge theory ideas, have been central to the emergence of new kinds of robots, as described in Chapter 7.

Flow Networks

Fluids flow not only outside the organism, but also inside. In humans and many other animals, life depends on the circulation of blood. This is exquisitely well controlled, especially in the brain. Magnetic resonance imaging allows us to visualize these flows directly and see how they are modulated by demands on brain activity as people interpret what they see, plan movements, recall memories, and so forth (Chapter 6). In plant leaves, a similarly complex network of veins effectively connects every cell to the source of water in the stem (see Figure 1.7, left). Both brains and leaves devote considerable resources to their vasculature, and the biological physics community has explored whether there might be general physical principles governing the distribution of these resources. For example, are there networks that minimize (in the leaf) the pressure drop from the stem to the tips, or equivalently the power dissipation in the flow, on the assumption that segments of veins have a cost related to their conductance? An idealized version of this problem would predict that the network has no loops, but if we search for a network that functions in the presence of fluctuating loads or occasional damage, the optimal networks have loops, as do the real networks (see Figure 1.7, right). This theoretical work raises questions about how to characterize the "loopiness" of flow networks, and how such networks could develop.

Each leaf can have a different vascular structure, so structure is not specified by genetics. Rather, it has been suggested that the structure is determined by the

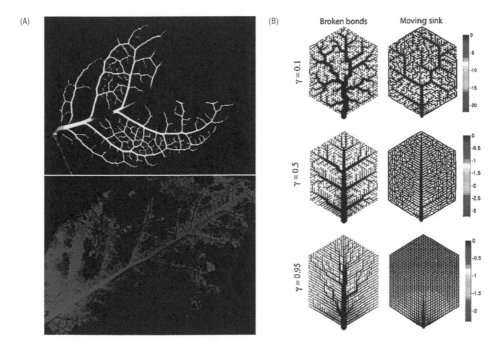

FIGURE 1.7 Fluids flow inside of organisms as well as outside. Plant leaves contain a complex network of veins that connects every cell to the source of water in the stem. (A) Visualization of fluid flow through the veins of a leaf, with a fluorescent dye. Fluid coming from the stem (at bottom) reroutes through loops in leaf's veins, avoiding the damage (black circle) to the main vein, and eventually arriving at the leaf tip. (B) Models of vascular networks that minimize power dissipation develop loops when random links are damaged (left column) or when there is water loss at a varying location in the leaf (right column). The thickness of each vein indicates its conductance. Models with different costs for vein conductance (top to bottom) show that the development of loops is robust. SOURCE: Reprinted with permission from E. Katifori, G.J. Szöllősi, and M.O. Magnasco, 2010, Damage and fluctuations induce loops in optimal transport networks, *Physical Review Letters* 104:048704, copyright 2010 by the American Physical Society.

process of growth of the leaf combined with biochemical cues that can, for example, impose the rule that each channel in the network grows according to the amount of fluid passing through it—the higher the flux, the more the channel widens. This rule leads to vascular structures "designed" for low dissipation, enhancing the function of the vasculature. In this view what is encoded genetically, and thus subject to selection, are the rules of growth, which ultimately determine the functional performance of the system. Closely related ideas about the interaction of flow and growth have arisen in thinking about the remarkable coordination of behavior over long distances in acellular slime molds.

Toward the Physics of Behavior

The interaction of organisms with their surroundings defines what is necessary for movement, and also sets the context within which this movement is controlled. A moth hovering near a flower to extract nectar, for example, confronts a collection of problems. Hovering near the flower requires seeing the flower itself and compensating for the wind, and it is surprising how well this works even as a bright blustery day turns into dusk, connecting to the problems of sensing (Chapter 1). Notably, visual responses slow down as light levels drop (Chapter 4), increasing integration times to reduce noise, but this is constrained by the speed needed to control the mechanics of flight itself. The wing beats are generated by nonlinear dynamics of neural circuits and muscles, with connections to nonlinear dynamics more generally (Chapter 5), while rhythm-generating circuits in other systems have been important testing grounds for ideas about how organisms navigate the high-dimensional parameter space of possible circuits (Chapter 4). Describing the aerodynamics of the flight itself is a challenging problem, as is defining the algorithm that the moth uses to stabilize itself. In moths it is possible to monitor almost all the neural signals that control the flight muscles, providing an opportunity to test ideas about how information is represented (Chapter 2). This gives a sense for how just one seemingly simple pattern of movement provides paths into many questions in the physics of life. This example also illustrates how the questions asked by the biological physics community connect to questions asked by neurobiologists, engineers, control theorists, and others, as explored more fully in Part II of this report.

Faced with the wide range of questions associated even with one movement, many physicists and biologists have made progress by constraining animal behavior so that some more limited set of movements could be studied more precisely. This is in the reductionist spirit (noted in the "Introduction and Overview" chapter), and has been extremely productive. But there is the worry that constraining movements misses something that is essential to the organism. Within the broad biological community, this point has been emphasized by ethologists, who are interested in the often complex behaviors exhibited by organisms in their natural environments, such as the dance that bees use to communicate to their hive about the location of food sources. In many ways, the challenge ethologists raise is paradigmatic for modern biological physics: Can the complexity of a living system be tamed in its functional context? In the spirit of physicists' approaches to other complex problems, the goal is not just to build better tools for characterizing behavior, but to discover some underlying principles that govern these complex dynamics. In recent years, the community has taken up this challenge, and many biological physicists now describe themselves as working on the physics of behavior. Over the course a decade these efforts have laid a foundation for a vastly more complete view of naturalistic behaviors, with the discovery that these behaviors themselves are more structured and hence simpler than they might have been. In some cases, we see the emergence of the next generation of physics questions about how these dynamics are organized on longer time scales.

The starting point for work on the physics of behavior is the effort to collect more complete data on what organisms are doing in more complex contexts. Classical approaches to this involve attaching large numbers of probes to the animal, for example, lights to track the angles of joints as people walk through the world. A more modern approach is to start with high-resolution, high-bandwidth video. One can search these high-dimensional data in an unsupervised way, looking for low-dimensional structure, or provide hand-labeled examples of the locations of cardinal points, which can then be used to train an artificial neural network (Chapter 7) to find these points efficiently in the video stream (pose estimation). These methods can be used when observing small organisms under a microscope, large animals in the wild, or interactions among individuals in small groups (see Figure 1.8).

 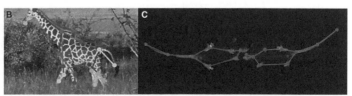

FIGURE 1.8 High-resolution, high-bandwidth video is a modern approach to collecting more complete data on what organisms are doing in more complex contexts, informing our understanding on the physics of behavior. Combining high-resolution video imaging with machine learning extracts the posture of animals during natural movements. Deep networks are trained to identify cardinal points on the body, including joint positions and angles, reducing images with tens of thousands of pixels to a handful of intrinsic coordinates. (A) Flies during courtship. (B) A giraffe. (C) Mice in a social interaction. SOURCE: Reprinted from S.R. Datta, D.J. Anderson, K. Branson, P. Perona, and A. Leifer, 2019, Computational neuroethology: A call to action, *Neuron* 104:11, copyright 2019 with permission from Elsevier.

The combination of video and machine learning has created the opportunity to study more complex, naturalistic behaviors in many systems, and generated considerable excitement, but this is only a start. It is an accomplishment to reduce high-resolution videos of walking flies automatically to more than 40 coordinates describing the movements of relatively rigid body parts, as in Figure 1.8A, but surely flies do not wander randomly in this 40+ dimensional space. Many groups are searching, with different methods, for further simplification; examples include the projection onto lower dimensional spaces that still capture most of the variability in movements, and the discovery of stereotyped segments of the organism's trajectory through the high-dimensional space that can be identified as behavioral states. With this further reduction, it becomes possible to reconstruct the dynamics of movements. This is a literal "physics of behavior," because it results in equations of motion for the organism, an analog of Newton's equations.

A recent example shows that apparently random components of the crawling behavior of the worm *Caenorhabditis elegans* can be traced to deterministic chaos in the underlying dynamics (see Figure 1.9A and 1.9B). More detailed analysis shows that each mode of behavioral variation that grows in time is paired with a mode along which variability decreases, quantitatively, so that the dynamics exhibit a symmetry analogous to the symplectic symmetry of Hamiltonian mechanics (see Figure 1.9C). Efforts along these lines are developing rapidly, in many different systems, from the simple and slow movements of plants to the rapid transitions among multiple gaits as animals move over complex terrains. The example of *C. elegans* illustrates how raw movies of unconstrained animal movements can lead to discoveries of hidden symmetry principles.

Beyond observing complex movements and trying to infer the underlying dynamics, it is possible to perturb these movements and study how the system responds. In thermal equilibrium, spontaneous fluctuations are related precisely to the respond functions, but of course this is not true in actively moving systems. Indeed perturbation experiments connect to very different points of view on animal movement, notably ideas from control theory. An example that brings physics and control together is the response of fly flight to mechanical perturbations (see Figure 1.10). By attaching small ferromagnetic pins to the fly, one can apply forces during free flight and observe the responses with multiple high-speed cameras to allow three-dimensional reconstruction of body and wing movements. These measurements reveal highly stereotyped responses on time scales of tens of milliseconds, which can be understood in terms of feedback from gyroscopic sensors called halteres.

Perspective

The motion of organisms through fluids—from swimming bacteria to soaring birds—has long provided inspiration for the physics community, pushing our

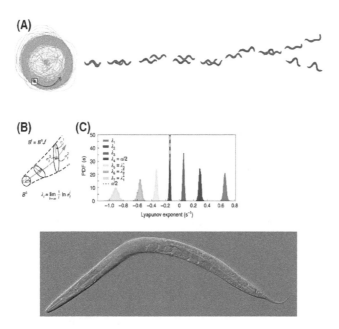

FIGURE 1.9 The worm *Caenorhabditis elegans* is a widely used model organism, shown at bottom. Recent work has uncovered chaos and hidden symmetries in the worm's crawling movements. (A) Measurement of the worm's shape can be projected into lower dimensional spaces. As the worm crawls forward, a wave passes along its body, and this oscillation corresponds to the counterclockwise rotations in the two-dimensional projection at left. Two trajectories (red and blue), which begin close together, both correspond to forward crawling, but gradually diverge. This can be seen in the full images of the worm, evolving in time at right: Although starting in almost identical configurations, the red and blue worm go their separate ways. (B) Embedded in slightly higher dimensional space, one can find all the trajectories that start within a ball B^0 and then follow the stretching or compression of this ball along different dimensions, defining the Lyapunov exponents λ_i. (C) The distribution of Lyapunov exponents from multiple experiments, illustrating the near symmetry around a central negative value. SOURCES: (A–C) Reprinted by permission from Springer: T. Ahamed, A.C. Costa, and G.J. Stephens, 2020, Capturing the continuous complexity of behaviour in *Caenorhabditis elegans*, *Nature Physics* 17:275, copyright 2020. Image of *Caenorhabditis elegans* from Zeynep F. Altun, editor of www.wormatlas.org, https://en.wikipedia.org/wiki/Caenorhabditis_elegans, Creative Commons license CC BY-SA 2.5.

understanding into new regimes and far eclipsing what human-made machines can accomplish. Conversely, insights from physics are indispensable for understanding the mechanics of movement and the special requirements that must be met in movement through fluctuating environments. The transport of fluids by and within organisms has likewise challenged us; for example, problems posed by vascular networks can be mapped in their simplest incarnation into electrical resistor networks, but the latter networks traditionally studied by physicists are not

(A)

(B)

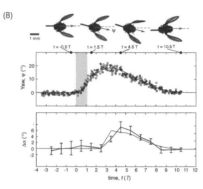

FIGURE 1.10 Fly flight control in response to controlled perturbations is an example of physics and control theory working together. (A) Three-dimensional reconstruction of the fly (*Drosophila*) body (2.5 mm long) and wing kinematics during an aerial "stumble" during transit through a Helmholtz coil; the perturbation is induced by a brief magnetic field pulse which couples to a small ferromagnetic pin attached to the fly's thorax. Images are captured 35 times per wing beat cycle, but shown only every four beats. (B) Top view images of the fly before, during, and after the perturbation, beneath which are shown body yaw (heading), wing relative attack angle over time (normalized by wingbeat period, $T = 4.5$ milliseconds), and aerodynamic/control model estimated torque. SOURCE: L. Ristroph, A.J. Bergou, G. Ristroph, K. Coumes, G.J. Berman, J. Guckenheimer, Z.J. Wang, and I. Cohen, 2010, Discovering the flight autostabilizer of fruit flies by inducing aerial stumbles, *Proceedings of the National Academy of Sciences U.S.A.* 107:4830, Creative Commons License CC BY-NC-ND 4.0.

required to adapt to minimize dissipation, remain robust to enormous damage, or send extra current to different locations on demand. In the last decade, physicists have confronted animal behavior in its full complexity, showing, for example, that one can reconstruct the effective equations of motion for naturalistic movements, even to the point of discovering underlying symmetries hidden in these dynamics. There are many frontiers where progress is expected in the coming decade: better connections of microscopic with macroscopic understanding; better understanding of movements through materials that can behave either as fluids or solids such as sand and soil, where our understanding of the materials themselves is still evolving; and the search for physical principles governing more complex movements in their natural context. The beautiful collective motions of flocks and swarms are discussed in Chapter 3.

SENSING THE ENVIRONMENT

In order to do the right thing, organisms must sense their environment. Sense organs are the instruments that organisms, including humans, use in making measurements on the world, and perceptions are the inferences that we draw from these data. As such, physicists have been especially fascinated by sensation and perception at least since the 19th century. Organisms and even single cells also sense their internal states, and generate signals. Physical principles that govern sensing the environment thus should also govern much of the everyday business of all cells as they exchange and process information, as described in Chapter 2.

Photon Counting

Thinking about the nature of light has been bound up with our understanding of vision for millennia. By the late 1800s it was understood that light was an electromagnetic wave, and the new unification of electricity, magnetism, and optics explained many important aspects of vision, including the irreducible level of blur that comes from diffraction. But in 1900, the quantum revolution began, and by 1905 it was proposed that the interaction of light with matter is described by the absorption and emission of discrete particles of light, called photons. Just a few years later, it was suggested that the dimmest lights humans can see, on a dark night, deliver just a handful of photons to our retina. The idea that the limit to vision is set by the quantum nature of light set an agenda that would unfold over the course of a century, with broad implications.

It now is established that, even under controlled conditions, our perception of very dim lights fluctuates, not because our attention is wandering but because normal sources of light deliver photons at random. Only recently has the full power of quantum optics been used to make light sources that eliminate much of this randomness, and our perception of these light sources is more reliable and deterministic, as predicted. There was a decades-long march from observations on human behavior to recording the electrical responses of receptor cells to single photons (see Figure 1.11A and 1.11B), and these now have been seen in animals across the tree of life—from butterflies to mice, and from horseshoe crabs to monkeys whose visual systems are very much like our own.

Inside the receptor cell, photons are absorbed by the rhodopsin molecule. A typical receptor has roughly 1 billion of these molecules, packed so densely that the cell is almost black. The absorption of one photon triggers a change in the structure of one rhodopsin molecule. As in photosynthesis, the first molecular events that follow photon absorption happen within trillionths of a second, so fast that these events compete with the loss of quantum mechanical coherence. Subsequent structural changes unfold on longer times scales, until the rhodopsin molecule reaches a metastable state that can trigger biochemical events through interactions with other molecules. These events form a cascade that serves as an amplifier, so that one rhodopsin molecule at the start of the cascade results in the degradation of many thousands of cyclic guanosine monophosphate (cGMP) molecules at the end of the cascade (see Figure 1.11C). The cGMP molecules in turn bind to ion channel proteins in the receptor cell membrane, regulating the flow of electrical current into the cell. Arrival of a single photon results in a pulse of current roughly one picoAmpere in size, well above the background of random fluctuations.

Although details vary, the molecular components of the amplification cascade that enables photon counting have direct analogs in a wide variety of processes throughout the living world. When a hormone molecule circulating in our blood binds to a cell surface receptor, this receptor interacts with a protein that belongs to the same "G-protein" family as the transducin (T) molecule in the visual cascade of Figure 1.11C. Following the hormonal response through its cascade, the G-protein activates an enzyme that changes the concentration of a different cyclic nucleotide, playing the role of cGMP in Figure 1.11C. These mechanisms are so widespread that it was possible to find the receptor molecules in our sense of smell by searching for G-protein coupled receptors encoded in the genome. Different parts of the amplification cascade have been discovered first in different systems,

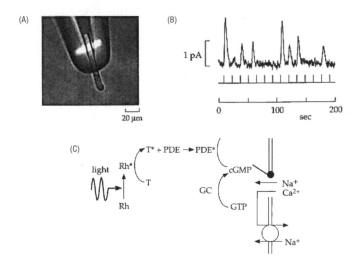

FIGURE 1.11 Normal sources of light deliver photons at random, causing our perception of very dim light to fluctuate. This is demonstrated by photon counting in single rod cells from the eye. (A) A single rod cell is drawn into a pipette and stimulated by light. A tight seal insures that current flowing across the cell membrane flows up the pipette, where it is measured. (B) Current trace (top) in response to a series of dim, brief flashes of light (below). Responses are quantized and probabilistic, as expected in the regime where individual photons are being counted. (C) Absorption of one photon by a single molecule of rhodopsin (Rh) triggers a molecular cascade. One molecule at the input results in the degradation of many molecules of cGMP at the output, which in turns changes the number of open ion channels in the membrane, producing the currents in (B). SOURCE: Reprinted with permission from F. Rieke and D.A. Baylor, 1998, Single-photon detection by rod cells of the retina, *Reviews of Modern Physics* 70:1027, copyright 1998 by the American Physical Society.

with the universality of the mechanisms emerging only gradually. Some measure of the importance of these mechanisms is their recognition in multiple Nobel Prizes: the discovery of rhodopsin itself (1967); the discovery of cyclic nucleotides as internal signaling molecules (1971); the discovery of G-proteins (1994) and olfactory receptors (2004); and the elucidation of the structural basis for the interaction between the receptors and G-proteins (2012). Photon counting is the example in which our understanding can be tested in the greatest quantitative detail, in the physics tradition, and has provided a touchstone throughout these developments.

Despite its importance, the amplification cascade is not enough to explain the ability of the visual system to count photons. A single rhodopsin molecule will continue to drive changes in the cGMP concentration so long as it is in its active state. But a single molecule makes transitions between states at random times, and this randomness would be passed through the cascade, ultimately resulting in a highly variable current across the cell membrane. Such variability would make it impossible for cells to report reliably that different numbers of photons had been counted; in fact the current pulses in response to single photons are stereotyped and reproducible. Part of the answer to this problem is that the active rhodopsin molecule does not just spontaneously switch off; rather, it is actively turned off by another protein that attaches multiple phosphate groups to the rhodopsin. Experimentalists can manipulate the genome so that cells produce rhodopsin molecules that are missing the sites at which these phosphate groups are added, and even deleting one out of six sites results in noticeably more variable responses to single photons; variability increases as more sites are deleted, and this pattern follows theoretical predictions. This is an inspiring example of how the complexity of biological molecules can be understood, quantitatively, as a response to the physics problems that organisms must solve.

More subtly, the multiple steps involved in turning off the activity of rhodopsin all dissipate energy, and this is essential. If no energy were dissipated, each step would be reversible and the molecule would take a random walk from its active to inactive state, restoring the original randomness of the transition. Simple models show that there is a tradeoff between energy dissipation and the reduction of variance in the time the molecules spend in the active state, anticipating the "thermodynamic uncertainty relations" that are part of recent progress in non-equilibrium statistical mechanics.

Looking carefully at Figure 1.11B, there is one example of a single photon response that seems to come before the flash of light. This is not a violation of causality, but an example of the "dark noise" that one finds in all photodetectors. In this case, there is some probability that rhodopsin will change its structure as the result of a thermal fluctuation rather than the absorption of a photon. In a single molecule this transition occurs roughly once every thousand years. But a single photoreceptor cell is packed with 1 billion rhodopsins, so there is one event per minute. Remarkably, these random events provide a dominant source of noise, limiting the organism's ability to be sure it has seen very dim flashes of light. In cold-blooded animals, one can lower the temperature, reducing the dark noise and increasing the reliability of seeing, just as is done with photodetectors in the physics lab.

The problem of photon counting in vision does not end with a current pulse in the photoreceptor cell. This current drives a change in voltage across the cell membrane, which in turn drives the flow of current carried by calcium ions, and increases in calcium concentration trigger the release of vesicles into the synapse onto cells in the next layer of the retina. Vesicle release is a central feature of signaling between cells, not just in the retina but throughout the brain and at the connection between nerve and muscle. Indeed, the phenomenon is much more general, encompassing many processes where cells need to export materials, from hormones to waste products. The idea that transmission across a synapse involves discrete vesicles, each containing thousands of neurotransmitter molecules, emerged from the discovery of quantization in high-resolution recordings of the electrical signals at the neuromuscular junction, and eventually it was possible to measure directly the added capacitance of the cell membrane as single vesicles fuse with it; this classical chapter in the interaction between physics and biology was recognized with a Nobel Prize in 1970. The commonality of vesicle release mechanisms was crucial in the identification of the key protein molecules involved in the process, which was recognized by a Nobel Prize in 2013. Today, vesicle release is studied with the full range of experimental methods from the biological physics community, down to the single molecule level. Variations in molecular properties tune different vesicle release systems to different requirements, from the slow release of hormones to the transmission of signals with near microsecond precision in the auditory system. A major theoretical question is whether there are unifying principles that govern the molecular events across these many orders of magnitude in time scale.

The synapse connecting the photoreceptor to the next layer of the retina has been an important example of how sensory signals are processed. In the fly, this synapse holds the record for the highest rates of information transmission seen in neurons, approaching the physical limit set by counting every single vesicle with millisecond resolution. In animals more like us, the synapse acts as a filter, helping to separate the single photon responses from the background of dark noise. The synapse also is nonlinear in its response, further enhancing this separation and making it possible for individual neurons to sum the signals from many receptor cells without being swamped by summed noise. The structure of this filtering and nonlinearity can be derived, quantitatively, from a common principle of maximizing the signal-to-noise ratio for detecting these dim flashes of light, and in this way aspects of these first steps in visual signal processing can be understood as solutions to the underlying physics problem. A challenge for the coming decade is to determine whether these physical principles can predict the dynamics of signal processing at synapses in the retina more generally.

Molecule Counting

Photon counting is not the only example where biological signaling systems encounter fundamental physical limits to performance. As bacteria swim, propelled by the rotation of their flagella (see Figure 1.4), they move toward sources of food and away from noxious chemicals. A breakthrough came in the early 1970s, with the construction of a tracking microscope that could follow the trajectories of single bacteria, demonstrating that their motion consists of relatively straight "runs" interrupted by "tumbles" that select a new direction almost at random; illustrations related to these measurements are shown in Figure 1.12. Runs correspond to counterclockwise rotation of the cell's multiple flagella, which allows them to come together in a bundle, while clockwise rotation causes tumbles as the bundle flies apart. Runs typically last for a few seconds, but in mutant bacteria that are incapable of chemical navigation, runs last much longer. Experiments show that when cells are swimming up toward increasing concentrations of attractive molecules, runs are prolonged, and this is the basis of navigation: These bacteria do not "steer" toward a source of food, but rather take a random walk that is biased toward the source. This is an algorithm for finding the maximum concentration that is similar to Monte Carlo optimization.

The chemical navigation system of bacteria is called chemotaxis, and it is extraordinarily sensitive: Significant changes in run length occur when the concentration changes by just a few parts in a million across the length of the cell. But molecules arrive at the cell surface by random motion, and this randomness obscures the small differences between the front and back of the cell. The only possibility is that cells monitor how concentrations change as they move, integrating over the several seconds of a run which carries them dozens of body lengths through their surroundings. Even so, they must effectively count every single molecule that arrives at their surface. These theoretical inferences from the sensitivity of chemotaxis were confirmed, for example in experiments where a bacterium is tethered to a glass slide by a single flagellum and then the rotation of the motor causes the whole cell to rotate; changes in the probability of clockwise versus counterclockwise rotation can then be monitored in response to brief pulses of attractive or noxious molecules. These and other experiments show that cells are responding to changes of concentration over time, not space; that there are significant responses when one extra molecule is bound to a receptor on the cell surface; that the cell can ignore the overall concentration of molecules and respond only to changes; and that the response averages these changes over times long enough to suppress noise but not

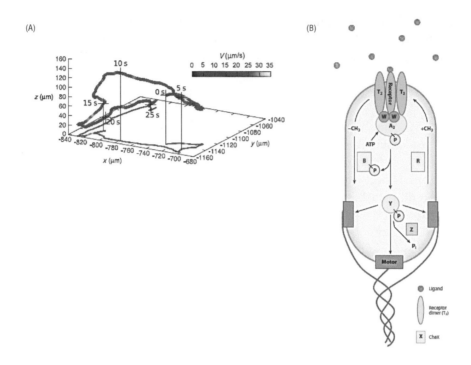

FIGURE 1.12 Bacteria swim toward sources of food and away from noxious chemicals, their motion characterized by relatively straight "runs" interrupted by "tumbles" that select a new direction almost at random. (A) Chemotaxis in the bacterium Escherichia coli. Three-dimensional trajectory of a single cell, with velocity coded by color. Runs with high, relatively constant velocity are interrupted by tumbles with low velocity. (B) A modern schematic of chemotaxis system, showing the flow of information from ligand molecules outside the cell through receptors, the addition of phosphate groups to proteins, and finally control of the flagellar motor's rotation. The label X marks the key proteins CheA, CheB, CheR, CheW, CheY, and CheZ, and the phosphate group is represented by the purple circle labeled P. SOURCES: (A) N. Figueroa-Morales, R. Soto, G. Junot, T. Darnige, C. Dourache, V.A. Martinez, A. Lindner, and É. Clément, 2020, 3D spatial exploration by *E. coli* echoes motor temporal variability, *Physical Review X* 10:021004, Creative Commons License Attribution 4.0 International (CC BY 4.0). (B) Y. Tu, 2013, Quantitative modeling of bacterial chemotaxis: Signal amplification and accurate adaptation, *Annual Review of Biophysics* 42:337.

so long that cells would be disoriented by their own rotational Brownian motion. In this way, much of the chemotactic behavior of bacteria can be understood as solutions to the underlying physics problems.

Ideas about the physical limits to molecular signaling that have their roots in thinking about chemotaxis have reappeared in connection with experiments on many different systems, from the control of gene expression to axon guidance during the development of the brain. The problem also continues to inspire new theoretical developments, notably generalizations to dynamic signals and to the case where there are many species of molecules to be detected and many different kinds of receptors. It is encouraging to see how the scientific community takes these steps toward more complex and biologically realistic formulations while remaining grounded in general physical principles.

In the case of chemotaxis, generations of scientists have connected the overall strategies for solving the underlying physics problems to detailed molecular mechanisms (see Figure 1.12). The enormous sensitivity of the system has contributions from multiple components: cooperative interactions among neighboring receptor molecules in the cell membrane; a cascade of molecule multiplication not unlike that found in photon counting; and cooperative interactions of the final signaling molecule in controlling the direction of the flagellar motor. This detailed mechanistic understanding was built using a combination of experimental methods from biology and physics, for example using genetic engineering to make fluorescent analogs of crucial molecular components and monitoring their interaction through measurements of energy transfer. These developments have gone hand in hand with increasingly precisely theoretical descriptions of the system, which have been used, for example, to address questions about the relations between energy dissipation and signal-to-noise ratio, which are much more general. Nearly 50 years after the first tracking microscope measurements, this system continues to inspire new developments.

Mechanical Sensing

Cells that move along solid surfaces or through the spaces of porous materials can sense and respond to the mechanical properties of their surroundings. Cells crawl using molecular motors (Chapter 1) and exert forces on their substrates via adhesive connections. The same mechanisms of force transduction are at play in the differentiation of stem cells placed on substrates of varying rigidity; stem cells placed on soft gels differentiate into cells belonging to soft tissues such as the brain, while those placed on very stiff gels differentiate into cells belonging to bone. These phenomena illustrate that mechanical cues can play as strong a role as chemical ones in biological processes, and can often work in tandem with chemical cues to influence behavior. Another example of this interplay occurs at the organ level in the early embryonic heart. The heart is the first organ to function, beating and pumping fluid, in animals. Each heartbeat consists of a wavefront of cell contraction that traverses the heart from one end to the other. The mechanism for cell signaling that coordinates the heartbeat has long been understood to be electrical—hence electrical defibrillators. It was recently shown, however, that the early embryonic heart does not use electrical signaling to coordinate its heartbeat. Rather, the mechanical strain that contracting cells exert on other cells is instrumental in signaling them to contract.

Mechanical sensing and extraordinary sensitivity come together in the specialized cells of the inner ear. In a quiet room, we can hear sounds that cause our eardrums to vibrate by less than the diameter of an atom. Several different mechanical sensors are embedded in the bones behind the human ear—the cochlea, which responds to sounds that move the eardrum; the semicircular canals, which respond to fluid motions caused by the rotation of our head; and otoliths, which respond to movements of tiny calcium carbonate crystals caused by linear accelerations, including gravity. At the heart of all these organs are the hair cells, which generate electrical signals in response to displacement of their "hairs" or stereocilia (see Figure 1.13). To be clear, these "hairs" are quite different from the hairs on our head at the molecular scale, but early microscopists were struck by the appearance of these cells, and the name persists. Mechanical sense organs with hair cells are found in all animals with backbones, including the lateral line in fish, which senses motion of the water as the fish swims, and the frog sacculus, which senses ground vibration. In one species of tropical frog, ground vibrations of just 10 trillionths of a meter are sufficient to trigger responses.

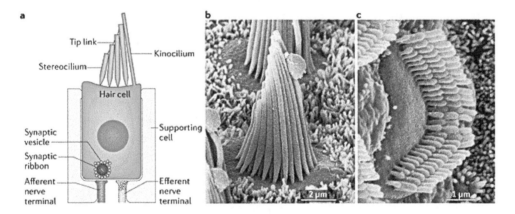

FIGURE 1.13 The hair cells of the inner ear generate electrical signals in response to displacement of their "hairs" or stereocilia. (A) Schematic of a hair cell surrounded by supporting cells. A ribbon synapse at the base of the cell releases vesicles that drive electrical activity of the afferent neuron, which carries information to the brain. The efferent neuron carries signals from the brain that modulate the sensitivity of the hair cell. Stereocilia vary systematically in length across each hair bundle and are connected by "tip links" that transmit forces to ion channels in the membrane. In many hair cells, there is a true cilium, the kinocilium, which has a role in organizing and orienting the bundle. (B) A scanning electron micrograph shows an individual hair bundle from the frog's sacculus, an organ that detects gravity and ground-borne vibration. (C) A scanning electron micrograph shows the specialized hair bundle of an outer hair cell from the bat's cochlea, which is V-shaped and has only three ranks of stereocilia. SOURCE: Reprinted by permission from Springer: A.J. Hudspeth, 2014, Integrating the active process of hair cells with cochlear function, *Nature Reviews Neuroscience* 15:600, copyright 2014.

A dramatic development in our understanding of mechanical sensing in living systems is the realization that hair cells are not just sensors, but active sensors, so that the inner ear—and presumably most other systems based on hair cells—are mechanically active. Dramatic qualitative evidence for this comes from the fact that ears can emit sound. These acoustic emissions are very pure tones, unique to each individual, and quite common. Presumably, these result from minor pathologies that allow too much of the mechanical activity to couple back into the macroscopic mechanics of the ear, much as in the instability of a microphone pointed at a loudspeaker. Quantitative evidence is based on interferometric measurements of the spontaneous hair bundle movements, and the demonstration that these violate the fluctuation-dissipation theorem.

These observations suggest a model in which elements of the inner ear are active filters, poised close to their instability. As this critical point or bifurcation is approached, the frequency range of the mechanical response narrows, suppressing the effects of thermal noise, and the magnitude of the response to external forces increases. Independent of the underlying mechanism, the behavior in the neighborhood of the bifurcation is universal. An example of this universal behavior is nonlinear mixing of nearby frequencies, with a strength that is nearly independent of amplitude. These essentially parameter-free predictions are consistent with classical perceptual observations on "combination tones," and with direct measurements on hair cells.

Perspective

This discussion of just three of the very many sensing systems that organisms possess has revealed deep principles. Sensors in living systems are precise despite having to function with signals that can vary by many orders of magnitude. This precision in many cases approaches fundamental physical limits to their performance, which belies the commonly repeated claim that living systems are irreducibly noisy and messy. In some cases the observation of near-optimal performance can be turned into a theoretical principle, from which aspects of system function and mechanism can be derived, following the classic example of the insect eye. Molecular components that implement these mechanisms have a surprising universality, and sensing itself plays a role in the behavior of all cells. The physics of sensing continues beyond the receptor cell into signal processing within the sensory organs and then on into the brain. A growing understanding of these architectures is blurring the distinction between sensing, filtering, and de-noising so that an organism can extract the most useful signals from the environment, which will be discussed further in Chapter 2.

STRUCTURES IN SPACE AND TIME

Organisms act on spatial and temporal scales that are removed by orders of magnitude from the natural scales of their molecular components. These phenomena are reminiscent of many pattern-forming systems in the inanimate world, and this connection has provided a path for many physicists to begin their exploration of the living world. At the same time, pattern formation in living systems poses qualitatively new challenges, driving the search for new physical principles that can explain these beautiful phenomena.

Allometry

The first spatial structures of plants and animals that attracted human attention were the most macroscopic, and it would take into the 20th century to state the problem of how these are encoded by the underlying molecules. The macroscopic structures have a logic of their own, often grounded quite directly in physical principles: The cross-sections of bones scale with overall body size so that animals do not collapse under their own weight, stems and roots scale with plant size to be sure that water and nutrients can be transported to the leaves, and so on. The study of such scaling or allometric relationships has a long history in the biological literature, and extends to the scaling of metabolic rates, lifespan, population density, and even brain size with body size.

Allometric scaling defines power-law dependences, for example of metabolic rate on body mass across species. These relations often are quite precise, and extend over many decades for each variable. In many cases, the scaling exponents are simple rational numbers, but not what would be expected from dimensional analysis. For example, animals should lose heat at a rate proportional to their surface area, while body mass is proportional to volume and area $\sim(\text{volume})^{2/3}$, suggesting that to keep constant temperature requires heat production scaling with the 2/3 power of body mass. In fact mammalian heat production scales as the 3/4 power of body mass, across a factor of 100,000 from mice to elephants. Theory, in a similar spirit to ideas about fluid flow in leaf vein networks (see Figure 1.7), predicts that this scaling emerges from the properties of resource distribution networks that carry nutrients through the body. As usual in physics, the leading scaling behavior is derived for systems that are asymptotically large, and more recent work emphasizes how finite size correction can lead to observed departures from simple scaling. It remains to be discovered whether any allometric relations exhibit anomalous scaling dimensions, as in more familiar physics problems.

Self-Assembly and Physical Virology

At the opposite extreme, an important example of structure formation is the assembly of viruses. Viruses typically consist of a protein shell, the capsid, which in the simplest case is made up of many copies of a single protein, which envelopes the viral genome in the form of either single- or double-stranded DNA or RNA. For viruses that infect mammalian cells, there is generally also a lipid bilayer coat. It is an extraordinary fact that viruses often will reassemble in a test tube from purified protein and nucleic components. Historically, this was important in convincing the scientific community as a whole that one could, in practice, study the building blocks of life with concepts and methods from physics and chemistry. In modern times, the self-assembly of viruses has emerged as a physics problem, and the interplay between the biological physics community and the larger biology community interested in viruses has led to a field called physical virology, now the subject of regular conferences.

X-ray diffraction images of crystals of viruses revealed, long ago, that many of them have icosahedral symmetry. More recently, electron microscopy has provided high-resolution reconstructions of viral capsids. An icosahedron can be folded from a flat sheet of hexagons by replacing some 6-fold vertices by 5-fold defects; this realization led to the classification in the 1960s of icosahedral virus shells (capsids) in terms of triangulation numbers, which characterize the distances between neighboring 5-fold defects—a classification still used today. While isolated 5-fold defects lead to icosahedral viruses, it is now understood that a line of 5-fold defects is responsible for conical capsids, as in the HIV virus. A modern example of these ideas is from recent work on the Brome Mosaic Virus (BMV), shown in Figure 1.14. In this system the nearly spherical capsid shell, composed of 180 identical proteins, has axes of 5-, 3-, and 2-fold symmetry. The BMV virus packages three different RNA molecules separately into different virions that are all present during an infection. Because it needs to package all genomes separately but simultaneously, it uses non-specific electrostatic interactions between the RNA and the capsid, resulting in much more disordered RNA in the interior. This is in contrast to other viruses in which the RNA acts as a template for assembly of the protein components and can serve as a "molecular ruler" to set the size of the virus as a whole.

Studies of viral self-assembly have been pushed forward by observations and manipulation of single viruses. This approach from the biological physics community parallels single molecule experiments aimed at understanding force generation in muscle (see Figure 1.3), the synthesis of ATP from the flux of protons across a membrane (see Figure 1.5), the readout of information encoded in the genome (see Figure 2.1), and the flow of electrical current through ion channels (see Figure 2.8). These methods reveal how the mechanisms of self-assembly address the physics problems that arise: Weak interactions among small numbers of capsid proteins allow the system to avoid kinetic traps, while the fully assembled structure is strong enough to resist the osmotic pressure generated by the long polymer of the genome packed inside. In bacteriophages—viruses that infect bacteria—direct measurements show that these pressures are on the order of 10 atmospheres, and special packaging motors are required to stuff the genome into the capsid; this pressure in turn drives the genome into the host cell during infection. Viruses thus provide examples of both spontaneous self-assembly and the active construction of stable far-from-equilibrium structures.

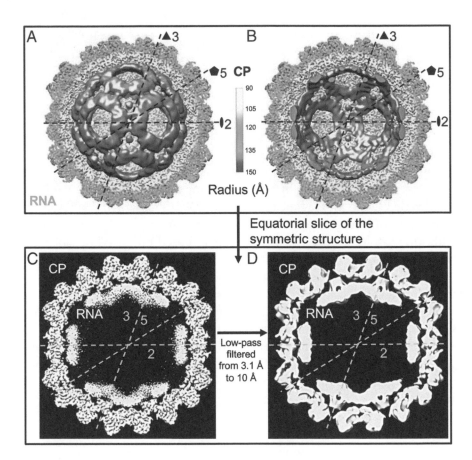

FIGURE 1.14 The Brome Mosaic Virus (BMV) is a modern example of the 1960s classification of icosahedral virus shells in terms of triangulation numbers characterizing the distance between neighboring 5-fold defects. The BMV system has axes of 5-, 3-, and 2-fold symmetry. (A) and (B) Interior view of the virus, reconstructed from electron microscope images, showing the back half of the capsid protein (CP) shell and either the entire (A) or the back half of (B) the RNA genome. Protein components are color-coded based on their radius from the center of the virus, and symmetry axes are shown as a guide. RNA sits near the 2- and 3-fold axes but not near the 5-fold; no RNA density is resolved at the center. (C) and (D) Slices through the reconstruction without (C) and with (D) low-pass filtering to 10 Å. The symmetry axes have been indicated, and it is clear that the RNA is situated preferentially near the 2- and 3-fold axes and away from the 5-fold axes. SOURCE: C. Beren, Y. Cui, A. Chakravarty, X. Yang, A.L.N. Rao, C.M. Knobler, Z.H. Zhou, and W.M. Gelbart, 2020, Genome organization and interaction with capsid protein in a multipartite RNA virus, *Proceedings of the National Academy of Sciences U.S.A.* 117:10673, Creative Commons License CC BY-NC-ND 4.0.

Bacterial Growth, Shape, and Division

Bacteria are larger than viruses, typically several microns in length. In many cases, including the well-studied *Escherichia coli*, the overall structure of the cell is supported in part by a polymer of protein molecules that wraps the cell, underneath its membrane, with a helical structure. This helix has a radius essentially equal to the radius of the cell itself, many hundreds of times larger than the diameter of the constituent protein molecules. As in the assembly of viruses, this long length scale results from a small angle between proteins along the polymer, and this small angle is determined by the structure of the protein–protein interfacial surface, pointing toward a more general physical principle in the construction of living systems. The possibility of creating structures with long internal length scales by assembly of microscopic components inspires a broader exploration of self-assembly, as discussed in Chapter 5.

Many important examples of structure formation are connected with cell division. In many bacteria, the process is triggered by polymerization of a single protein in a belt or ring around the middle of the cell. But this middle position is determined by a surprisingly dynamic process. It had been known for some time that a set of proteins called "Min" were essential for division; the name derives from the fact that mutations in these proteins cause the appearance of miniature cells. The surprise was that dynamic measurements on the concentration of these proteins inside the cell revealed an oscillation, with protein accumulating at one end of the cell and then the other, periodically, with a cycle of less than a minute (see Figure 1.15); the middle of the cell is where concentrations are persistently low, and this is consistent with the role of Min proteins as an inhibitor of ring formation. Theory suggested that these oscillations could arise if ATP binds to one protein (MinD) and increases its affinity for the membrane, then a second protein (MinE) binds to the first and triggers ATP hydrolysis. Diffusion, coupled with these reactions, is enough to generate the observed spatiotemporal oscillations. If this is correct, it should be possible to reproduce the essential behavior in a purified system. Indeed, this works, but it is essential to mimic the geometry of the cell (see Figure 1.15E and 1.15F). In different geometries, this simple system can produce many different patterns, consistent with theoretical predictions. Also consistent with theory, in cells where cell division is blocked the pattern of oscillations includes several periods along the length of the cell.

The simple picture of Min oscillations as the mechanism by which cells define their middle is incomplete. In some species of bacteria, the division ring can be placed precisely in the middle even in the absence of Min proteins. In other species, the formation of miniature cells is inhibited by Min, but the patterns are static, and targeting of the protein to the ends of the cell requires other factors. It has never been clear whether Min is sufficient, even in *Escherichia coli*, to explain the preci-

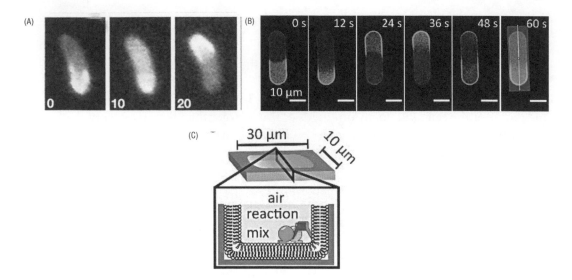

FIGURE 1.15 In many bacteria, a set of "min" proteins are essential for cell division. The concentrations of these proteins oscillate, accumulating first at one end of the cell then the other. Min protein oscillations, in cells and in a reconstituted system. (A) Time-lapse images of MinD protein, fused with green fluorescent protein (GFP), in live *Escherichia coli* cells. Time is noted in seconds and scale bar is 1 μm. DM Raskin and PAJ de Boer, Rapid pole-to-pole oscillation of a protein required for directing division to the middle of *Escherichia coli*. (B) MinD oscillations in a purified system containing only two proteins, MinD and MinE, with a supply of ATP. The solution containing the proteins is confined in a chamber (C) that is lined with lipids to mimic the cell membrane. SOURCES: (A) D.M. Raskin and P.A.J. de Boer, 1999, Rapid pole-to-pole oscillation of a protein required for directing division to the middle of *Escherichia coli*, *Proceedings of the National Academy of Sciences U.S.A.* 96:4971, Creative Commons License CC BY-NC-ND 4.0. (B–C) Adapted from B. Ramm, P. Glock, and P. Schwille, 2018, In vitro reconstitution of self-organizing protein patterns on supported lipid bilayers, *Journal of Visualized Experiments* 137:e58139, https://doi.org/10.3791/58139.

sion with which cells divide in half. What certainly is missing is an understanding of what physical principles determine the advantages and disadvantages of these different molecular mechanisms.

Reactions, Diffusion, Scaling, and Size Regulation

Despite open questions, the Min system illustrates the power of coupling reactions and diffusion to generate patterns. The general mathematical structure of such systems allows for different kinds of patterns that can be fully classified. These patterns have been recognized in many different living systems (e.g., Figure 1.16), even if it can be difficult to identify the particular molecular components that implement these dynamics. The study of reaction–diffusion systems has a remarkable history (see Box 1.4).

In coupled reaction–diffusion systems, there are characteristic length scales that correspond, roughly, to the distance that a molecule can diffuse before it reacts. These length scales are intrinsic to the dynamics, and set the size of pattern elements, such as the width of stripes or the distance between spots. In a larger system, there would be more stripes or spots. While this happens, there also are systems where the patterns scale to the size of the organism. An example is the segmented body plan of a caterpillar or maggot, where individuals of different size have the same number of segments and the segment size or spacing changes in proportion to overall body size; in maggots (larval flies) this can be traced to scaling in the patterns of gene expression that drive these patterns (as in Figure 2.7). It seems fair to say that there is no general understanding of how this scaling is achieved. While the inanimate world provides many examples of pattern formation, some of which may remind us of patterns in living systems, these patterns do not scale. Perhaps this is one more example of life finding new physics.

Related to the problem of scaling is the problem of size regulation. What sets the size of an organism? What sets the size of an organ, or a single cell? Which of these are tightly regulated, and which are fluctuating widely across individuals? Within a single cell, what sets the size of organelles? There is a classical example of size regulation in the algal cell *Chlamydomonas reinhardtii*, which has two flagella of equal length, and this is crucial for its swimming. If one flagellum is broken or removed, the other will shorten, and the two flagella will lengthen together only once they are of equal length. This problem has come back into focus because of a new generation of quantitative experiments and mathematical analyses that exclude many classical models. The problems of size regulation and scaling are simple to state, but may provide hints of deeper principles.

FIGURE 1.16 Patterns in reaction-diffusion models and in animal skin coloration. (A) Two molecular species diffuse and react. Increased concentrations of the activator promote its own (auto-catalytic) synthesis, and the synthesis of an inhibitor. (B) Patterns formed with different parameter settings of the model in (A). (C) Patterns in nature, left to right: hybrid fish (*Salvelinus leucomaenis* x *Oncorhynchus masou masou*); marine beta (*Calloplesiops altivelis*); cheetah (*Acinonyx jubatus*); bengal cat (*Felis catus*); and giraffe (*Giraffa camelopardalis reticulate*). SOURCE: H.C. Metz, M. Manceau, and H.E. Hoekstra, 2010, Turing patterns: How the fish got its spots, *Pigment Cell and Melanoma Research* 24:12, © 2010 John Wiley & Sons A/S.

BOX 1.4
Reaction, Diffusion, and Turing

The idea that patterns form in a developing embryo due to the interplay of biochemical reactions and diffusion goes back to a remarkable paper[a] by a remarkable historical figure, Alan Turing.[b] In 1952, he presented a strikingly modern analysis of how instabilities in such a system could lead an initially homogeneous tissue to develop spatially varying structures on scale relevant for real embryos.

The exploration of reaction-diffusion models was a major theme in the interface of mathematics and biology for a generation. While there were efforts to make Turing's model more realistic, a deeper question is whether the embryo really faces the problem of making patterns from a homogeneous initial condition. In many cases, the mother breaks the symmetry of the egg during its construction, and in other cases, fertilization plays this symmetry-breaking role. But Turing patterns clearly are relevant to a wide variety of living systems, notably the diverse patterns of animal coloring (see Figure 1.16). It is striking that this range of phenomena can be captured in a single mathematical framework.

In presenting his work, Turing also gave voice to an approach that resonates strongly with many members of the biological physics community even today:

> A mathematical model of the growing embryo will be described. This model will be a simplification and an idealization, and consequently a falsification. It is to be hoped that the features retained for discussion are those of greatest importance in the present state of knowledge.

---

[a] A.M. Turing, 1952, The chemical basis of morphogenesis, *Philosophical Transactions of the Royal Society of London B* 237:37.

[b] A. Hodges, 1983, *Alan Turing: The Enigma*, Burnett Books, London, UK.

Perspective

As with energy conversion, the building of structures in space and time is a prerequisite for many other functions of living systems. There has been a very productive exchange between the exploration of particular living systems and the synthesis of artificial systems that operate under similar or perhaps even the same principles of pattern formation and self-assembly. At the same time, there is something different about the living systems, and the community has struggled to articulate this difference. There is tension, for example, between the idea of patterns emerging spontaneously out of homogeneous backgrounds and the idea that information about position in a developing embryo is passed through a cascade of molecular signals, starting with some initial symmetry-breaking event (Chapter 2). In a different direction, we have the idea that information for self-assembly is encoded in molecular structures and especially in the energetics of contacts between the assembling subunits, but we do not really know how to measure this information or relate it to the matrix of contact energies. As in many examples of biological function, an essential part of the problem is not just to stabilize the correct outcome, but to avoid the vastly more numerous incorrect outcomes. It will be exciting to see how the examples of self-assembly and pattern formation guide the search for more general physical principles that address this challenge.

2

How Do Living Systems Represent and Process Information?

A traditional introduction to physics emphasizes that the subject is about forces and energies. This might lead us to think that the physics of living systems is about the forces and energies relevant for life, and certainly this is an important part of our subject. But life depends not only on energy; it also depends on information. Organisms and even individual cells need information about what is happening in their environment, and they need information about their own internal states. Many crucial functions operate in a limit where information is scarce, creating pressure to represent and process this information efficiently. Understanding the physics of living systems requires us to understand how information flows across many scales, from single molecules to groups of organisms. From the theoretical side, these explorations often have reinforced the deep connections between statistical physics and information theory, and on the experimental side we have seen the development of extraordinary new measurement techniques. The search for the physical basis of information transmission in living systems has led to foundational discoveries, pointing to new physics problems. A sampling of these issues is provided in Table 2.1.

INFORMATION ENCODED IN DNA SEQUENCE

The fact that humans look like their parents is evidence that information is transmitted from generation to generation. This is perhaps the most fundamental example of information flow in living systems, underlying the persistence of life itself and its evolutionary change. The fact that this information is encoded in DNA,

TABLE 2.1 Physics of Information in Living Systems

Discovery Area	Page Number	Broad Description of Area	Frontier of New Physics in the "Physics of Life"	Potential Application Areas
DNA sequences	88	Representation and readout of information in the genetic code, and beyond.	Non-equilibrium physics of proofreading; physical encoding of regulatory signals.	Synthetic biology; personalized medicine.
Molecular concentrations	97	Extracting information from the environment, sensing and controlling internal states, all through changing concentrations.	Physical limits to sensing and signaling in realistic contexts; information processing in networks; information about spatial patterns.	New tools and models for genetic networks in health and disease.
The brain	104	Understanding the principles of electrical signal transmission within cells and across synapses; understanding how these signals embody codes for perception, action, memory, and thought.	Theories of coding in single neurons and large networks; physical limits to information transmission; connecting molecular dynamics to macroscopic functions.	New tools for exploring the brain; new ingredients for neural networks in artificial intelligence.
Communication and language	112	Exploring physical representation of information in communication systems from bacteria to bird song, and organizational principles across these scales.	Discovery of new communication modes; communication and self-organization in bacterial communities; searching for information; long-ranged correlations in "meaningful" signals.	New algorithms for robotic search; tools for understanding language models in artificial intelligence.

and that the transmission of this information is enabled by the double helical structure, are among the most profound results of 20th-century science. As described previously, these results emerged from an interplay among physics, chemistry, and biology, and these basic facts about DNA and the encoding of genetic information are now taught to high school students. But the search for physical principles of genetic information transmission did not stop with the discovery of the double helix. Efforts to understand how this information is copied so reliably from one generation to the next, how is it "read" by the cell, and how it can be rewritten all have driven the development of new experimental tools and new theoretical ideas in the physics community.

The Genetic Code

The structure of DNA immediately suggested that information encoded in the sequence could be transmitted reliably through base pairing: In the four letters or bases of the DNA alphabet, A pairs with T and C pairs with G, and the structure of the molecule makes the "wrong" pairings much less favorable. This is an example of "complementarity," in which molecules find their correct partners because their structures literally fit together. This is important not only for A to find T as DNA is copied, but for many other molecules to find their partners in the complex environment of the cell.

But the fidelity with which DNA is copied is extraordinary. In humans, there are a few billion letters in the DNA sequence, and when a cell makes a copy of its DNA there are only a handful of mistakes. The structure of DNA favors the correct AT and CG pairings, strongly, but not strongly enough to explain that mistakes occur with a probability of only one in a billion.

In more formal terms, the molecular structure determines an energy difference between the correct and incorrect pairings. The correct pairing has lower energy, and thus is favored, much as a ball will roll downhill to lower its energy in the gravitational field of Earth. But unlike the ball, a single molecule is jostled, significantly, by the random motions of neighboring molecules. As a result, there is some probability that the system will find itself in higher energy, less favorable states, and this is determined precisely by the energy difference and the temperature of the surroundings. For base pairing in copying DNA, this probability is roughly one in 10,000. This is small, but nowhere near the level of one in a billion found in real organisms.

The discrepancy between error probabilities determined by molecular structure and the observed error probabilities observed in living cells exists at every step in the processing of information encoded in the DNA: in the copying of the DNA itself; in the transcription of DNA into mRNA; in the connection of transfer RNA (tRNA) molecules to amino acids; and in the binding of tRNAs to mRNA during the translation of the mRNA sequence into the amino acid sequence of proteins. In each case living cells achieve a sorting of molecular components that is vastly more accurate than would be expected from energy differences alone. In the 1970s, it was proposed that these very different biochemical processes all face a common physics problem.

In order to drive error probabilities below the levels expected from energy differences, cells must perform a function very much like the hypothetical "Maxwell's demon" that was introduced in the 1800s as a challenge to our molecular understanding of thermodynamics. Briefly, the demon could sort molecules, so that for example it could arrange for all the fast-moving molecules to be collected on one side of a container. But a collection of fast moving molecules is hotter than a collection of slow-moving molecules, so the demon would produce a temperature difference, and in this way could build an engine—out of nothing. While the original description of the demon focused on molecular speeds, sorting by any molecular property would be sufficient to power some kind of engine. Thus the seemingly innocuous sorting of molecules, if it could be done, would make it possible to build a perpetual motion machine, violating the second law of thermodynamics. The solution to the problem of Maxwell's demon drove a much deeper understanding of the connections among energy, entropy, and information.

The short answer is that Maxwell's demon cannot sort molecules reliably without itself expending energy, and the minimum energy expenditure is enough to compensate for the energy that could be extracted by an engine after the molecules were sorted. Similarly, in order to lower the error probabilities in processing information encoded in DNA, the cell expends extra energy in the processes of DNA replication, transcription, tRNA charging, and translation. Although the details vary, all of these processes involve steps that dissipate energy, sometimes in apparently wasteful ways, but these futile steps serve to increase precision. These mechanisms are called "proofreading," analogous to the correction of spelling errors in text. Proofreading is an important example of how diverse mechanisms of life can be understood as addressing a common physics problem, and these molecular events provided important inspiration for understanding the thermodynamics of information and computation.

Optical trapping and single molecule manipulation, as described in Chapter 1, created the opportunity to observe proofreading in action, step by step. Figure 2.1 shows a schematic of experiments that probe the transcription of DNA into mRNA, in a minimal system that includes just the one essential protein, RNA polymerase (RNAP). In order to "read" the DNA sequence and produce the corresponding mRNA, RNAP moves along the DNA strand, trailing the growing mRNA behind it. After several generations of improvement, it was possible to measure this motion with a precision comparable to the diameter of an atom, and thereby resolve the discrete steps from one letter to the next along the DNA sequence.

Watching RNAP walk, base by base, along the DNA, reveals that there are occasional pauses, and even reversals. These reversals are enhanced when the polymerase is forced to make mistakes by providing an excess of incorrect bases. It is known that errors in transcription increase if a discrete subunit of RNAP is removed, and when this is done the reversals disappear. It seems very likely that these experiments thus have observed directly the proofreading steps in transcription of mRNA. Single molecule experiments on the ribosome have given similar insight into the proofreading processes involved in translation.

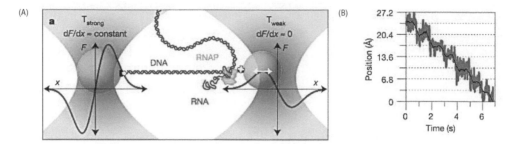

FIGURE 2.1 Optical trapping and single molecule manipulation allow us to observe directly as genetic information, encoded in a DNA sequence, is "read out" to produce the corresponding messenger RNA (mRNA). Watching a single molecule read the genetic code. (A) A single molecule of DNA is between two beads, each of which is held in an optical trap. The right-hand bead is coated with RNA polymerase (RNAP) molecules which "walk" along the DNA and synthesize mRNA. (B) Displacement of the bead versus time, showing the first signs of discrete steps. Step size is 3.4Å, the spacing between bases along the DNA double helix. SOURCE: Reprinted by permission from Springer: E.A. Abbondanzieri, W.J. Greenleaf, J.W. Shaevitz, R. Landick, and S.M. Block, 2005, Direct observation of base-pair stepping by RNA polymerase, *Nature* 438:460, copyright 2005.

The sequence of bases along DNA encodes two very different kinds of information, genes and their regulatory elements. In genes, the DNA sequence represents the amino acid sequence of a protein molecule that the cell will synthesize, and this mapping from DNA to amino acids is "the genetic code." The genetic code is almost universal across the entire tree of life, although there are important variations. One of the questions that has intrigued both physicists and biologists is whether the code(s) used were selected for some functional property, or whether it is just an accident of history. Since cells expend energy to minimize the errors in reading the code, it might be that evolution has selected codes that are more tolerant to errors, or perhaps even error correcting, as in the codes used in engineered communication and data transmission systems. The standard genetic code does have the property that the most common errors in translation lead to amino acids that have very similar physical properties, thus minimizing the impact of errors. Direct theoretical calculations indicate that only a tiny fraction of all possible codes could be more error tolerant in this sense.

In order to make use of encoded information, one needs the codebook. In living cells, the codebook for the genetic code is embodied in the tRNA molecules that physically connect the bases along mRNA to the corresponding amino acids. But these molecules themselves need to be synthesized by the cell, and there is a separate protein molecule (enzyme) that attaches each of the 20 amino acids to the corresponding tRNAs. Thus, while genetic information is localized in DNA, the codebook is distributed across this family of proteins. Analysis of the sequence of chemical reactions catalyzed by these enzymes provided the first convincing evidence for proofreading.

Regulatory Sequences

Beyond the DNA sequences that encode the amino acid sequences of proteins, there are sequences that carry information about how the synthesis of proteins is regulated. This reading out of genetically encoded information is referred to as "expression" of the corresponding genes. In bacteria, regulatory sequences often are physically close to the genes that they control. Proteins called transcription factors (TFs) bind to these sequences and can interact directly with RNAP or act through an intermediate protein. The geometry can be such that TF binding gets in the way of RNAP binding at the site where transcription begins, in which case the transcription factor inhibits or represses the expression of the gene. Alternatively, TF binding can stabilize the binding of RNAP to the start site, enhancing or activating transcription.

Even in a bacterium such as *Escherichia coli*, there are more than 200 different transcription factor proteins; in human cells, there are more than 1,000. Classical studies focused on the interaction of one transcription factor with one gene. But early work made clear that a single TF could bind to many different short sequences of DNA, either to provide more complex regulation of a single gene or parallel regulation of multiple genes. From the earliest data, there was an effort to build models of TF binding to DNA based on equilibrium statistical mechanics, and also effective statistical mechanics for the ensemble of target sequences. These theoretical ideas linked gene regulation, protein/DNA interactions, and the evolution of sequence variation in ways that continue to influence our thinking.

In the 21st century, it became possible to explore regulatory sequences on a much larger scale, exploiting tools from genetic engineering and concepts from physics. An example of such an experiment is to insert the gene for a fluorescent protein (Chapter 6) into *Escherichia coli* and place a known regulatory sequence close enough to this that it will control expression. As a result, when an environmental signal is read by the regulatory sequence leading to the expression of the gene under its control, this will trigger the synthesis of the fluorescent protein, so that the bacteria literally will glow in response to light. But rather than inserting a single regulatory sequence into all the bacteria in a population, one can insert tens of thousands of different regulatory sequences into different bacteria. The binding of the TF will be stronger or weaker depending on the sequence, and this will change the amount of the fluorescent protein that is synthesized and hence the brightness of the fluorescence. Measuring the fluorescence from each individual bacterium and sequencing makes it possible to measure quite precisely how much information each individual letter along the DNA carries about the fluorescence signal.

The "information footprint" in Figure 2.2 corresponds directly with what is known about the contacts between these proteins and DNA from their molecular structures. RNAP and the transcription factor (here called CRP) bind at particular positions along the DNA, measured from the point at which RNAP starts to transcribe messenger RNA (position 0). Strikingly, there are positions where the DNA base carries zero information, and thus can be chosen at random without changing the level of transcription. Although the contribution of any single base is small, combinations of bases carry much more information. This can be captured by a model in which every base makes an additive contribution to two numbers, one of which defines the interaction of CRP with DNA and one of which defines the interaction of RNAP with DNA. This is exactly what is expected from the equilibrium statistical mechanics models—these two numbers are the free energy

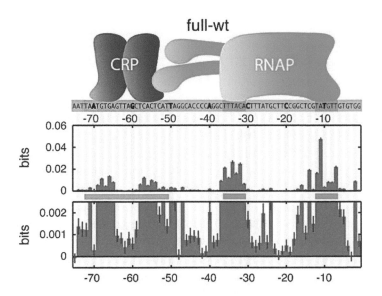

FIGURE 2.2 How much information, in bits, do DNA sequences provide about gene expression? A transcription factor protein (CRP) binds to a specific site along DNA where it can interact with the RNA polymerase (RNAP) and influence transcription and hence the level of gene expression. As described in the text, experiments probe many DNA sequences and estimate the information that each letter (A, C, T, G) provides about the resulting level of gene expression, shown in the bottom two panels. This "information footprint" mirrors the known structural contacts between the proteins and DNA. Zooming in shows that the information is really zero at sites where there is no contact. SOURCE: J.B. Kinney, A. Murugan, C.G. Callan, Jr., and E.C. Cox, 2010, Using deep sequencing to characterize the biophysical mechanism of a transcriptional regulatory sequence, *Proceedings of the National Academy of Sciences U.S.A.* 107:9158, Creative Commons License CC BY-NC-ND 4.0.

for binding of CRP and RNAP to DNA. But statistical mechanics predicts that the numbers combine in a very specific way, with one more parameter describing the interaction energy between the two proteins when they are both bound. This combination captures all the available information, and the inferred value of the CRP/RNAP interaction agrees with independent measurements.

The information footprinting experiments provide strong support for relatively simple statistical physics models of protein/DNA interaction and the regulation of transcription, at least in bacteria. These models predict a wide range of phenomena: the dependence of TF effectiveness on DNA sequence, the dependence of gene expression levels on TF concentration, and more. With the advent of techniques for measuring gene expression in single cells, many of these predictions have been tested in quantitative detail, providing the kind of dialogue between experiments and theory that is the hallmark of other areas of physics.

Sometimes the regulatory sequence is more distant from the gene and in order for it to influence the expression of the gene, it has to contact the distant RNA polymerase, thereby looping out the intervening DNA. Since DNA is a polymer, this looping costs (free) energy, which can be incorporated in statistical physics models of gene expression. Such models have been successful in explaining how the amount of gene expression depends on the distance between the regulatory sites and the gene. Because DNA is a double helix, bending is not enough to bring two sites into contact; the molecule must also twist. The twist entails an additional energy cost, and this depends periodically on the distance between sites, with the periodicity matching the periodicity of the helix itself. One can engineer the genome to vary distances with single base pair resolution, and it is dramatic to see how these sub-nanometer rearrangements are propagated, quantitatively, to order of magnitude variations in the macroscopic level of gene expression. It now is possible to measure directly the looping of DNA in response to transcription factor binding, although it remains challenging to connect these single molecule experiments to the macroscopic behavior of gene expression, quantitatively.

The relatively simple picture of these processes in bacteria, where a small number of transcription factors regulate the expression of nearby genes, stands in contrast to what has been learned about the control of gene expression in higher organisms. In these cases, a single gene can be regulated by transcription factor binding to dozens of enhancer sites, spread over a length of DNA covering tens of thousands of bases. Recent work tracks these individual molecular components in several systems, showing that activation of transcription requires an occupied enhancer site to come into close proximity to the transcriptional start site, but super-resolution microscopy suggests that proximity is not contact. Independent measurements show that there is condensation of a droplet of protein molecules around active transcription sites, and this could mediate the interactions that are thought to occur by direct contact in bacteria. Interestingly, similar droplets now have been observed in bacteria. All of these results point toward a view in which the regulation of gene expression is more of a collective effect, emerging from interactions among a large number of individual protein molecules. This connects to other examples of protein condensation (Chapter 3) and raises theoretical questions about how the specificity of individual molecular interactions is preserved in the presence of these collective effects.

CRISPR

Finally, bacteria can edit their own DNA, storing memories of their experiences. The edited sequences are CRISPR—clustered regularly interspaced short palindromic repeats. These sequences include segments that are extracted from bacteriophages that infect the bacterium, and are inserted into the bacterial DNA by the enzyme Cas9 (see Figure 2.3). The CRISPR/Cas9 apparatus has been adapted into a tool for modifying the genomes of other organisms, and this has had a revolutionary impact on biology, on many areas of biological physics, and on prospects for gene therapy. This work was recognized with a Nobel Prize in 2020. In the life of the bacterium, however, CRISPR/Cas9 serves as a kind of immune system, carrying a memory of previous infection and allowing more rapid recognition and response to future infections. Our immune system also carries a memory, but humans do not pass this information on to their offspring. Theorists in the biological physics community have tried to understand how the different dynamics of environmental challenges drive the emergence of these different strategies in different classes of organisms.

Bacteria have limited resources that can be devoted to the CRISPR system. If the cell tries to keep a memory of too many different phages in the environment, then there simply will not be enough copies of the Cas proteins to do the job. On the other hand, in an environment with diverse challenges, keeping track of too few phages can leave the cell vulnerable. With reasonable assumptions, one can calculate the probability of survival as a function of the number of stored memories and the number of different phages in the environment, and the result is that the optimal number of memories is close to the size of the CRISPR systems in real cells. This is just the start of an ambitious program to understand features of this remarkable system as responses to the environment, quantitatively.

Perspective

Information encoded in DNA has been the source of deep questions about the physics of life for nearly 70 years. This section highlights some of the major results and points to several frontiers where progress is expected in the coming decade. A

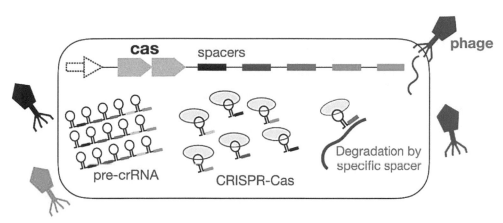

FIGURE 2.3 Bacteria can edit their own DNA into clustered, regularly interspaced, short palindromic repeats (CRISPR), which include segments that are extracted from bacteriophages that infect the bacterium. A bacterium with CRISPR machinery encounters a diverse set of phages (colors). The CRISPR-Cas locus is transcribed and then processed to bind Cas proteins (gray ovals) with distinct spacers (colors), thus producing CRISPR-Cas complexes. The complex with a spacer that is specific to the injected phage DNA (same color) can degrade the viral material and protect the bacterium from infection. SOURCE: S. Bradde, A. Nourmohammad, S. Goyal, and V. Balasubramanian, 2020, The size of the immune repertoire of bacteria, *Proceedings of the National Academy of Sciences U.S.A.* 117:5144, Creative Commons License CC BY-NC-ND 4.0.

common theme is a shift from thinking about isolated bits of information to seeing this information in context, particularly important since DNA forms the substrate on which evolution takes place (Chapter 4). This context may be provided by the whole population of tRNA molecules that embody the genetic codebook, by the large number of molecules that are involved in controlling the expression of even single genes in higher organisms, or by the ensemble of phages that challenge the health of individual bacteria. In each of these cases and more, it is reasonable to expect progress in the coming decade both from the introduction of new experimental methods and from the formulation of sharper theoretical questions about how these systems function.

INFORMATION IN MOLECULAR CONCENTRATIONS

Throughout the living world, information crucial to life's functions is represented or encoded in the concentrations of specific molecules. This happens on a macroscopic scale, as when insects find their mates by following the odor of pheromones, or when cells in our body secrete and respond to hormones that circulate in the blood. It also happens on a microscopic scale as cells control their internal states by changing the concentrations of signaling molecules. While such molecular signaling is ubiquitous in the living world, it is less connected to the traditional subject matter of physics, where information most often is carried by electrical or optical signals. On the theoretical side, this raises new physics problems: Which physical principles of information transmission are universal, and what new principles emerge for the case of molecular signaling? On the experimental side, methods from physics have combined with those from chemistry and biology to provide an unprecedented ability to observe and manipulate molecular signals in living cells. These foundational developments set the stage for today's excitement in this crucial area of biological physics.

Signaling, Growth, and Division

In describing the ability of the visual system to count single photons, and the ability of bacteria to count the molecules arriving at their surface (Chapter 1), we have encountered signaling via changing molecular concentrations as intermediate steps along the path from input to output (see Figures 1.11 and 1.12). In vision, the input is light, but the absorption of one photon triggers a structural change in one rhodopsin molecule, and it is useful to think of the cell as having to "smell" this one molecule out of 1 billion others. The output of the cell is an electrical voltage or current across the membrane, but this current flows through channels whose state is controlled by the concentration of a small signaling molecule, cGMP. In this sense, light intensity is represented internally by the concentration of cGMP, and this concentration in turn is the result of a cascade of molecular events. Similarly, in bacterial chemotaxis there is a cascade from the cell surface receptors to the phosphorylation of the CheY protein, which ultimately controls the flagellar motor much as cGMP controls ion channels. In both cases, prolonged inputs generate adaptation, which reduces the response, but this requires the accumulation of another internal molecular signal (Chapter 4). Amplification via molecular cascades is widespread, and phosphorylation is both a common signal and a central component of many cascades. Proteins that act as enzymes to phosphorylate other proteins are called kinases, and there are kinase kinases as well as kinase kinase kinases, testimony to the ubiquity of these cascades.

As cells grow and divide, their size, structure, and timing are encoded in the concentrations of several different molecules. One example are the Min proteins discussed in Chapter 1. The discovery of molecules that control the cycle of cell division in eukaryotic cells was recognized by a Nobel Prize in 2001, but controversies remain about the connection of these core molecules to the phenomenology of growth. It has been proposed that there are master molecules which trigger cell division at a threshold concentration, so that the accumulation or dilution of these molecules is analogous to the accumulation of evidence for a decision. Candidates for these molecules are not universal, and it is not clear whether this is the right picture. Many groups in the biological physics community have gone back to the macroscopic features of growth and division in bacteria, discovering for example that fluctuations in growth rate can be inherited and correlations maintained across a dozen generations; the molecules whose concentrations represent this information have not been identified.

From Transcription Factors to Genetic Networks

An important class of molecules that convey information through their concentrations are transcription factors. As explained in relation to Figure 2.2, these proteins bind to particular DNA sequences and regulate the synthesis of mRNA from nearby genes. At the start of the 21st century, the study of information flow in transcriptional control was revolutionized by the introduction of new experimental methods. The discovery and engineering of green fluorescent proteins, described in Chapter 6, opened a path to genetically engineering organisms so that TFs could control the synthesis of fluorescent proteins, or so that the TFs themselves could be fused with fluorescent proteins. In addition, DNA sequences of interesting genes could be modified so that the resulting mRNA molecules attract other fluorescent proteins engineered into the genome, providing a readout of transcription at its start. Improved optical microscopes then allow visualization of these molecules in the living cell. Even classical ideas about tagging mRNA molecules with fluorescent labels could be pushed, with better microscopes, to the point of counting each individual molecule, one by one.

Measurement with fluorescent proteins in single cells allowed the first measurements of noise in transcriptional control, separating intrinsic noise in the control mechanism from extrinsic variations in cellular conditions. This led to a flurry of theoretical work, trying both to understand the precise physical origins of this noise and to explore the implications of noise for information transmission. A second generation of experiments, such as that in Figure 2.4, follows the dynamics of multiple proteins to resolve the direction of information flow from the transcription factor to its target gene. Careful measurements of correlations in the fluctuations of these protein concentrations provide a detailed test for models of the regulatory interactions, which in turn provide a foundation for engineering new genetic circuits (Chapter 7).

Transcription factors are proteins, and their expression is regulated by other TFs, resulting in networks of genetic control. There is considerable interest in understanding information flow through these networks, and the possibility that they generate emergent, collective states. Interest comes both from the biological physics community (e.g., in the spirit of Chapter 3) and from the biology community (Chapter 6). Because fluorescent proteins have broad absorption and emission profiles, however, it is difficult to adapt these methods to studying many genes simultaneously in live cells. As a result, many investigators have looked to counting mRNA molecules rather than proteins. One approach uses microfluidic methods developed in the biological physics community to manipulate large numbers of single cells, ultimately breaking them open and identifying all of their mRNA molecules using biochemical sequencing methods (Chapter 6). These single cell sequencing methods have swept through the biology community over the last decade. A different strategy, based again on optical methods, is to fix the cell so that one can look for longer times, perhaps at many genes in sequence. An example of this is in Figure 2.5, where bacteria are fixed and then labeled by fluorescently tagged DNA molecules that are complementary to the mRNA molecules of different genes. With modern microscopy methods, one can count these molecules, one by one, and do this for each of many genes to build up a snapshot of the state of a single cell.

The comparison of Figures 2.4 and 2.5 highlights an important point about experimental methods. In one case (see Figure 2.4), genetic engineering turns the concentration of a handful of proteins into fluorescence signals with different colors, and this allows measuring molecular concentrations over time in live cells. While there are many technical issues, in favorable cases this really provides a readout of the information at several nodes of a genetic network, in a live, functioning cell. A basic limitation of these methods is that the absorption and emission spectra of fluorescent proteins are broad, so that if we try to monitor too many nodes simultaneously the signals will be corrupted by crosstalk. In the other case (see Figure 2.5), the action is stopped, providing time for multiple measurements that eventually probe the concentrations of several thousand different molecular species, with single molecule precision. While these observations on fixed cells cannot provide dynamical information, we can sample the distribution of signals across large populations of cells. Both approaches still are developing rapidly, and we expect to see deeper insights emerging from the analysis of these measurements in the coming years.

Positional Information

It has long been appreciated that developing embryos are a system in which ideas about information and the control of gene expression come together. Embryos begin as one fertilized egg cell, and then go through many rounds of cell division. For several cycles, the cells are functionally equivalent, and could in principle grow to become any part of the organism's body. But at some time in the course of development, cells need to adopt distinct identities. Cellular identities are defined, in part, by their patterns of gene expression, and these need to be matched to their spatial location in the body. In this sense, cells need to "know" their positions.

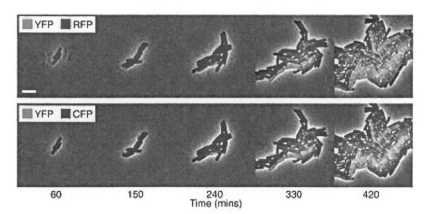

FIGURE 2.4 Measurements of correlations and dynamics in a synthetic gene circuit provide detailed tests for models of regulatory interactions. Time-lapse images of a bacterial colony growing from a single cell, in the red/green (*top*) and blue/green (*bottom*) channels. As the green protein rises, the red protein declines, while green and blue are weakly correlated. This pattern of temporal correlation reflects the underlying architecture of the circuit. SOURCE: Reprinted by permission from Springer: M.J. Dunlop, R.S. Cox III, J.H. Levine, R.M. Murray, and M.B. Elowitz, 2008, Regulatory activity revealed by dynamic correlations in gene expression noise, *Nature Genetics* 40:1493, copyright 2008.

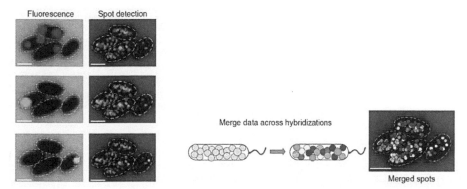

FIGURE 2.5 There is considerable interest from both the biological physics community and the biology community in understanding information flow through networks that control transcription of genes. Toward this end, methods have been developed for counting individual transcripts (i.e., messenger RNA [mRNA] molecules from multiple genes in single bacteria). Short DNA molecules complementary to mRNA sequences are labeled in different colors and applied sequentially to the same sample. In each cycle, a new set of secondary readout probes are introduced. Raw fluorescence data is shown on the right, and the detected local spot maxima are shown in the spot detection image. Merged spots for many genes are shown in shuffled colors. SOURCE: D. Dar, N. Dar, L. Cai, and D.K. Newman, 2021, In situ single-cell activities of microbial populations revealed by spatial transcriptomics, *bioRxiv* 2021.02.24.432972, Creative Commons CC BY-NC-ND 4.0 international license.

Cells in a developing embryo could acquire positional information in many ways. One common theme is that this information is carried by the concentrations of specific molecules—morphogens—as schematized in Figure 2.6. But how do the necessary spatial variations in molecular concentration arise? One possibility that we have encountered in Chapter 1 is that patterns can arise out of instabilities in the dynamics of some underlying biochemical or genetic network, operating in an otherwise homogeneous embryo. Another possibility is that the homogeneity or symmetry is broken not spontaneously but by some external event, perhaps at the moment of fertilization or even in the construction of the egg itself. These two different scenarios lead to different physics problems, and real living systems almost certainly combine elements of both.

A dramatic step forward was the identification of these morphogens in particular systems. In the fruit fly, which had been a favorite experimental testing

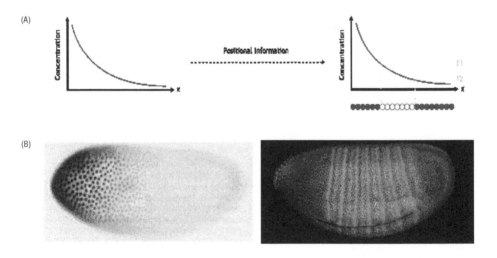

FIGURE 2.6 How do cells "know" where they are in a developing embryo? (A) The concentration of a single "morphogen" molecule carries information about the position *x* of cells along the length of the embryo. This positional information is transformed and finally interpreted, in this simple case to make three different cell types that form segments of the body. (B) Instantiation of these ideas in the fruit fly embryo. *Left*, an embryo stained with antibodies to the protein Bicoid, a transcription factor that accumulates in the nuclei. Darkness of the stain is proportional to concentration, which is larger at the end of the embryo that will become the head. The embryo is ~1/2 mm in length. *Right*, embryo stained for two of the "pair rule" genes; striped patterns parallel the segments of the larval fly that crawls away once the egg hatches. SOURCES: (A) Republished with permission of The Company of Biologists Ltd. from G. Tkacik and T. Gregor, 2021, The many bits of positional information, *Development* 148:dev176065; permission conveyed through Copyright Clearance Center, Inc. (B) *Left*, Reprinted from W. Driever and C. Nüsslein-Vollhard, 1988, A gradient of *bicoid* protein in *Drosophila embryos*, *Cell* 54:83, copyright 1988, with permission from Elsevier. *Right*, courtesy of E.F. Wieschaus, Princeton University.

ground for ideas about genes since the early years of the 20th century, there had long been indications that mutations in a single gene could cause macroscopic changes in the pattern of development, leading to the deletion, rearrangement, or duplication of body parts. In the 1970s, there was a systematic search for all the genes that control pattern formation in the fly embryo, and this led to the startling conclusion that there are only about 100 such molecules, and even fewer involved in patterning the long axis into body segments. Furthermore, these molecules are organized into a layered network, starting with molecules that are placed in the egg by the mother. These primary maternal morphogens control the expression of a collection of gap genes, so named because mutations in these genes lead to gaps in the body plan, and then the gap genes control the expression of pair rule genes that form striped patterns, an approximate blueprint for the segments of the fully developed organism.

The demonstration that just a small number of molecules were responsible for pattern formation in the fly embryo came at an opportune time. In molecular biology, new tools were making it possible to sequence these genes, to synthesize the proteins that they code for in the laboratory, and thus to make probes that could measure their concentration in the embryo. Almost all of the relevant molecules are transcription factors, so these earliest steps in pattern formation involve only the control of gene expression, rather than changes in cell structure or overall geometry of the embryo. These macroscopic, mechanical manifestations of pattern formation come later. The whole process, however, is quite rapid: The egg hatches in 24 hours, and the striped patterns of pair rule gene expression are visible just 3 hours after the egg is laid.

At the start of the 21st century, several groups had the idea that the fly embryo could be turned into a physics laboratory. These experimental efforts were launched in a background of theoretical work on noise and information transmission in molecular networks, and ideas about the robustness of these networks to parameter variation. The new generation of physics experiments on the fly embryo revealed the system to be much more precise than anyone had anticipated. Concentrations of just a handful of the proteins encoded by the gap genes provide enough information to specify the position of cells to 1 percent accuracy. Primary morphogens have absolute concentrations that are reproducible from embryo to embryo with $\sim$8 percent accuracy. This precision occurs despite the fact that relevant molecules are present at low concentrations, as with other transcription factors, and these notions of precision could be formalized in the language of statistical physics and information theory.

Results on the precision of positional information processing in the fly embryo have stimulated the exploration of many theoretical questions. What are the physical limits to information transmission through genetic networks given the limited number of molecules and low concentrations involved? What strategies can cells use to reach these limits, allowing the construction of more complex and reproducible body plans? Is there something special about the fly that allows all these events to happen so reliably and so quickly? Alternatively, what new possibilities emerge for organisms that use a longer time scale for development? Approaches to these theoretical questions are connecting quite abstract considerations to the quantitative details of experiments on particular systems, including the fly embryo. It is especially interesting that ideas about information flow in these genetic and biochemical networks have parallels with ideas about information flow in the brain.

Perspective

Physicists are accustomed to the idea that information can be carried by electrical currents and by light. More abstractly, correlations in the fluctuations of any field can be thought of as carrying information across space and time. But living systems' use of changing molecular concentrations to convey information poses new challenges. Basic physical limits to this mode of communication have been known for decades, but there still are new discoveries being made as the community generalizes these ideas to contexts more relevant to real cells. In some systems, foundational work in biology has provided a nearly complete list of the relevant molecules, so there is a closed system within which to ask about the representation of information and the physical principles that govern life's choice of this representation. In other cases, the set of relevant molecules expands as experiments probe more widely, and in the limit one can ask about information contained in the patterns of expression across the entire genome. These developments hint at a more collective view of signaling and information flow, connecting to the ideas of Chapter 3.

INFORMATION IN THE BRAIN

When we run our fingers across a textured surface, receptor cells in our fingertips convert the pressure on our skin into electrical currents that flow across the membrane of these cells, much as happens in other sensing systems (Chapter 1). But it is a long way from our fingertips to our brain, or even to the spinal cord. For this long distance communication of information, the continuously varying currents are converted into discrete electrical pulses, called action potentials or spikes. Spikes can propagate without decaying or changing shape, at an essentially constant speed, until they reach the synapse where one cell connects to another. This conversion to spikes is essentially universal, happening in all our senses, and in almost all organisms. What humans see and hear, smell and feel thus are not light and sound or odors and textures; instead, what is sensed are the patterns of action potentials over time arriving from our sensory neurons. When we move, the brain sends action potentials in the other direction, along the motor nerve cells out to the muscles. Spikes also are what neurons inside the brain use in sending signals to one another. Patterns of action potentials are the language of the brain, and all the information that the brain uses—thoughts and perceptions, memories and actions—is at some point encoded in spikes.

Action Potentials and Ion Channels

The uncovering of the mechanisms by which nerve cells generate action potentials is another great success story in the interaction between physics and biology. There was a long path from quantitative observations on the relatively macroscopic electrical dynamics of single neurons to a precise mathematical and physical description of the underlying molecular events, including the structures of the relevant molecules, the ion channel proteins that are embedded in the cell membrane. Along the way were the very first direct measurements of dynamics in single molecules, the observations of electrical current flow through these channels. This work in the biological physics community was contemporary with the first measurements of electrical current flow through nanoscale devices in the condensed matter physics community. For decades, this research program was in the reductionist spirit, finding the elementary building blocks that generate the electrical behaviors of cells. We describe this largely complete program here because it provides a model for understanding the phenomena of life, quantitatively, with our understanding summarized in mathematical terms, as expected in physics. These discoveries also provide a foundation for many currently exciting questions in the physics of living systems, from understanding the energetic costs for reliable information processing to the way in which life explores the parameter space of microscopic mechanisms (Chapter 4).

All cells are bounded by a membrane that separates inside from outside. The membrane by itself is a good electrical insulator, allowing very little current flow. Interesting electrical dynamics in cells happen because there are specific protein molecules inserted into the membrane. Currents that do flow are carried by ions, not by electrons as in familiar electronic devices. All cells have pumps to maintain concentration differences for these ions between the inside and outside of the cell, effectively acting as batteries, and the inside of a cell typically is 0.03–0.05 volts negative relative to the outside.

The action potential is a brief pulse, lasting a few thousandths of a second, during which the voltage difference between the inside and outside of the cell changes by roughly one tenth of a volt. The first major clue to the dynamics of the action potential was that during the spike, the resistance of the membrane is reduced, or equivalently the conductance is increased (see Figure 2.7A). A series of beautiful experiments and mathematical analyses showed that these changes in conductance could be dissected into one component specific to the flow of sodium ions, and one specific to potassium ions. These conductances have their own dynamics in response to the changing voltage across the cell membrane, and the coupled dynamics of voltage and conductances have a remarkable mathematical structure—in a long cylinder shaped cell, as with the axons that extend outward from most neurons, the solutions to these equations converge to stereotyped pulses that propagate at a constant speed, and this provides a nearly perfect quantitative description of the action potential.

A natural interpretation of the equations describing the sodium and potassium conductances is that the membrane has "channels" that allow these ions to pass, and that the dynamics of current flow are the dynamics of these channels opening and closing. If this is correct, then small patches of membrane will have small numbers of these channel molecules, and since single molecules behave randomly, the resulting current flow will have measurable randomness or noise, and it does.

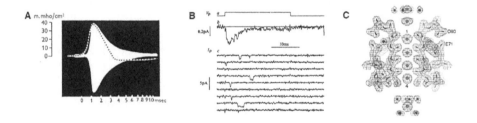

FIGURE 2.7 The macroscopic electrical dynamics of cells results from the underlying dynamics of specific protein molecules—ion channels—that are embedded in the cell membrane. (A) The action potential consists of a voltage pulse, shown as a dashed line (without units). These experiments imposed very high frequency voltage oscillations on top of the action potential, and the resulting excursions in current (white) measure the conductance of membrane, which increases as the action potential rises and then decays slowly. Note the nonlinear time axis. (B) A small step in voltage across the membrane (a) produces, on average, a proportionally small current (b). Responses to individual pulses (c) are composed of square pulses of current, corresponding to the opening of single ion channels. With these small voltage changes, channel opening is rare. The average current traces the probability that a channel is open as a function of time. (C) The structure of a potassium channel, from X-ray crystallography. Slice through the density map of the molecule, cutting through the entrance to the conduction pore, with potassium ions (green spheres) and water molecules (red spheres) visible. SOURCES: (A) K.S. Cole and H.J. Curtis, 1939, Electric impedance of the squid giant axon during activity, *Journal of General Physiology* 22:649. (B) Reprinted by permission from Springer: F.J. Sigworth and E. Neher, 1980, Single Na+ channel currents observed in cultured rat muscle cells, *Nature* 287:447, copyright 1980. (C) Reprinted by permission from Springer: Y. Zhou, J.H. Morais-Cabral, A. Kaufman, and R. MacKinnon, 2001, Chemistry of ion coordination and hydration revealed by a K+ channel-Fab complex at 2.0 Å resolution, *Nature* 414:43, copyright 2001.

In sufficiently small patches of the membrane, with sufficiently sensitive amplifiers, one can see individual channel molecules opening and closing (see Figure 2.8), and reconstruct the original macroscopic description of current flow by averaging over these random molecular events. Looking more closely at the original equations, the dynamics of the channels are described by multiple "gates" that open and close. For gates to open and close in response to voltage, basic statistical mechanical principles require that opening and closing is associated with structural changes in the channel that move charge across the membrane; this small gating charge movement was eventually measured and agrees quantitatively with theory. Once it was possible to isolate the channel proteins, a major effort to determine their structure by X-ray diffraction resulted in a clear view of how ions pass through the channel (see Figure 2.7C).

It took nearly 30 years to go from the mathematical description of conductances and voltages to the observation of current flow through single channels, and another 20 years to reveal the structure of the channels. As always, between these milestones many important results accumulated. An interesting feature of the history is that decades of work were driven by the search for the physical embodiment of individual terms in the equations—channels, gates, and more—taking the mathematical description literally. Although the results of this search have become part of the mainstream of biology, this style of exploration is very much grounded in physics. Several of the milestones in our understanding of the action potential were recognized with Nobel Prizes: the mathematical description of membrane conductances (1963), the observation of single channel currents (1991), and the structure of ion channels (2003).

The reductionist program that brought us from action potentials to ion channels is largely complete, though there are open questions about the physics of ion transport through the channel. But while the squid axon that was the subject of early studies has dynamics dominated by two types of channels, it now is known that the genomes of many animals, including humans, encode over one hundred different kinds of channels, and a typical neuron in the brain might use seven of these different types. Our precise understanding of individual channel dynamics sets the stage for asking questions at the next scale of organization. In particular, how do cells choose the number of each type of channel to insert into the membrane? This provides an accessible example of a larger question about how organisms navigate the large parameter space accessible to them, as described in Chapter 4. In a different direction, electrical signaling through action potentials, or through smaller amplitude graded changes in voltage, provide concrete examples where we can understand the energy costs of coding and computation in the nervous system. Our understanding of the inherently stochastic molecular dynamics of the channel molecules also means we can characterize the reliability or fidelity of information transmission and processing, and relate these measures of performance to the energy costs. Can we connect these concrete ideas to more abstract ideas from non-equilibrium statistical physics about dissipation and noise? Are there lessons for engineering, in the choice of analog versus digital computation? Finally, while it is tempting to think that the details of molecular dynamics are erased once we mark the occurrence of relatively macroscopic action potentials, there are hints that some features of these dynamics, which determine the sharpness of the threshold for triggering spikes, may influence even the collective dynamics in large networks of neurons (Chapter 3).

Coding in Single Neurons

Beyond the mechanisms of action potential generation, one can ask how these signals represent information. The earliest observations on spikes in single sensory neurons showed that constant sensory input, such a steady light shining into the eye or a constant weight on the mechanical sensors in muscles, resulted in spikes generated at a rate that increased with the strength of the stimulus. This idea of "rate coding" has had considerable influence, moving attention away from the discrete nature of action potentials. But under natural conditions, sensory inputs are seldom constant, and there is considerable evidence that many sensory neurons generate roughly one spike before the inputs change. Confirming that these spikes are not just noise, experiments have demonstrated the reproducibility of the response across multiple experiences of the same sensory inputs, down to a time resolution of a few thousandths of a second. With care, this reproducibility is visible in the responses of neurons much deeper in the brain. Moving away from the sensory inputs, however, neurons receive input from many places; a typical cell deep in the cortex receives thousands of inputs from other neurons, and it becomes more difficult to separate signal and noise convincingly. Approaching the brain's output, one can once again see the correlation of single spikes and patterns of spikes with particular trajectories of muscle activity.

The biological physics community has been keenly interested in the more abstract properties of the code by which sensory signals and motor commands are represented by sequences of action potentials. Information can be conveyed only if the sequences of spikes vary. This variability can be characterized by an entropy, the same quantity that arises in statistical mechanics, and the information (in bits) that spikes represent cannot be larger than this entropy. In several systems there is evidence that the information carried by spikes comes close to the physical limit set by their entropy, down to millisecond time resolution; that this coding efficiency is higher for sensory signals with statistics more like those that occur in the natural environment; and that this efficiency is achieved through adaptation processes that match neural coding strategies to the statistics of their inputs (see also Chapter 4). There are more ambitious efforts, with roots in the 1960s and continuing until today, to derive these strategies directly from optimization principles, maximizing information transmission subject to physical constraints. As an example, these theories predict that on a dark night the neurons in the retina average over space and time to combat the noise of random photon arrivals, while as light levels increase they respond to more rapid spatial and temporal variations in image contrast, to remove redundancy in the signals transmitted to the brain. These predictions are qualitatively correct, and there are similar ideas about the nature of filtering in the auditory system. There is a continuing effort to push these theories into a regime that includes the full dynamics of real neurons, and to connect abstract models of coding to the known dynamics of ion channels.

The abstract measure of information in bits may seem mismatched to the concrete tasks that organisms need to accomplish in order to survive. Which bits are relevant for life? There have been efforts to define relevance by reference to other signals, or to the animal's behavioral output. Alternatively, many tasks require organisms to make predictions, and perhaps it this predictive information which is almost always relevant. These ideas have deep connections to many problems in statistical physics and dynamical systems, and have even led to experiments that estimate the amount of information that small populations of neurons carry about the future of their sensory inputs.

Coding in Populations of Neurons

Although the spike sequences from individual cells carry surprisingly large amounts of information, it is clear that most functions of the brain require the coordinated activity of many neurons. This coordination can be so extensive that currents flowing into and out of neurons add up to generate macroscopic current flows, leading to voltage differences that are measurable even outside the skull—the electroencephalogram, or EEG. The coordinated current flows also generate magnetic fields that are detectable outside the skull, and the analysis of these signals has been called magnetoencephalography, or MEG. Modern MEG often is done with arrays of superconducting quantum interference devices (SQUIDs), connecting the frontiers of precision low-temperature physics techniques to the study of brains and minds. While these methods certainly have limitations, it remains striking to see how the electrical activity of the brain is correlated with our internal mental life. It is a classical demonstration, for example, that humans can change the structure of their own EEG simply by thinking about different things, even in the absence of immediate sensory inputs or motor outputs.

The representation of information by populations of neurons connects naturally to the search for collective dynamical behavior of these populations, as described in Chapter 3. One way in which collective phenomena emerge, across many physical systems, is through symmetry and the breaking of symmetry. In the context of a population or network of neurons, symmetry would be manifest as a range of possible, reasonably stable states for the network that are all equivalent to one another; the breaking of symmetry occurs when the network chooses one of these states. If the possible states, or attractors, form a discrete set, then each of these states—which correspond to different patterns of activity across the network—can represent an object, a person, or a category of things. The set of states, taken together, represent a list of things or people that the brain can identify, a list that must have been learned (Chapter 4). The network may be driven into one of these states, breaking the symmetry, by an external stimulus, for example, when the brain recognizes an object from its appearance, or a person from the sound of her voice. But the network may break symmetry spontaneously, recalling the memory of an object or person from the most feeble of reminders. Models of neural networks with discrete symmetry breaking thus provide ideas for how the brain solves the problems of object recognition, memory, and more. Over the last two decades these seemingly abstract models have been connected, at increasing levels of detail, to experiments on real brains.

A more subtle possibility is that the set of attractors is continuous rather than discrete. Such states in a network of neurons could then represent the orientation of an object, the position of an animal in space, the direction in which an animal is looking or moving. It is known that neurons in particular brain regions provide animals with a "map" of the world in which they are moving; this discovery was recognized by a Nobel Prize in 2014. Some of the most successful models for the origin of these maps rely on the ideas of a continuous underlying symmetry in the network dynamics. As in other statistical physics problems, signatures of the interactions that allow the emergence of these collective states are found not in the mean behavior of the individual neurons but in the correlations among fluctuations in their behavior, and this has opened new directions for the analysis of these networks. Correlations also can limit or enhance the transmission of information by neurons, and this has been a central theme in experiments that probe the relation of activity in populations of neurons to the reliability of decisions.

The patterns of correlation among activity in large populations of neurons also provides hints of more general collective behaviors in the network, as discussed in Chapter 3. Theories of collective behavior make predictions for the structure of measurable correlations, but one can also turn the argument around and ask for the simplest collective states that are consistent with the measured correlations. These ideas are deeply grounded in statistical physics, and have connections to other examples of emergent behaviors of living systems, as discussed in Chapter 3. Throughout these developments, the example of neural networks has been a touchstone, providing some of the earliest and deepest examples of new physics.

A network of neurons that represents orientation or direction would be very special, because the stable states of activity must map to a closed circle or ring. This structure could be implicit in a very densely interconnected network, so that there is no ring of neurons even if there is a ring of attractors. It came as a great surprise, then, when it was discovered that neurons in a genuinely ring shaped structure deep in the fly's brain behave very much as predicted for the class of ring attractor networks (see Figure 2.8). This system has fewer than 100 neurons, so it is now feasible to map the connections among all these cells, and some of their sensory inputs, part of the larger efforts to map the full patterns of connectivity (the "connectome") of substantial pieces of the brain or even the whole brain (Chapters 3 and 6).

Perspective

The search for the physical principles underlying the representation and transmission of information in the brain led to the discovery of ion channels. This effort is important on its own, but also provides perhaps the most successful mathematical description of the interactions among many different species of protein. As such, ion channels in neurons constitute a laboratory for many more general questions about the physics of life. The exploration of information flow in the brain also has led to remarkable experimental methods for monitoring the

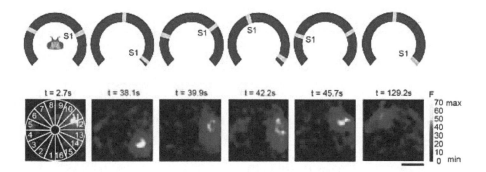

FIGURE 2.8 The stable states of activity maps to a ring, making a network of neurons that represent orientation or direction of special interest. An example of a ring of neural states exists in the brains of flies. The fly experiences virtual reality while electrical activity of neurons in the ellipsoid body is monitored through calcium-sensitive fluorescent proteins (Chapter 3). Snapshots of the virtual arena (*top*), and the fluorescence images (*bottom*) over the course of a minute. There is a localized "blob." With improved data and more quantitative analysis, one can reconstruct the fly's orientation from the pattern of activity, moment by moment, with high precision. SOURCE: Reprinted by permission from Springer: J.D. Seelig and V.V. Jayaraman, 2015, Neural dynamics for landmark orientation and angular path integration, *Nature* 521:186, copyright 2015.

electrical activity of neurons, and to a broad range of theoretical ideas about the way in which information is represented, and why. These developments highlight the gap between the reductionist view that ends with molecules and the functional view that ends with abstract descriptions of neural codes and network dynamics. Measurement of simultaneous activity in large numbers of neurons already has been transformational, and the growing data deluge—while presenting its own challenges for analysis—is poised in the coming years to sharply refine theoretical models, both for the representation of information and for collective behavior in neural networks (Chapter 3). The problem of connecting molecules to more phenomenological characterizations of function reappears in many contexts, and is likely to emerge as a theme in biological physics research over the coming decade.

COMMUNICATION AND LANGUAGE

Although it is tempting to "simplify" the phenomena of life by thinking about organisms in isolation, in fact many organisms communicate actively with others. As humans, we are especially fascinated by our own abilities to communicate through language, and search for analogs in other species such as songbirds. But even bacteria communicate, secreting and sensing a collection of molecules that allow individual bacteria to know about the density of other bacteria in the environment, and many insects find their mates across long distances though odor cues alone. Physicists have engaged with the full range of these problems, working to understand the mechanisms involved in each case but also searching for new unifying physical principles.

Bacterial Communication

Communities of bacteria communicate with one another and form beautiful patterns as they grow (see Figure 2.9). These patterns have fascinated physicists and biologists alike. As the 20th century drew to a close, these patterns attracted renewed interest from the biological physics community. This was driven in part by progress in understanding pattern formation in inanimate systems, such as snowflakes. Bacterial communication could be a byproduct of more basic processes: As bacteria grow and move through a medium, they consume resources, and other bacteria navigate the resulting concentration gradients. But it had been known since 1970 that single celled organisms can communicate more directly, and in parallel with the biological physics community's interest in pattern formation, microbiologists were exploring the molecular basis of this communication and realizing that it is widespread among bacteria. In particular, there are "quorum sensing" mechanisms in which individual cells excrete particular molecules, and

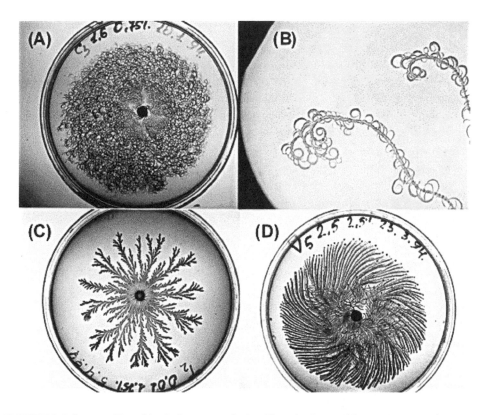

FIGURE 2.9 Communities of bacteria communicate with each other and form patterns as they grow. Here, colonies of the bacterium *Paenibacillus dendritiformis* exhibit multiple heritable patterns termed morphotypes. For scale, Petri dishes are 8 cm in diameter. (A) Chiral morphotype. (B) Zoomed-in view of (A). (C) Branching or tip-splitting morphotype. (D) Vortex morphotype. SOURCE: E. Ben-Jacob, I. Cohen, and D.L. Gutnick, 1998, Cooperative organization of bacterial colonies: From genotype to morphotype, *Annual Review of Microbiology* 52:779.

these accumulate, so that the concentration provides information about the local cell density.

Quorum sensing is vital because cells make decisions based on their local density. Luminescent bacteria only generate light when in a group large enough to make a difference for their symbiotic partners. Infectious bacteria only carry out their nefarious program when present in high enough density not to be overwhelmed by the immune system. More modestly, cells make decisions to attach to surfaces and grow communally only when there are enough compatriots in the neighborhood. As with modern work from the biological physics community on flocks and swarms (Chapter 3), the experimental frontier is to monitor each of the thousands of individuals in one of these communities, as in Figure 2.10.

With bacteria one can do more than track their movements as they engage in collective behaviors. By genetically engineering the cells one can induce them to report on the concentration of crucial internal signaling molecules that respond to the quorum sensing signals. This connects the problem of communication among cells to the problem of representing information through changing concentrations (Chapter 2) and even the problem of chemical sensing itself (Chapter 1), illustrating how one system embodies many different physics problems, each of which arises in many systems. Much is now known about the molecular mechanisms of information flow through the quorum sensing system, and there are even efforts to understand aspects of these mechanisms as solving the problem of maximizing information flow with limited molecular resources.

A different kind of communication among single celled organisms occurs in communities of *Spirostomum ambiguum* (see Figure 2.11). Living in aqueous environments, a single cell can contract its body by more than 50 percent on millisecond time scales, creating accelerations reaching 14 times that of gravity. These contractions create hydrodynamic flows that trigger other cells to contract, resulting in a wave that passes through the entire community. Cell contractions release toxins, and it is possible that the cells use this fast form of communication to repel predators or immobilize prey. The physics of flow and mechanical sensing is very different from diffusion and detection of molecules, and comparing these

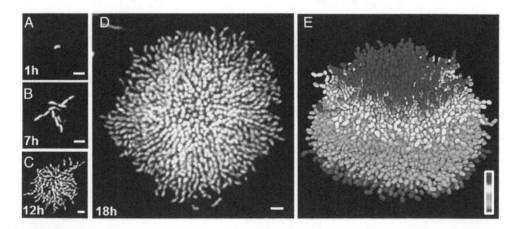

FIGURE 2.10 Bacteria communicate through quorum sensing to organize their growth in a community. (A) Cross-sectional images of the bottom cell layer at 1 hour, (B) 7 hours, (C) 12 hours, (D) 18 hours; scale bars: 3 μm. (E) Segmenting the three-dimensional cluster in (D) into 7,199 cells, color-coded according to z position (0–21 μm). SOURCE: J. Yan, A.G. Sharo, H.A. Stone, N.S. Wingreen, and B.L. Bassler, 2016, Vibrio cholerae biofilm growth program and architecture revealed by single-cell live imaging, *Proceedings of the National Academy of Sciences U.S.A.* 112:E5337, Creative Commons License CC BY-NC-ND 4.0.

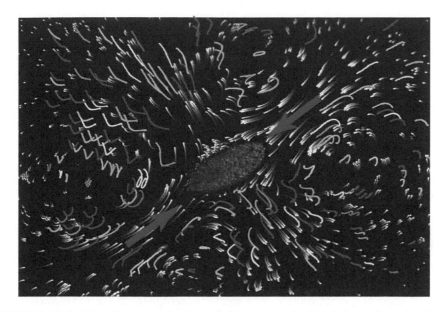

FIGURE 2.11 An interesting case of communication among single celled organisms can be seen in *Spirostomum ambiguum*. A single cell contracts its body by ~60 percent within milliseconds in order to create the flow pattern revealed by the trajectories of tracer particles (colored and white lines) in order to communicate with other cells nearby. SOURCE: Reprinted by permission from Springer: A.J.T.M. Mathijsen, J. Culver, M.S. Bhamla, and M. Prakash, 2019, Collective intercellular communication through ultra-fast hydrodynamic trigger waves, *Nature* 571:560, copyright 2019.

different modes of communication may provide novel ways of testing ideas about the organizational principles of communication in biological systems.

Searching for Sources

Chemical communication is central not only in the lives of bacteria, but also in the lives of insects. In particular, many species of insects find their mates by following the odor of pheromones over extraordinary distances, up to 30 miles in the case of silkworm moths. In some ways this is similar to the ability of bacteria to swim toward sources of nutrients by sensing the concentration of the relevant molecules along the way (chemotaxis, Chapter 1), but the physics in these two cases is very different. On the micron scale of bacterial life, molecules move through water by diffusion, and as a result, concentrations become smooth functions of position. For insects, molecules move through the air carried by the wind, which is turbulent.

In a turbulent flow, odors are carried by plumes. Standing in one place, the odors are intermittent, as plumes pass by, providing only very limited cues about the direction of the source. Classic experiments, however, show that moths "know" about this challenging physics problem, and actually fail to find the source of odors when turbulent flows are replaced by smooth (laminar) patterns of airflow in a wind tunnel. In the wild, insects searching for a source fly into the wind, occasionally casting sideways, and these sideways motion become more frequent when they lose track of the odor plumes (see Figure 2.12A).

The full problem of insect flight control involves synthesizing many different cues—from wind, odors, and vision—and connecting to the aerodynamics of flight itself. But the strategy behind this control might be simpler. In order to fly toward the source of an odor, the insect must make some inference about the location of that source from the limited data it has collected. Although "information" can appear as an abstraction, the minimum search time is related directly to the amount of information, in bits, that the insect has about the location of the source. In moving through the world, it might thus make sense to use a strategy that collects as much information as possible. By analogy with chemotaxis, this strategy has been called infotaxis. Infotaxis generates flight trajectories that are quite similar to the trajectories of real insects (see Figure 2.12B); the strategy works well enough that it can be used to guide robots in real world search problems, and infotaxis has revitalized the discussion of search and foraging strategies in animals more generally. More deeply, it is an example where the abstract goal of gaining information can be used in place of more detailed descriptions of underlying mechanisms, unifying our understanding of diverse animal behaviors.

Vocal Communication

Our human preoccupation with vocal communication leads to special interest in other animals that use the same modality. Frog calls, bird songs, and the mysterious sounds of whales and dolphins all attract our attention, and the attention of the physics community. Thus, the inner ear of frogs has been an important experimental testing ground for ideas about the mechanics of hearing (Chapter 1). Songbirds share with us the fact that vocalizations are learned, and thus provide an important test case for theories and experiments on learning (Chapter 4), as well as interesting examples for the neural coding of complex, naturalistic signals (Chapter 2). Bird song also provides an opportunity to explore the statistical structure of complex behaviors (Chapter 1) across time scales and species. Zebra finches sing individual songs that are stereotyped to millisecond precision, while these songs are linked together in more complex and variable structure on longer time scales. In other species, such as Bengalese finches, even individual songs are variable, and sequences of song elements are strongly non-Markovian.

And what of human language itself? Physicists have long been fascinated by the evidence for scale invariant correlations in written texts and speech. These observations are reminiscent of scale invariant behaviors near critical points, which

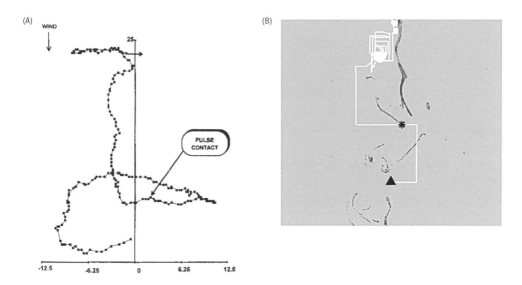

FIGURE 2.12 Chemical communication is central to the lives of insects, including silkworm moths that find their mates by following the odor of pheromones, and this concrete task can be reformulated as the abstract problem of gathering information. (A) Trajectory of a male moth *Cadra cautella* in response to a single pulse of the pheromone released by a potential mate. The moth is "casting" until coming into contact with the odor pulse, then turns upwind with a delay; after some time without contact it begins casting again. (B) Theoretical trajectory (white) of an organism that maximizes the information gathered about source location. Colors denote the concentration of the odor, measured in a naturalistic plume carried by turbulent flow; light blue is a very low concentration, darker blue through red correspond to higher and higher concentrations. The black triangle is the starting point, and the black asterisk is the point of first contact with the plume. SOURCES: (A) Reprinted by permission from Springer: A. Mafra-Neto and R.T. Cardé, 1994, Fine-scale structure of pheromone plumes modulates upwind orientation of flying moths, *Nature* 369:142, copyright 1994. (B) Reprinted by permission from Springer: M. Vergassola, E. Villermaux, and B.I. Shraiman, 2007, "Infotaxis" as a strategy for searching without gradients, *Nature* 445:406, copyright 2007.

have a deep meaning, but it has been controversial whether this connection is more than a metaphor. What it not controversial is that our linguistic interactions with machines—from automatic speech recognition to colloquial queries in search engines and machine translation of text from one language to another—have been revolutionized in just the past few years by new computational models. These models, including "long short-term memories" and transformers, are dynamical, recurrent versions of the deep neural network models that have had such a huge impact on artificial intelligence more generally. As discussed in Chapters 4 and 7, these models have their roots in statistical physics models for networks of real neurons in the brain, and many people see statistical physics as a natural language within which to understand why such systems work so well. Among other features, the new language models capture correlations over much longer time scales than previous models, and this certainly is part of what gives them their power.

In many ways, the frontier of this subject has moved from the examination of natural language to the problem of understanding why artificial neural networks are so successful at language processing tasks. Although this effort involved contributions from many disciplines, there has been special interest from the biological and statistical physics communities. As emphasized in Chapter 3, neural networks are a continuing source of new physics problems, and the language processing systems push these questions into new and unexplored regimes.

Perspective

The biological physics community's exploration of communication spans from hydrodynamic trigger waves to language, from songbirds to information-based search, and more. From the biological point of view, these are vastly different systems, the subjects of quite separate literatures. It is too soon to know if the search for common physical principles will succeed, but this search certainly is motivating many exciting developments. The comparison of these different systems can seem like a primarily theoretical question, but these comparisons also highlight opportunities for new and more precise experiments that will be realized in the coming decade. The idea that concrete tasks can be solved by methods that refer only to abstract goals, such as gathering information, provides a hint for how we may able to generalize away from microscopic details in a wider range of problems, connecting macroscopic functions more directly to new physical principles. The connections to artificial intelligence strengthen one of the most important paths for biological physics research to have impact on technology and society (Chapter 7).

3

How Do Macroscopic Functions of Life Emerge from Interactions Among Many Microscopic Constituents?

One of the great triumphs of science in the 20th century was the enumeration and characterization of the molecular components of life. But much of what strikes us as most interesting about living systems emerges from interactions among many of these molecular components. Much of our behavior as humans happens on the scale of centimeters or even meters. For a single cell, this behavioral scale is on the order of microns, something visible only through a microscope but still a thousand times larger than the nanometer scale of individual molecules. A major thrust of biological physics is to understand how to bridge these scales, from microscopic to macroscopic.

The emergence of macroscopic behaviors from microscopic interactions is not a question uniquely about the physics of life. In an ice cube, for example, individual water molecules interact only with their near neighbors, over very short distances. But if one pushes on an ice cube, touching only the molecules on the surface, molecules on the other side of the cube also move, even though they are separated from our finger by hundreds of millions of other molecules. If we raise the temperature by just a few degrees, the ice melts, leaving liquid water; now our finger passes through the liquid, leaving distant molecules unperturbed. The same molecules behave very differently with just a slight change in conditions, and this difference emerges clearly only with a very large number of molecules.

An important part of physics is devoted to understanding emergent phenomena in matter. There are solids and liquids, but also different kinds of magnets; a chunk of metal can conduct electricity or act as an insulator, and when made very cold it can become a superconductor; complex molecules can form the liquid crystals that were a mainstay of computer displays and electronic watches a generation ago and are still widely used in digital cameras. All of these phenomena, which happen in matter at thermal equilibrium, are described in a single unifying language, statistical mechanics. A profound result of statistical mechanics is that macroscopic phenomena often can be described, quantitatively, by models that are much simpler than the underlying microscopic mechanisms. This emergent simplification empowers us to explore much more complex systems, and it has long been hoped that statistical mechanics would provide a language for describing emergent phenomena in living systems. The past decade has seen extraordinary progress toward realizing this dream. Crucially, this has involved much more than applying what is known from physics to the living world. Rather in each case, the biological physics community has uncovered exciting new questions, focusing attention on how the emergent phenomena of life are different from their counterparts in the inanimate world. A sampling of these efforts is given in Table 3.1.

PROTEIN STRUCTURE, FOLDING, AND FUNCTION

Even single biological molecules are so large that one can think of their structure and function as emerging from the interactions among their many component parts. Proteins, for example, are polymers of amino acids. There are 20 possible amino acids at each site along the polymer chain, and proteins vary from a few dozen to several thousands of amino acids in length. The typical polymer does not have a well-defined structure, behaving instead like a randomly crumpled string. A large class of proteins have the remarkable property of folding into compact structures, and these structures are linked intimately to the functions that these molecules carry out in the living cell.

It is useful to identify at least three different questions about emergence of protein structure from the interactions among amino acids. First, what are the structures of these complex molecules? Second, what is the nature of the mapping between amino acid sequences and protein structures? Finally, and perhaps most deeply, what is it about these molecules that makes it possible for them to fold into well-defined structures, so unlike typical polymers? There has been enormous progress on all these problems, driven both by theory and experiment.

Protein Structures

The determination of protein structures is one of the great success stories in the interaction between physics and biology. These developments trace back to the early years of the 20th century, when it was discovered that X rays scatter from crystals to form a pattern of diffraction spots, and that this pattern can be

TABLE 3.1 Emergence of Macroscopic Functions from Microscopic Interactions

Discovery Area	Page Number	Broad Description of Area	Frontier of New Physics in the "Physics of Life"	Potential Application Areas
Protein structure, folding, and function	120	Proteins represent a different organizational state of matter than found in the non-living world, evolved for particular functions but also for the more general task of folding efficiently into compact structures.	Revolutionary tools for structure determination; energy/entropy tradeoffs; avoiding frustration and glassiness; statistical mechanics of sequences.	Protein design.
Chromosome architecture and dynamics	129	Understanding how 10 feet of DNA is packaged into nucleus just one-thousandth of an inch across; dynamics in the packed state.	Combining equilibrium polymer dynamics with non-equilibrium drive; chromosomes as a state of matter.	New tools for exploring chromosomal rearrangements in cell function and disease.
Phases and phase separation	134	Understanding when and how structural organization inside cells and membranes can emerge spontaneously, as with the formation of oil droplets in water.	New phases and droplet configurations in multicomponent systems; active droplets; departures from generic behaviors.	Self-assembly.
Cellular mechanics and active matter	139	Understanding how molecular motors and filaments organize to generate coherent movements on the scale of cells and tissues.	Hydrodynamic theories for "active matter;" reconstitution of more and more complex examples of self-organization for building a model cell; cell mechanosensing and feedback.	Cell and environmental mechanics in tumors and cancer treatment.
Networks of neurons	144	Perceiving the world, moving in response, and remembering what we did all involve the coordinated electrical activity of thousands of neurons in the brain.	Neural networks as a source of statistical physics problems; new experimental tools to monitor large numbers of neurons; data analysis to connect theory and experiment; the connectome.	Brain imaging; artificial intelligence.
Collective behavior	152	Understanding the beautiful, coordinated movements of birds in a flock, insects in a swarm; emergence of "construction projects" in social insects.	New universality classes; direct inference of statistical physics models from data; non-classical modes of ordering.	Active matter; coordination of autonomous vehicles.

analyzed to reconstruct the positions of individual atoms in the crystal. Not long after this discovery, it was shown that protein molecules could be crystallized, and in some cases they could still carry out their functions as enzymes while in the crystalline state. It was even possible to crystallize viruses and show that they retained their virulence once the crystal was dissolved. These results raised the possibility that one could use X-ray diffraction, or crystallography, to determine the positions of atoms in proteins in the same way that it was used to determine the positions of atoms in a salt crystal. But in salt crystals, the repeating units of the crystal have only a handful of atoms, while a single protein molecule has thousands of atoms, and established methods for the analysis of X-ray diffraction patterns did not generalize to this scale. Using X rays to help understand life required not just the application, but a dramatic expansion, of the methods of physics.

It would take decades until the first protein structures were revealed through the analysis of X-ray diffraction patterns, in the late 1950s and 1960s. These structures were not so precise as to reveal the positions of individual atoms, but showed clearly that the polymer of amino acids folded into local or "secondary" structural elements, helices and sheets, as had been predicted theoretically; these elements then pack into the overall globular structure of the protein. Structures of different proteins accumulated slowly, and the resolution of these structures improved (see Figure 3.1). The community of physicists, chemists, and biologists interested in protein structure realized that more open exchange of these results would accelerate progress, and in 1971 founded the Protein Data Bank (PDB) with just seven structures. Fifty years later, the PDB holds more than 170,000 structures, and has provided a model for open science. This exponential expansion of structural data was enabled in large part by the advent of synchrotron light sources.

The analysis of protein structures by X-ray crystallography had a revolutionary impact on our understanding of many processes in living cells, providing a literal scaffolding on which to build explanations of mechanism, but the constraint of crystallizing the proteins remained significant. Not long after the discovery of nuclear magnetic resonance (NMR) it was realized that resonances are sensitive to the structural and chemical environment and thus NMR spectra have a fingerprint of structure, but this is hard to extract. But when the magnetic moment of one nucleus is excited by radio waves, it "relaxes" by transferring energy to nearby nuclei, and this transfer is very sensitive to the distances between atoms. Thirty-five years after the first theoretical and experimental explorations of these relaxation dynamics, understanding had developed to the point where they were used to determine the structure of a small protein free in solution. An important aspect of this analysis is that, from the start, it provided not a single structure but an ensemble of structures, focusing attention on the flexibility of proteins.

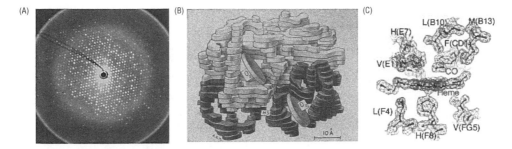

FIGURE 3.1 The analysis of X-ray diffraction patterns began in the mid-20th century, and has been developing ever since to allow us a clearer view of protein structures. (A) Early X-ray diffraction pattern from a hemoglobin crystal. The pattern fades out near the rim of the picture, which corresponds to a spacing of 1.8 Å. (B) Reconstruction of the protein structure from data in (A), with heme groups shown as gray discs with oxygen bound. MF Perutz, X-ray analysis of hemoglobin. (C) Modern high-resolution structure of one hemoglobin subunit, focusing on the heme and nearby amino acids. Carbon monoxide (CO) was bound to the heme, and the bond was broken with a flash of light, but at low temperatures in the crystal the CO remains trapped inside the protein. SOURCES: (A–B) From M.F. Perutz, 1963, X-ray analysis of hemoglobin: The results suggest that a marked structural change accompanies the reaction of hemoglobin with oxygen, *Science* 140:863, reprinted with permission from AAAS. (C) S. Adachi, S.-Y. Park, J.R.H. Tame, Y. Shiro, and N. Shibayama, 2003, Direct observation of photolysis-induced tertiary structural changes in hemoglobin, *Proceedings of the National Academy of Sciences U.S.A.* 100:7039, Creative Commons License CC BY-NC-ND 4.0.

Most recently, electron microscopy has taken its place alongside X-ray diffraction and NMR as a method for protein structure determination. Electron microscopes have their roots in the early days of quantum mechanics, with the realization that electrons have wavelike properties. There were steady improvements in resolution throughout the 20th century, and many important discoveries about the internal structures of cells. In the early days, viewing biological samples under an electron microscope involved using heavy metal stains to improve the contrast of the images, but the 1970s brought a combination of better microscopes and samples that were hydrated—that is, surrounded by water as in the living cell—and frozen. This began with proteins that naturally form regular arrays, such as the neurotransmitter receptors at a synapse, where the quality of the image was enhanced by averaging over the many repeating units, much as in X-ray diffraction. In the 2010s, there was a "resolution revolution," driven by better electron sources, more sensitive detectors, and improved analysis methods. Today it is possible to take electron microscope images of hundreds of thousands of individual protein molecules at extremely low (cryogenic) temperatures, combining these to resolve the underlying structure with a precision sufficient to trace the protein chain, as in Figure 3.2.

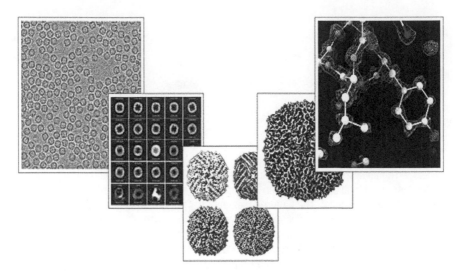

FIGURE 3.2 Combining electron microscope images of many single protein molecules provides enough information to define the molecular structure completely, resolving the positions of each individual atom. Shown here is a simplified Cryo-EM data processing workflow. From left to right, motion-corrected and summed image of mouse apoferritin in vitreous ice, two-dimensional class averages of apoferritin particles, three-dimensional sorting and subsequent "GoldStandard" refinement of the "best" particles, and zoomed in view of the apoferritin EM density (gray mesh, Electron Microscopy Data Bank ID1638) with the atomic model shown as sticks (Protein Data Bank ID 7A4M); carbons colored white, nitrogens colored blue, and oxygens colored red. SOURCE: Courtesy of Mark Herzik, University of California, San Diego.

Being able to see the structures of proteins at atomic resolution has completely changed how the scientific community thinks about the molecular mechanisms of life; many of these implications are explored in Part II of this report. Although rooted in physics, X-ray diffraction, nuclear magnetic resonance, and cryogenic electron microscopy have combined into a field known as structural biology and absorbed into the biological sciences more broadly. Some measure of the impact of these developments is in the stream of Nobel Prizes recognizing many of the breakthroughs: the first structures of proteins and other complex biological molecules (1962, 1964); the use of electron microscopy to solve structures with repeating units, as with the many proteins in certain viruses (1982); the development of NMR methods for structure determination (1991, 2002); and the development of cryogenic electron microscopy (2017). Alongside the recognition of methods, there is recognition of what has been learned from these methods about particular mechanisms, including photosynthesis (1988), the function of ion channels in the cell membrane (2003), the transcription of DNA into messenger RNA (2006), the translation of messenger RNA into proteins at the ribosome (2009), and amplification in cellular signaling (2012).

Folding

The extensive knowledge of protein structures provides a solid foundation from which to ask questions about how and why these structures emerge. The 20 types of amino acids come in two broad classes, the "hydrophilic" ones that interact favorably with water and the "hydrophobic" ones, which (like oil) do not. Protein structures pack the hydrophobic amino acids into the core, leaving a shell of hydrophilic amino acids to interact with the surrounding water. Although the interactions are complex, this suggests that the driving force for protein folding into compact structures is the hydrophobicity of the core amino acids.

While hydrophobic amino acids end up packed in the interior of proteins, much as a drop of oil separates from surrounding water, these amino acids are not all close to one another along the polymer chain. If the polymer is crumpled at random, there are many ways for hydrophobic and hydrophilic amino acids to end up as neighbors. More subtly, if the wrong pairs of hydrophobic amino acids come into contact, this favorable interaction could prevent the rest of the polymer from folding into a compact structure. This suggests that if hydrophobic and hydrophilic amino acids appeared in random order along the polymer chain, it would be very hard for the protein to fold into a well-defined structure because of competition among the many different possible contacts.

The intuition that random sequences cannot fold was made precise using theoretical approaches to the statistical mechanics of disordered systems (Chapter 5). These methods allow us to predict the behavior of the typical "random heteropolymer." In such random systems, the competition among many different possible contacts becomes a genuine frustration in which not all amino acids can find favorable neighbors. Imagining the energy of the system as a function of the protein structure, the picture is like that of a rough landscape, with many valleys inside of valleys. As molecules try to move on this landscape, they get stuck in local valleys, unable to find their way over the mountain pass to some ultimately more favorable structure. In a sense that can be made mathematically precise, almost all randomly chosen sequences of amino acids behave in this way. Experimentally, one can synthesize a random sequence of amino acids, and typically it does not fold. But real proteins do.

Evolution must have selected amino acid sequences that avoid the rough landscapes that are typical of random sequences. Certainly, the sequences found in today's organisms are a tiny fraction of the possible sequences, but it might have been that this is just a consequence of evolution not having had enough time to try out more possibilities. Instead, the physics of the folding problem teaches us that only a tiny fraction of sequences are allowed if proteins are to be functional. This shift from a historical view to a functional view is profound, and is encapsulated in the image of a funnel-like landscape for protein folding, as in Figure 3.3.

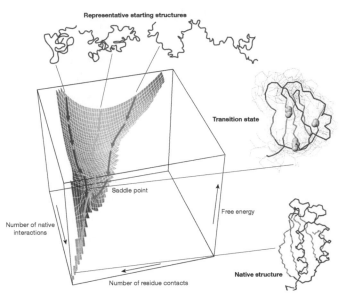

FIGURE 3.3 If proteins are to be functional, only a tiny fraction of amino acid sequences are allowed. Schematic energy landscape for protein folding, showing how the ensemble of unfolded structures is "funneled" to the unique native structure. Importantly, the molecule never visits states in which large numbers of amino acids form incorrectly, as opposed to "native" contacts. This behavior is atypical of polymers, and correspondingly only a tiny fraction of all possible amino acid sequences can become folded, functional proteins. SOURCE: Reprinted by permission from Springer: C.M. Dobson, 2003, Protein folding and misfolding, *Nature* 426:884–890, copyright 2003.

One can idealize the problem of avoiding rough landscapes by saying that evolution selects for sequences that minimize frustration, sculpting the energy landscape into a funnel that guides proteins to their final folded configuration. Energy landscape theory has created a formalism that quantifies the funnel picture and provides a candidate principle for the dynamics of folding and, importantly, makes detailed predictions, in quantitative agreement with experiments. While X-ray crystallography, NMR, and cryo-EM are the methods needed to determine the final folded structures of proteins, probing the dynamics of folding requires a wider variety of methods involving single molecule experiments and microfluidic devices, atomic force microscopy, several optical spectroscopies including fluorescence resonance energy transfer (FRET) and circular dichroism, and a variety of imaging approaches.

This discussion has emphasized the conceptual problems of protein folding, how and why well-defined structures emerge from interactions among amino acids. But there also are practical versions of these problems. If a new protein is discovered and its amino acid sequence is determined, is it possible to predict the structure? This is the usual formulation of the "protein folding problem." Conversely, if there is a function that would be useful to implement, is it possible to design a sequence that will realize the required protein structure? This is the "protein design problem." These questions are discussed in Chapter 6, which describes the connections among biological physics and molecular and structural biology. Applications of these ideas in the search for proteins with engineered functions are discussed in Chapter 7. As will be emphasized in Part II, these more practical formulations of the folding problem have had recent and dramatic input from artificial intelligence. This seems a good place to note the continuing importance of experiments on protein structure, both to explore uncharted territory in the universe of possible structures and to probe structural fluctuations and their dynamics, especially as we gain more appreciation for the functional importance of partially disordered proteins.

Statistical Mechanics in Sequence Space

Although the typical random sequence does not fold, neither does every single amino acid along the polymer chain need to be chosen correctly in order for the protein to fold into a particular structure. Looking across the tree of life, and sometimes even within a single organism, there are many proteins with related but not identical sequences, and the different proteins in these families fold into very similar structures. If these structures are stabilized by contacts among particular amino acids, and by the need to avoid competing interactions, then when mutations occur, changing the identity of one amino acid along the chain, it is reasonable to expect that evolution will select for compensating mutations in other amino acids. This leads to statistical patterns of amino acid covariation that encode physical and functional constraints.

During protein folding, the structure of the molecule changes, while the sequence of amino acids stays fixed. During the evolution of proteins that belong to a well-defined family, the amino acid sequence changes while the structure stays approximately fixed. While for many proteins folding really does correspond to the physical process of coming to equilibrium, and hence will be described by statistical mechanics, the dynamics of evolution in the presence of functional constraints is more complex. Nonetheless, many physicists have explored the idea that there can be a "statistical mechanics in sequence space." There is a path from the measured covariation in amino acid identities to construct the simplest models that capture these correlations, and these models are equivalent to statistical mechanics problems known as Potts glasses.

The analysis of the Potts glass models, which are determined entirely by the observed sequence variations in protein families, has produced two startling results. First, having constructed a statistical mechanics in sequence space allows the simulation of new sequences in much the same way that one can simulate the positions of molecules in a liquid. But with the modern tools of genetic engineering, these new sequences can be synthesized, and experiments show that, with high probability, they fold and function as do other members of the protein family. Second, the models can be interpreted as describing interactions between the amino acids, and these interactions turn out to be much more spatially restricted than the directly measurable correlations. In many cases, this spatial restriction is enough to identify the amino acids that are in contact, and hence infer the three-dimensional structure of the protein from the patterns of sequence variation alone.

A more general lesson from statistical mechanics in sequence space is that the mapping from sequences to structures is a many-to-one mapping. There is not one sequence that allows a given structure and function, but a whole ensemble of sequences. This idea that functions emerge from microscopic details in a many-to-one mapping is a theme in biological physics, and appears more explicitly in thinking about how organisms navigate the large space of possible parameters that is available to them (Chapter 4). The example of protein folding allows us to see very clearly that functional, living systems do not emerge by setting parameters at random, nor does function require every parameter to be set precisely. Variation in this view is not biological messiness, but rather an exploration of what is allowed by physics. Indeed, looking more closely at the sequence to structure mapping in simplified models, one can show that this level of allowed variation itself is extremely inhomogeneous. There are structures that could be reached by only one sequence, but there are structures even of short proteins that can be reached by thousands of sequences. In this sense, some structures are easier to "design," and it is plausible that evolution selects for maximally designable structures.

Perspective

Proteins represent a different organizational state of matter than is found in the non-living world, selected by evolution for particular functions but also for the more general task of folding efficiently into compact structures. The great expansion of experimental methods for determining protein structures, now largely exported from biological physics into the structural biology community, encourages us to think more globally about the mapping between sequence and structure. It now is almost possible to state how physical constraints of folding and function shape the ensemble of allowed protein sequences, and it is reasonable to expect that this problem will be solved fully in the coming few years. Beyond compact structures, there are families of intrinsically disordered proteins, whose role in many cellular functions is being appreciated more deeply. All of these developments support a view of life's functional mechanisms as belonging to ensembles, as in statistical physics. Success in understanding the physics of life requires us to construct these functional ensembles, and progress on protein structure and folding provides a model for the export of this view to function on larger scales.

CHROMOSOME ARCHITECTURE AND DYNAMICS

The human genome is divided among 23 pairs of chromosomes. If the DNA in these chromosomes, from just one cell, were stretched to its full length, it would be nearly 10 feet long. Along this length, there are more than 20,000 genes and millions of shorter sequences that have been identified as regulatory elements, helping to control which genes are "expressed" as proteins. Thanks to generations of experiments, one can point to the physical location of these many functional pieces along the genome. But despite this detailed information, it still is not known how these regulatory elements, which are often separated from their target genes by thousands to millions of base pairs along the DNA, manage to find and activate their specific target genes. Part of what is missing is the three-dimensional arrangement of these elements in space, as opposed to their one-dimensional arrangement along the genome. Ten feet of DNA is packaged into a nucleus just a thousandth of an inch across, smaller than the thickness of a human hair, and this packing is not random. Genomic regions that are separated by large distances along the DNA may end up in spatial proximity and thus become interacting neighbors. Their interactions, in turn, may activate the transcription of a gene that changes the fate of the cell, enable genomic recombination that leads to the production of an antibody, or result in a recombination event that leads to cancer.

While the structure of proteins is more dynamic than one might expect from simple pictures, this is even more true for the genome, because different segments of the genome can move relative to one another over relatively long distances. Nonetheless, some organizational rules are obeyed. The past decade has brought increasing appreciation for the role of chromosomal structure and dynamics in cellular function. In parallel we have seen glimpses that chromosomes organize themselves into a state quite unlike that of inanimate, polymeric materials, and that this new physics is important for function.

Architecture

How does 10 feet of DNA fit into a box less than one thousandth of an inch across? As shown schematically in Figure 3.4, the double helix of DNA is spooled around protein assemblies known as histone octamers to form nucleosomes. The nucleosome encompasses roughly 150 DNA base pairs, and is small enough that its structure now is known from X-ray crystallography, to atomic resolution. These "beads on a string" condense into fibers, and this can be seen with purified materials in a test tube but it has been challenging to connect these observations to what happens in the living cell. New methods combining chemical labels with electron microscopy, as in Figure 3.4, hold promise, but there still is no full picture connecting the nanometer scale of the nucleosome to micron scale of the nucleus.

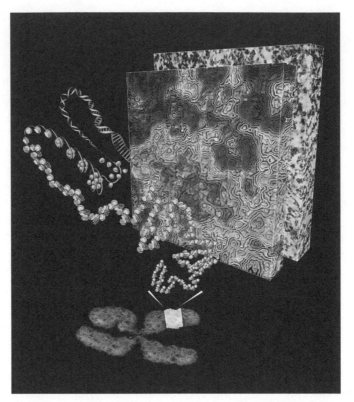

FIGURE 3.4 The double helix of DNA is packed into the cell nucleus through a series of higher order structures, first looping around proteins, then organizing into fibers and folding. Here, chromatin fibers in the three-dimensional (3D) space of the mammalian cell nucleus are visualized with methods that combine chemical labeling and electron microscopy. Large 3D sampling volumes (rear block) reveal that the vast majority of chromatin in the nucleus is a disordered polymer of 5 to 24 nm in diameter, shown schematically. The polymer is packed at different densities in interphase nuclei and mitotic chromosomes (front block): high density (red); medium density (yellow); and low density (blue). SOURCE: H.D. Ou, S. Phan, T.J. Deernick, A. Thor, M.H. Ellisman, and C.C. O'Shea, 2017, ChromEMT: Visualizing 3D chromatin structure and compaction in interphase and mitotic cells, *Science* 357:370, reprinted with permission from AAAS.

A very different experimental approach uses chemical methods to crosslink elements of DNA, which are close in space, and then cuts the whole chromosome into small pieces. By sequencing these pieces and comparing the results with the full DNA sequence, one can identify segments of the chromosome that are separated by long distances along the chain but folded to be close in space. Progress in sequencing methods makes it possible to do this in a "genome-wide" survey mode, sampling essentially all of the DNA in an ensemble of cells. An alternative approach attaches fluorescent labels to particular locations along the DNA, and then locates these labels with nanometer precision using super-resolution optical microscopy, building up a map of the chromosome as shown in Figure 3.5. These maps, here for human chromosome 21, illustrate the domains or long segments of the chromosome that are in close proximity, with these domains having relatively sharp boundaries. In addition, the maps are different in different cells. The chromosome is a dynamic structure, and the experiments on single cells capture snapshots of these dynamics.

Some sense for the frontier of this exploration comes from the fact that both the optical methods of Figure 3.5 and the chemical/genomic methods currently are limited to locating segments of the chromosome that are tens of thousands of base pairs in length. This scale is roughly 100× larger than the nucleosome, leaving a substantial gap in our understanding. Importantly, the missing scales overlap the scale of DNA regulatory elements in higher organisms, so what is missing is very relevant for how information flows through genetic networks (Chapter 2) and how cells navigate their parameter space (Chapter 4).

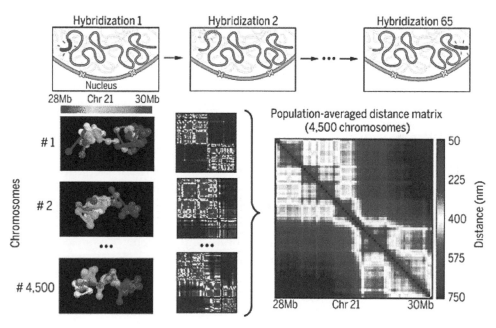

FIGURE 3.5 Super-resolution tracing optical microscopy makes it possible to map the three-dimensional distances between small segments of DNA along a chromosome, one cell at a time. (*Top*) Many rounds of hybridization label 30 kilobase segments of human chromosome 21, and the labels are then located with super-resolution fluorescence microscopy. (*Left*) The pseudocolor images of the positions of individual chromatin segments in single cells and the corresponding matrices of intersegment distances reveal domain-like structures with a globular conformation. (*Right*) The population-average matrix reveals domains at the ensemble level. SOURCE: From B. Bintu, L.J. Mateo, J.-H. Su, N.A. Sinnott-Armstrong, M. Parker, S. Kinrot, K. Yamaya, A.N. Boettiger, and X. Zhuang, 2018, Superresolution chromatin tracing reveals domains and cooperative interactions in single cells *Science* 362:419, reprinted with permission from AAAS.

Dynamics

Chromosomes are not static occupants of the nucleus; they constantly wiggle around, and therefore their spatial organization changes over time. Recent developments in microscopy make it possible to follow these movements of individual chromatin regions on sub-second time scales, in live cells and in real time. As chromosomal organization becomes susceptible to these sorts of quantitative experiments, it has become conventional to analyze and interpret these experiments with reference to sophisticated models from physics. More profoundly, these physics approaches to chromosomal organization in specific systems are beginning to reveal general theoretical principles across different systems. Chromosomes have been found to exhibit a highly nontrivial dynamical behavior, whose properties are similar across different species (see Figure 3.6). Ideas from statistical physics have been used to identify the origin of these dynamics and to demonstrate that the robust scaling laws observed for chromosomal motion in vivo can arise from physical principles rather than system-specific biological mechanisms. In particular, trajectories followed by a point along the genome are consistent with anomalous diffusion, governed by the viscoelastic nature of the cellular environment and strongly influenced by the spatial confinement imposed by the hierarchical folding of chromosomal DNA. Evidence is accumulating that these dynamics control crucial processes such as the editing of the genome in the generation of antibody diversity.

While geneticists have known for decades that the genome forms loops, and that the loops bring regulatory elements into close proximity with genes that they control, it was unclear how these loops formed. A scenario of loop extrusion executed by molecular motors, mathematically described by physicists nearly two decades ago, recently began to gain experimental evidence as one of the governing mechanisms of chromosomal compaction.

A long polymer such as the chromosome, which can be crosslinked by other molecules, has multiple possible phases. Ideas from polymer physics have been used to show that recent experiments on chromosomal dynamics are consistent with the chromosome being poised at a special point in its phase diagram, near a sol-gel transition. There are also hybrid theoretical descriptions that treat the chromosome as a flexible polymer but add constraints consistent with measured contacts (as in Figure 3.5). These theories also point to the proximity of chromosomal DNA to phase transitions and hint at their functional role.

Perspective

Understanding chromosomal structure and dynamics would feed into understanding a wide range of life's essential functions. The advent of powerful new experimental methods is providing unprecedented looks at snapshots and even

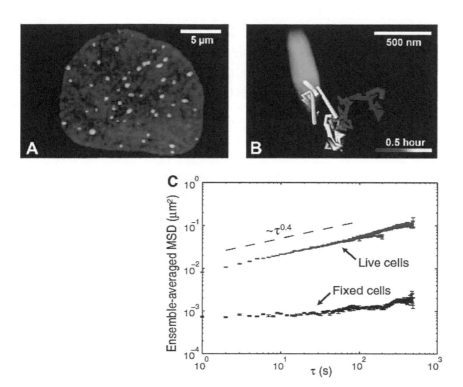

FIGURE 3.6 The dynamics of chromosomes. (A) Snapshot of chromosome ends (telomeres) in a mammalian cell nucleus, shown in a two-dimensional projection. (B) Three-dimensional motion of a single telomere. (C) Mean-square displacement versus time, averaged over an ensemble of trajectories, with fixed cells as a control. There is clean scaling over a 100-fold range of times, from seconds to minutes, with an anomalous exponent. SOURCES: (A–B) Reprinted with permission from I. Bronstein, Y. Israel, E. Kepten, S. Mai, Y. Shav-Tal, E. Barkai, and Y. Garini, 2009, Transient anomalous diffusion of telomeres in the nucleus of mammalian cells, *Physical Review Letters* 103:018102, copyright 2009 by the American Physical Society. (C) Reprinted with permission from S.C. Weber, A.J. Spakowitz, and J.A. Theriot, 2010, Bacterial chromosomal loci move subdiffusively through a viscoelastic cytoplasm, *Physical Review Letters* 104:238102, copyright 2010 by the American Physical Society.

dynamics of chromosome structure on multiple spatial scales. Gaps in these experiments likely will be filled over the next decade. Existing theories capture different aspects of the existing data, in some cases fitting the data and in other cases trying to derive at least global features of the data from more general principles. A more fully integrated, and more principled, theoretical understanding seems within reach, perhaps even within the next decade. Importantly, there are hints that the dynamic organization of chromosomes is not a simple or generic example drawn from the world of inanimate materials. Rather it seems likely that these crucial building blocks of the cell instantiate new forms of order, exploiting new physics to achieve their functions.

PHASES AND PHASE SEPARATION

Cells are the smallest basic structural, functional, and biological unit of all known organisms and are capable of independent self-replication. Cells consist of cytoplasm enclosed within a lipid bilayer membrane; the cytoplasm contains water soluble biomolecules such as proteins and nucleic acids that carry out the basic functions of energy production, nutrient uptake and processing, self-replication, and shape control and movement, as well as specialized functions unique to cells in specific tissues such as information processing and transmission, secretion of extracellular signals and structures, or detection and response to threats. To carry out specific activities within this diverse array of functions with spatial and temporal precision, cells organize their contents into specialized subunits known as organelles. Specific organelles were initially discovered by light or electron microscopy, techniques which reveal them as entities distinct from the surrounding cytoplasm. The revolution in chemical preservation ("fixation") of cellular ultrastructure in the 1960s led to the identification of most cellular organelles known today. The functional autonomy of organelles was validated by the fact that they could be isolated by biochemical fractionation and maintain their activities.

Although the classical textbook picture evolved, the 21st century brought revolutionary changes, grounded in new discoveries about the physics of these systems. It had long been known that purified versions of biological materials had interesting phases and transitions, but except in special cases—such as the behavior of proteins in the lens of the eye—it was never clear that this physics was relevant to the business of life. Over the course of a decade, this has changed dramatically, with novel phases, phase transitions, and phase separation becoming central to discussions of myriad processes in living cells.

Membranes

Physicists have long been interested in the cell membrane as an example of self-assembly. The lipid molecules in these membranes have "oily" tails and charged heads, so they are driven to organize themselves in ways that hide the tail from the surrounding water, exposing only the charged head groups; the bilayer is the simplest structure that does this. This can be reproduced in a test tube, even with just one species of lipid in water. But bilayers typically have at least two distinct phases, one liquid and one more rigid. Lipids in water have a much richer phase diagram, organizing into many different three-dimensional structures, including beautiful labyrinths. These phases, and the transitions between, provide fascinating examples of soft matter physics (Chapter 5), but it has been challenging to connect these phenomena to the life of the cell.

Real biological membranes have several different lipid components in carefully regulated proportions. These multicomponent systems have new axes to their phase diagram, and new phases where droplets or domains of different composition can condense in the membrane. These condensed domains were first observed in model systems made from mixtures of three naturally occurring lipids, then in vesicles formed from membrane extruded by live cells, and finally in fully natural cellular systems such as the yeast vacuole (see Figure 3.7). The transition into the phase with condensed domains is a liquid-liquid phase transition, and there is a critical point at a particular lipid composition. Near criticality, there are fluctuating domains on long length scales, and the spatial and temporal statistics of these fluctuations are predicted theoretically, by general statistical physics principles, with no free parameters; these predictions have been confirmed in detailed experiments on these membrane systems. The surprise is that real biological membranes have lipid compositions close to the critical point.

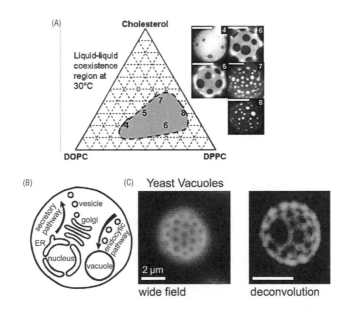

FIGURE 3.7 The membranes surrounding living cells are composed of many kinds of lipids, allowing for multiple phases and phase separation of droplets or domains. (A) Phase diagram of vesicles formed from mixtures of three lipids, DOPC, DPPC, and cholesterol, at 30°C. Gray region is where condensed domains are found, and images show the appearance of the vesicles near the transition into this region. Images are from a fluorescent probe molecule that partitions differentially between the coexisting phases. Vacuoles in living yeast cells, expressing a fluorescent version of the vacuole membrane protein Vph1. Images are taken with standard wide-field epifluorescence illumination (B) or wide-field illumination with z-sectioning followed by iterative deconvolution (C). SOURCES: (A) Reprinted from S.L. Veatch and S.L. Keller, 2003, Separation of liquid phases in giant vesicles of ternary mixtures of phospholipids and cholesterol, *Biophysical Journal* 85:3074, copyright 2003, with permission from Elsevier. (B–C) Reprinted from S.P. Rayermann, G.E. Rayermann, C.E. Cornell, A.J. Merz, and S.L. Keller, 2017, Hallmarks of reversible separation of living, unperturbed cell membranes into two liquid phases, *Biophysical Journal* 1113:2425, copyright 2017, with permission from Elsevier.

It has long been known that cell membranes have in-plane organization on long length scales, into domains and "rafts" that play a functional role in signaling and other processes; it was assumed that these structures are imposed on the membrane by other mechanisms. Discoveries about phase separation and criticality show that such large-scale organization will happen spontaneously, perhaps needing only to be stabilized by interactions with structure inside the cell. The proximity of the critical point also means that proteins embedded in the membrane will interact with one another over long distances, through the analog of Casimir forces known from other physics problems. The full implications of these results still are being explored, as is the mechanism by which cell membranes become tuned near their critical points. The observation of liquid-liquid phase separation in the two dimensions of a membrane prepares us for the possibility that something similar happens in three dimensions with proteins and nucleic acids in the cytoplasm.

Phase Separation in the Cytoplasm

Understanding the principles that drive the organization of biological molecules into function-specialized machines known as organelles has been largely undertaken by biologists, not physicists. The electron microscopy heyday of the 1960s, with the advent of glutaraldehyde fixation for ultrastructural preservation and staining methods dependent on composition, gave rise to the notion that there were two general classes of organelles: membrane-bounded and non-membrane bounded. Membrane-bounded organelles physically isolate and concentrate specific components in their interiors relative to the bulk cytoplasm, thereby forming reaction vessels containing all the necessary ingredients to perform their task. Phospholipids that make up biological membranes form bilayers in the aqueous cytoplasm by hydrophobic driving forces, and can enclose contents within vesicles, as described above. Some examples of membrane-bounded organelles include the nucleus, which contains the genome wherein genes are transcribed; the endoplasmic reticulum, in which much of RNA translation into protein and protein folding takes place; mitochondria, where oxidative metabolism occurs to generate ATP; and lysosomes, where proteins are degraded and processed into nutrients for cell growth.

Non-membrane-bounded organelles were identified by their lack of specific membrane staining. These include the highly ordered filamentous scaffolds of the microtubule, actin, and intermediate filament cytoskeletons; large semi-ordered macromolecular assemblies such as ribosomes and centrioles; and local concentrations of proteins that appear disordered, variously described by biologists as "bodies," "aggregates," "granules," "clusters," "plaques," or "osmophilic clouds." These include, for example, nucleoli, stress granules, the centrosome, Balbiani bodies, and focal adhesions. Identification of the protein and/or nucleic acid components of the more ordered non-membrane-bounded organelles such as the cytoskeleton and ribosomes, together with structural analysis and in vitro reconstitution, has led to a reasonably high level of understanding of how the assembly of these structures is driven by the same physical principles driving any protein-protein or protein-RNA interaction. However, although the components of the disordered non-membrane-bounded organelles were identified, the disordered nature of their structure and difficulty in achieving in vitro reconstitution made it difficult for biologists to decipher the physical principles driving their highly ordered formation.

It came as a surprise that structurally disordered, non-membrane-bounded organelles form by the process of liquid-liquid phase separation. Two key discoveries led to this idea. First was the direct observation of fluid behavior in one particular organelle, the P granule, through high-resolution microscope movies of living zygotes (see Figure 3.8). The second was the discovery that purified multi-valent proteins or repetitive RNAs that are made up of repeated, low affinity interaction motifs undergo liquid-liquid phase separation into organelle-sized droplets in aqueous solution in vitro (see Figure 3.9). These observations were supported by

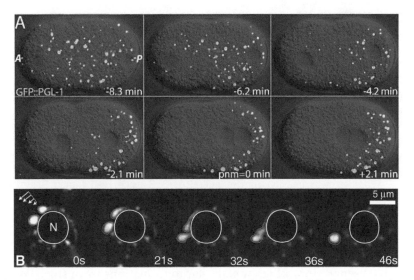

FIGURE 3.8 The direct observation of fluid behavior in the "p-granule" of a living embryo was one of two discoveries that some of the organelles identified in classical cell biology are really phase separated liquid droplets. (A) Images of a fertilized egg from the worm *Caenorhabditis elegans*. A single protein (PGL-1) has been tagged with the green fluorescent protein, and this protein localizes to "P granules," which eventually become germ line cells. Over the 10 minutes surrounding the meeting of the two pronuclei (pnm = 0), the P granules migrate from the anterior (left, marked A) to the posterior (right, marked P). The embryo is ~50 μm long. (B) P granules (outlined in red) dripping from a dissected cell, with the nucleus N (outlined in white). This is one of many liquid-like behaviors of the droplets. SOURCE: From C.P. Brangwynne, C.R. Eckmann, D.S. Courson, A. Rybarska, C. Hoege, J. Gharakhani, F. Jülicher, and A.A. Hyman, 2009, Germline P granules are liquid droplets that localize by controlled dissolution/condensation, *Science* 324:1729, reprinted with permission from AAAS.

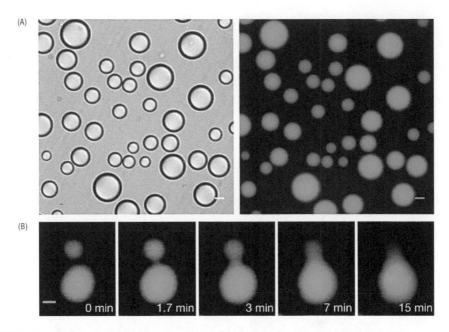

FIGURE 3.9 The observation that purified multi-valent proteins made up of repeated, low affinity interaction motifs undergo liquid-liquid phase separation into organelle-sized droplets in aqueous solution in vitro was the other key discovery leading to the finding that structurally disordered, non-membrane-bounded organelles form by the process of liquid-liquid phase separation "for highlighted part. Liquid droplets formed from solutions of weakly interacting proteins with repeating units. Differential interference contrast microscopy (A) and wide-field fluorescence microscopy (B), with a small fraction of the SH3 proteins carrying a fluorescent label. Concentrations are well below the affinity measured between individual SH3 and PRRM4 domains, so this is a collective effect. Scale bars: 20 μm. (C) Time-lapse imaging of merging droplets that were formed as in (A). Scale bar: 10 μm. SOURCE: Reprinted by permission from Springer: P. Li, S. Banjade, H.-C. Cheng, S. Kim, B. Chen, L. Guo, M. Llaguno, J.V. Hollingsworth, D.S. King, S.F. Banani, P.S. Russo, Q.-X. Jiang, B.T. Nixon, and M.K. Rosen, 2012, Phase transitions in the assembly of multivalent signalling proteins, *Nature* 483:336, copyright 2012.

a quantitative description of the relationship between valency, affinity, concentration, and phase separation, which was similar to transitions from small complexes to large, dynamic supramolecular polymers that had been described in non-living systems. Subsequent demonstrations that phase separation actually affects protein activity led to the notion that phase transitions may be used to spatially organize and biochemically regulate information throughout biology.

Perspective

In many of the examples described in this report, learning something essential about the physics of life required looking in places where nobody had looked before, often with new methods. In the case of intracellular condensates, on the other hand, many people had looked through a microscope at these organelles, over many decades. The paradigm-changing discovery that has launched immense progress was that these objects are not what they appear to be, and are governed by different physics. In little more than a decade, this has become a major focus of research not just in the biological physics community but in biology as well, with implications for medicine. These connections are discussed in Chapter 6. In the meantime, progress in understanding the phase behavior of real cell membranes continues. It took decades to realize that these membranes are not at generic points in their phase diagrams, and that this may have functional consequences. It remains to be seen if similarly non-generic behaviors are seen in cytoplasmic condensates. More generally, it will be exciting to learn if cytoplasmic condensates take advantage of novel phase separation behaviors that have not been uncovered in the exploration of inanimate systems. Finally, there are new physics questions about phase separation in the fundamentally non-equilibrium environment of the living cell.

CELLULAR MECHANICS AND ACTIVE MATTER

Living cells move. They change shape, divide, crawl over surfaces, and squeeze past each other even in dense tissues. The forces that drive these movements are generated by motor proteins (Chapter 1) such as myosin and kinesin, which act on filamentous proteins such as actin and tubulin. Filaments in turn can be bundled and cross-linked. The result is that the whole collection or proteins and filaments inside the cell, called the cytoskeleton, forms an active medium. These mechanical behaviors of the cell do not stop at the cell membrane. Instead, cells are responsive to the mechanical properties of their surroundings, which can affect their motility, shape and even decisions about which genes to express. The biological physics community has been interested in all these problems, and has had strong interactions with the larger community of cell biologists, as described more fully in Chapter 6.

Activity and Organization

A major effort in the cell biology community has been to purify the protein components of the cytoskeleton and reconstitute their behavior outside the cell. This makes possible very controlled and quantitative physics experiments, characterizing the mechanical behavior of the reconstituted material. As an example, one can put micron-sized beads into the medium, apply controlled forces with optical tweezers (see Box 1.3), and monitor the displacements of these beads. There will be random displacements, but also displacements in response to applied forces. In an equilibrium system, the random or Brownian motions are related, quantitatively, to the responses through the fluctuation-dissipation theorem (FDT). In an active system, one should see violations of the FDT.

Figure 3.10 shows tests of the FDT in a simple reconstituted system of actin and myosin. Remarkably, motors do not change the response of the medium to applied forces, but the spontaneous motions are an order of magnitude larger than thermal motions predicted by the FDT. On one hand, this is unambiguous evidence of non-equilibrium behavior. On the other hand, this shows that the active motions are not enormously larger than thermal motions, at least on micron length scales.

If the microscopic behavior of motors and filaments is understood, one can try to build a theory that averages over these details and describes the densities and flows of molecules on a scale of microns and larger. This is the same spirit as the derivation of fluid mechanics from molecular dynamics, with the difference that now the constituent particles are active. Pioneering efforts to derive these sorts of hydrodynamic theories for "active matter" were motivated by flocks and swarms (Chapter 3), but in the same way that fluid mechanics is the same for many different kinds of molecules, the hydrodynamics of active matter should be the same for all constituents that have the same symmetry properties. As discussed in Chapter 5, active matter now is a lively field of physics independent of its origins in the physics of living systems.

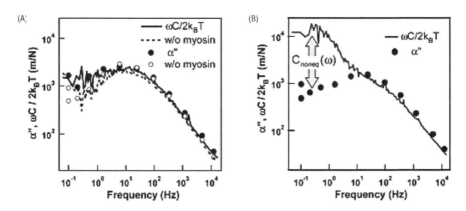

FIGURE 3.10 Observing fundamental signatures of irreversible, non-equilibrium behavior in simplified mixtures of motor proteins. Spontaneous movements of small beads are characterized by their power spectrum $C(\omega)$, which measures the amplitude of motion at each frequency ω. Displacements in response to applied forces also can be measured as a function of frequency, to give the response function $\alpha(\omega)$. The response function has an elastic component $\alpha'(\omega)$ and a viscous components $\alpha''(\omega)$; the fluctuation-dissipation theorem (FDT) connects $C(\omega)$ to $\alpha''(\omega)$ and the thermal energy $k_B T$ if the system is in equilibrium. (A) In the absence of myosin, or in the first few hours after myosin is added, the FDT is obeyed. (B) After a few hours, the activity of myosin molecules leads to a large violation of the FDT at low frequencies. SOURCE: From D. Mizuno, C. Tardin, C.F. Schmidt, and F.C. MacKintosh, 2007, Nonequilibrium mechanics of active cytoskeletal networks, *Science* 315:370, reprinted with permission from AAAS.

An important feature of filaments such as actin and tubulin is that they are polar: Particular species of motor molecules move primarily in one direction along the filament, and even the polymerization of the filament itself is directional. Early theoretical work showed that active polar fluids have defects analogous to those in liquid crystals, including asters and spirals or vortices. This is provocative because such organization of microtubules happens in cells, especially during the complex process of segregating newly copied chromosomes during cell division. These organized structures can be seen in simple reconstituted mixtures of tubulin and kinesin, conforming to the predicted phase diagram, as illustrated in Figure 3.11.

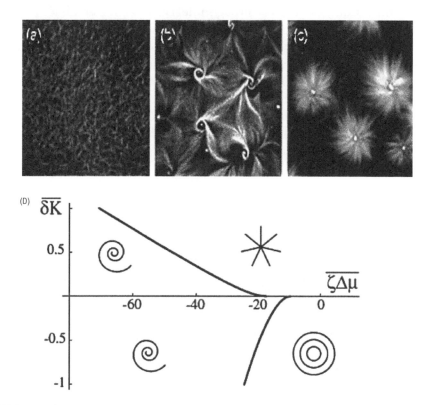

FIGURE 3.11 Protein filaments such as microtubules can organize in ways similar to what happens in liquid crystals, including forming defects with characteristic geometries. Here we see such self-organization in a mixture of microtubules with the motor protein kinesin. (A) Disordered array of microtubules. (B) Addition of modest amounts of the motor kinesin generates a spiral organization of the microtubules. (C) Higher concentration of motors generates asters. (D) Phase diagram for the hydrodynamics of an active polar medium. Axes are the elastic anisotrpy δK and the motor activity ζΔμ. SOURCE: Reprinted with permission from M.C. Marchetti, J.F. Joanny, S. Ramaswamy, T.B. Liverpool, J. Prost, M. Rao, and R.A. Simha, 2013, Hydrodynamics of soft active matter, *Reviews of Modern Physics* 85:1143, copyright 2013 by the American Physical Society.

One direction for this work is to try to reconstitute more and more complex examples of self-organization, perhaps to the point of building something that could be called a model cell. The other direction is to take these ideas back into real cells. Recently, quantitative theories and related experiments based on active matter ideas have addressed questions such as the size and shape of mitotic spindles and the cortical flow leading to polarization of worm embryos.

Connecting to the World

The cytoskeletal networks of filaments and motors inside the cell are linked to the environment outside the cell through integrin protein assemblies (see Figure 3.12). These have the unusual property of catch bonds—unlike most bonds, which weaken and detach more quickly when they are pulled apart, catch bonds strengthen and detach more slowly. Such bonds allow the cells to sense and respond to applied stresses and to the mechanical stiffness of their environments. The extracellular matrix that surrounds cells was long considered a simple passive scaffolding that simply houses and supports cells. The discovery of cell mechanosensing and feedback led to the recognition that the extracellular matrix—the cell microenvironment—has a profound role in regulating cell behavior and mediating interactions between cells.

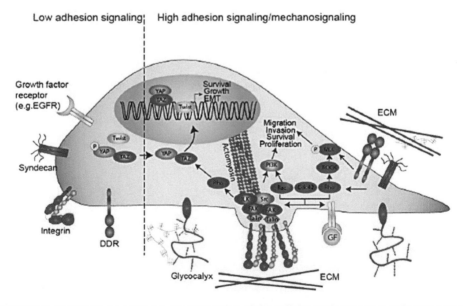

FIGURE 3.12 The many pathways for mechanics of the cellular environment to influence internal states via integrin molecules at the surface. SOURCE: Reprinted from F.B. Kai, A.P. Drain, and V.M. Weaver, 2019, The extracellular matrix modulates the metastatic journey, *Developmental Cell* 49:332, copyright 2019, with permission from Elsevier.

In particular, the stiffness of the microenvironment is important to how cells sense and respond to external forces, through a process called mechanotransduction.

As an example, stem cell proliferation and differentiation processes were long considered to be controlled solely by biochemical cues, but it is now recognized that these processes depend critically on mechanical properties of the constantly evolving microenvironment. The stiffness of the microenvironment helps determine the cell type during stem cell differentiation: Cells in stiff environments differentiate into stiff cells such as those in bone, while cells in soft environments differentiate into soft ones like neurons.

The understanding of cancer has been similarly transformed by the recognition that a tumor is intimately linked to its microenvironment and can be considered in itself as an organ—cells within tumors cannot be understood in isolation. Tumor cells can subvert their microenvironments to promote the tumor; conversely, targeting the microenvironment may be an effective way of inhibiting a tumor. Mechanics are critical to the interaction; breast tumors are found by palpation because they are stiffer than normal tissue. The tumors promote remodeling of the extracellular matrix that stiffens the matrix, further promoting tumor growth, which further promotes stiffening of the matrix in a downward cycle of malignant tumor progression. Even worse, stiffening of the extracellular matrix promotes tumor metastasis, perhaps through local force cues and increasing expression of proteins that promote cell migration, among other mechanisms.

These collective phenomena governing the behaviors of tumors and stem cells, as just two examples, arise from the many-body interactions that link cells to their microenvironments and thereby link cells to each other. Many of these phenomena are explicitly non-equilibrium phenomena. An important step in cancer progression is the epithelial-to-mesenchymal transition (EMT). During this process, interactions among cells and between cells and the extracellular matrix are modified, leading to detachment of epithelial cells from each other and from the underlying substrate membrane so that cells can migrate away and invade normal tissue, seeding metastasis. Thus, this process marks a transition from a solid state, in which cells do not change their neighbors, to a fluid state, in which cells migrate and change neighbors. Once in the fluid state, the system becomes a realization of a classic active matter system, namely a collection of self-propelled particles.

Perspective

Active matter provides a perspective on the emergence of structure and function from interactions among motile components. From this perspective, in looking at the cytoskeleton the "particles" are molecules, while in tissues the particles are cells, but the physical principles are the same. Thus, movements of cells in tissues recapitulate some of the phases and transitions seen for movements inside single cells, connecting to questions about active matter beyond the living world (Chapter 5). As recently as a decade ago, there was a substantial divide between (roughly) physicists interested in the mechanics of cells and biologists interested in the myriad pathways by which these mechanics are regulated. As experimental tools become more powerful, it becomes possible to explore the physics of the fully regulated mechanical system, with quantitative probes of force, displacement, and signaling molecule concentrations, simultaneously. It will be interesting to see whether these richer systems—as with flocks and swarms (Chapter 3)—have ways to generate behaviors outside the universality classes of conventional active matter theories. Concepts from active matter such as self-organization and self-healing are impacting thinking on engineered micromechanical systems, with much more to be explored.

NETWORKS OF NEURONS

Perceiving the world, moving in response to stimuli, and remembering past events all involve the coordinated electrical activity of thousands of neurons in the brain. It is an old dream of the physics community, dating back into the 1940s and receiving a major stimulus in the 1980s, that this coordination could be understood as an emergent phenomenon in the language of statistical mechanics. Some of the first theoretical ideas were prompted by experiments on the all-or-none nature of the action potential in single neurons, and on the coarse-grained behavior of large numbers of neurons seen in the electroencephalogram (EEG). Subsequent decades have seen great progress in both theory and experiment, as well as in our ability to bring theory and experiment together. Important successes often have become part of the mainstream of neuroscience, but the effort to understand collective behavior in networks of neurons continues to occupy a significant part of the biological physics community, as experimentalists develop new instruments for quantitative exploration of network dynamics and theorists use neural networks as a source of new problems in statistical mechanics.

Observing the Human Brain

Humans have a special interest in the dynamics of their own brains. In addition to the EEG, all of the methods for observing electrical activity in the human brain have had major contributions from the biological physics community. Magnetoencephalography (MEG) measures the magnetic fields that result from coordinated current flow among neighboring neurons in the brain, and the high sensitivity needed for these measurements has led to the use of superconducting quantum interference devices (SQUIDs) as field sensors. EEG and MEG both have high time resolution, but limited spatial resolution.

The electrical activity of neurons requires energy, and so the consequences of this activity are detectable in blood flow and metabolism. Positron emission tomography (PET) follows, for example, the uptake of radio-labeled glucose molecules that provide the (almost) unique energy source for the brain. For cells to extract energy from glucose requires oxygen, and oxygen binding to hemoglobin in the blood changes the nuclear magnetic relaxation behavior of protons in the surrounding water; this is the basis of functional magnetic resonance imaging (fMRI, Chapter 6). PET and fMRI have high spatial resolution, but the metabolic signals to which they are sensitive are slower than the electrical activity itself.

These physics-based experimental methods have provided dramatic glimpses of our brains in action. Experiments show that the pattern of brain activity when humans imagine an image is very similar to that when they see the image, and in some individuals this is true even in the primary visual cortex. When two people have a conversation, their brain activity becomes synchronized; while the listener's brain largely follows the speaker's brain with some delay, specific brain regions have activity that is predictive. These and other observations provide boundary conditions for theories of how the emergent behavior of neural networks underlies our experience of the world.

Functional magnetic resonance imaging has joined with PET scanning, EEG, and MEG to form the field of human brain imaging, and this has become part of the mainstream of neuroscience and psychology (Chapter 6), as well as playing a key role in the understanding of brain injuries and disease (Chapter 7). In many ways this parallels the merger of X-ray crystallography, NMR, and cryogenic electron microscopy into structural biology (Chapters 3 and 6). Physicists continue to improve these technologies—creating better scintillation detectors for PET studies and denser arrays of electrodes for EEG, and increasing the resolution of MRI through the use of stronger magnetic fields and more sophisticated pulse sequences.

Monitoring Many Single Neurons Simultaneously

At the opposite extreme from measuring the EEG is the measurement of electrical activity of single neurons in laboratory animals. The first such experiments in the 1910s strained the sensitivity of instruments in the physics laboratory, and there is continuing input from the physics community into these measurement techniques. Although much has been learned about the brain by studying the responses of single neurons, there was a gap between these measurements and ideas about emergent and collective behavior in networks of neurons. Closing this gap requires monitoring the electrical activity of many individual neurons, simultaneously. Some of the first efforts in this direction were aimed at the retina. Arrays of electrodes deposited on glass allow a relatively straightforward interface to a piece of retina dissected from the eye and kept alive in a dish. By the early 2010s, these methods had advanced to the point where one can monitor almost all of the hundreds of signals that the eye sends to the brain from a small patch of the visual world, giving new opportunities to study how visual information is represented (Chapter 2) and to search for collective behavior in this relatively simple network. More broadly applicable methods adapt semiconductor microfabrication techniques to build electrode arrays that can be inserted into three-dimensional brain tissue. Major efforts to make such tools available to the wider research community were mounted in the late 2010s. As with detector arrays in experimental particle physics, there are challenging problems in transforming the data from multiple electrodes into meaningful signals from multiple individual neurons; in parallel with hardware developments these problems have attracted attention from the physics, applied mathematics, and computer science communities. Today it is possible to resolve hundreds or even thousands of single neurons, and importantly one can track these signals continuously and stably over many months.

During an action potential, the electric field across the cell membrane changes by $\sim 10^7$ V/m. This large field is enough to generate large changes in the optical properties of molecules in the membrane, and there were efforts dating back to 1970 to use voltage-sensitive dye molecules that would dissolve in the membrane and literally make the electrical activity of neurons visible as a change in fluorescence. Although there were dramatic early demonstrations of activity in large populations of cells, the dyes suffered from various limitations that prevented them from becoming a viable alternative to electrodes.

The idea of recording electrical activity by optical imaging methods received a revolutionary push from the discovery of the green fluorescent protein, described in Chapter 6. These proteins have been modified to have their fluorescence depend on a variety of signals in the surrounding solution, including calcium. Because action potentials trigger an influx of calcium into neurons, which is pumped out (or into internal stores) more slowly, calcium concentration is a signal that traces a temporally smoothed version of electrical activity. Animals genetically engineered to express calcium-sensitive fluorescent proteins in their neurons thus make it possible to visualize electrical activity as a flickering movie of fluorescent signals. To make the most of these signals requires development and deployment of specialized optical methods, including scanning two-photon microscopy, microendoscopy, head-mounted miniature microscopes, and microscopes with adaptive optical capabilities for imaging deeper into brain tissue. The result of these developments is that one can monitor hundreds or even thousands of individual neurons simultaneously, with high signal-to-noise ratio, as illustrated by the experiments in Figure 3.13. These methods are undergoing continual development, with steady progress in the number of individual neurons that can be resolved and the quality of the recordings. There are also fluorescent proteins that insert into the membrane and respond directly to voltage. These methods are on the threshold of general use, which will realize a 50-year-old dream of directly visualizing electrical activity in the brain.

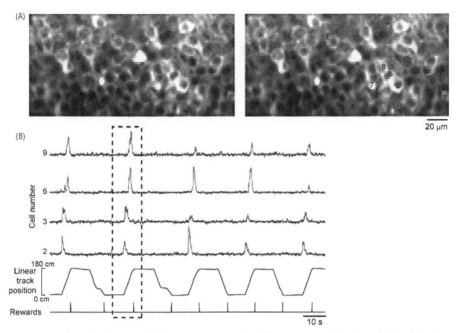

FIGURE 3.13 Genetically encoded fluorescent proteins allow us to monitor electrical activity in many neurons simultaneously, at high signal-to-noise ratio. (A) Image of neurons in the CA1 region of a mouse hippocampus (*left*). The cells express a protein whose fluorescence is sensitive to the calcium concentration, which changes in response to electrical activity. Cell bodies appear outlined because the protein is excluded from the nucleus. Fluorescence images are collected by scanning two-photon microscopy. Selected cells are outlined (*right*). (B) Fluorescence signals from four cells as the mouse runs along a (virtual) linear track, receiving rewards at the end. Note the low level of background noise. SOURCE: Reprinted by permission from Springer: D.A. Dombeck, C.D. Harvey, L. Tian, L.L. Looger, and D.W. Tank, 2010, Functional imaging of hippocampal place cells at cellular resolution during virtual navigation, *Nature Neuroscience* 13:1433, copyright 2010.

Theory

Recording the electrical activity of thousands of neurons creates the opportunity to search for collective, emergent behaviors in these connected networks. But such high-dimensional data cannot be explored without some guidance from theory. Theories of neural network dynamics date back to the 1940s, with the first efforts to understand what neurons could compute. This early work showed that arbitrary patterns of connections (synapses) between neurons can generate very complex dynamics. To make progress, two extreme simplifications emerged from the biological physics community. As often the case in the physics of interacting many-body systems, neither of these simplifications are literally correct for networks of neurons in real brains, but both have been powerful sources of ideas.

The first simplification is to imagine the neurons are organized into layers, and that synaptic connections carry signals from one layer to the next, with no feedback. This "perceptron" or feed-forward architecture was proposed around 1960, and reappeared in the 21st century as the foundation for the deep network revolution in artificial intelligence, as explained in Chapter 7. The second, alternative simplification is to imagine that all synaptic connections are symmetrical, in which case the dynamics of the network are equivalent to motion on an energy landscape. In both cases, ideas from statistical physics play a key role in the analysis; more deeply, these model neural networks have been the source of new statistical mechanics problems.

An essential step in the theoretical analysis of neural networks is to think not about a particular network, in which there is some specific pattern of connections among all the neurons, but rather about the behaviors that are expected in an ensemble of networks, where the patterns of connections are chosen from some distribution. In the limit that networks are large, there is self-averaging, so that a single network becomes typical of the whole ensemble. This approach connects the theory of neural networks to the statistical physics of disordered systems such as glasses and spin glasses (Chapter 5). Indeed, the symmetric model of neural networks maps exactly to a novel family of spin glasses.

In the symmetric model of neural networks, the dynamics in the absence of noise is just a downhill slide on an energy landscape. The network stops at local minima of the energy, or attractors. In the first proposal, these dynamics were envisioned as a model for the recall of a memory; a cue for recall would initialize the network in the basin of attraction for one memory, and the dynamics would recover that memory. The structure of the landscape, and hence the stored memories, depends on the detailed pattern of connections between neurons; the network can be "programmed" by changing the strengths of these synapses, sculpting the energy landscape. Importantly, in certain limits this programming can turn the current state of the network into an attractor by changing the synaptic connection between two neurons solely in relation to the activity of those two neurons. More generally, many problems that the brain has to solve—and many classical problems in the theory of computational complexity—can be formulated as minimizing a cost function and mapped into the dynamics of a network.

The symmetric model of neural networks thus connected brain dynamics, statistical physics, computational complexity, and the rules for synaptic plasticity. As the 20th century ended, many of these connections were solidified, for example, the use of statistical physics methods to identify phase transitions in large computational problems, and to understand the conditions under which these problems become hard (Chapter 5). For the collective behavior of neurons in real brains, perhaps the most important prediction of these models is that memories are stored in locally stable patterns of activity distributed across the whole network, patterns in which the activation of each neuron is maintained self-consistently by the activity of the other neurons. This gives us a mathematically precise version of classical ideas about reverberating activity, and connects directly to a large number of experiments that probe persistent activity of neurons under conditions in which animals remember and compare distinct sensory inputs. In both symmetric and feed-forward architectures, there is a notion of capacity for the network, and this capacity depends on the distribution of synaptic strengths. Maximizing capacity leads to nontrivial predictions for this distribution, which agree with experiment, including the large number of silent or nearly silent connections.

Beyond the symmetric and feed-forward simplifications, there have been substantial efforts to understand fully dynamical regimes of neural networks. Much has been learned from the study of relatively small networks, where it is possible to make nearly complete, microscopically realistic models and then analyze these models with all the tools of modern dynamical systems theory. These systems have been especially important for thinking about the mapping between microscopic parameters and macroscopic functions, as described in Chapter 4. There is a well-developed mean field theory for large networks with random synaptic connections, and these systems exhibit a transition from a quiescent to a chaotic state. While random connections may seem non-functional, these systems have rich dynamics, especially near the transition, which can be harnessed to generate or analyze temporal sequences on time scales much longer than the transient response times of individual neurons. There are several lines of evidence that real networks of neurons may be poised near critical points, although this remains controversial.

An important challenge in searching for collective behavior in networks of neurons, as in many other living systems, is the absence of the usual macroscopic, thermodynamic probes. Even in models that map to well-defined statistical physics problems, order parameters are complex combinations of activity across the network; available experimental manipulations do not couple naturally to these order parameters (as with applying a magnetic field to a ferromagnet), nor is it clear how to change the analog of the temperature. In the absence of such probes, there are efforts to infer a statistical or quasi-thermodynamic description directly from experiments on the activity of large numbers of neurons. Ideas along these lines include searching for low dimensional structure in the activity patterns; studying the behavior of these patterns under coarse-graining, inspired by the renormalization group; and constructing minimally structured or maximum entropy models for the distribution of activity patterns, matching measured correlations. The maximum entropy approach connects with the analysis of sequence variation in protein families (Chapter 3) and velocity fluctuations in flocks of birds (Chapter 3), and in some cases makes predictions that agree with experiment in quantitative detail. Still, it is not clear that the community has found the compelling theoretical framework to link the rapidly growing body of data on large populations of neurons with the ideas of emergent, collective behavior in these networks.

Connectomics

All theories of neural dynamics agree that the collective behavior of a network depends on the pattern of connections among neurons. There is a long history of probing these connections between pairs of individual neurons, but as the 21st century began many people started to take seriously the possibility of mapping connections—network architecture—on a much larger scale. Higher throughput methods of electron microscopy are being combined with machine learning to trace neurons and their connections through very densely packed brain tissue. There have been important successes in using these methods to study smaller systems, for example showing that the nearly crystalline, orderly structure of connections in the early stages of sensory processing in insects gives way to more nearly random connectivity deeper in the brain. These results have provided proof of principle, and there are now serious proposals to chart the full "connectome" of a mouse, a primate, or perhaps even a human. Such a large-scale project would provide a scaffolding on which to build a description of collective neural dynamics. The first such effort dates back to the mid-1980s, with the reconstruction of all the synaptic connections among the 302 neurons of the nematode worm *Caenorhabditis elegans*.

There remains considerable debate within the scientific community as to whether the enormous effort and funding needed to determine one or more truly complete mammalian connectomes by electron microscopy would constitute a prudent allocation of resources. Nonetheless, there is general agreement that the biological physics community has played, and will continue to play, a crucial role in the development of imaging techniques for the acquisition of data and in the development of analysis techniques for image processing and the elucidation of neural circuits. Extensive challenges concern the successful visualization and tracing of trillions of axons and their synaptic connections, key constituents of a complete connectome, as illustrated in Figure 3.14. As discussed in Chapter 9, the National Institutes of Health and the Department of Energy are now actively engaged in discussions about supporting the largest scale versions of such a project, building on the success of intermediate scale projects sponsored by the Allen Institute, the Howard Hughes Medical Institute, the Intelligence Advanced Research Projects Activity, and other agencies. Thus far, the impact of connectomics has been greatest when focused on smaller circuits where we have a substantial body of knowledge about the computational functions being carried out. An impending effort of national scale, however, could bring us to an inflection point at which the impact of large-scale connectomics research rises dramatically.

Perspective

The search for emergent behavior in networks of neurons has been enormously productive, but is far from over. The last decade has seen dramatic advances both

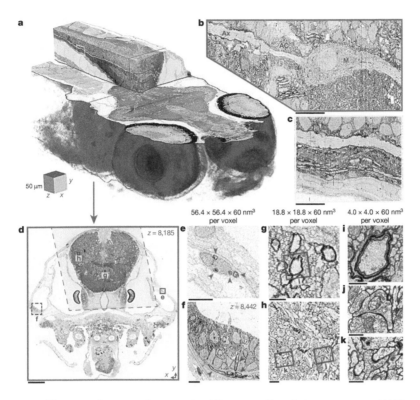

FIGURE 3.14 Electron microscopy to reconstruct the connections between neurons. (A) The anterior quarter of a larval zebrafish was captured at $56.4 \times 56.4 \times 60$ nm³ per voxel resolution from 16,000 sections. (B) The Mauthner cell (M), axon cap (AC), and axon (Ax) illustrate features visible in the $56.4 \times 56.4 \times 60$ nm³ per voxel image volume. (C) Posterior Mauthner axon extension. (D) Targeted re-acquisition of brain tissue at 18.8 18.8 60 nm³ per voxel (dashed) from 12,546 sections was completed after $56.4 \times 56.4 \times 60$ nm³ per voxel full cross-sections (solid). (E, F) Peripheral myelinated axons (arrowheads) recognized from $56.4 \times 56.4 \times 60$ nm³ per voxel imaging of nerves (E) and the ear (F). (G, H) Neuronal processes including myelinated fibers can be segmented at $18.8 \times 18.8 \times 60$ nm³ per voxel resolution. (I–K) Targeted re-imaging to distinguish finer neuronal structures and their connections. Scale bars: (B, C) 10 µm; (D) 50 µm; (E, F) 5 µm; (G, H) 1 µm; (I–K) 0.5 µm. SOURCE: Reprinted by permission from Springer: D.G.C. Hildebrand, M. Cicconet, R.M. Torres, W. Choi, T.M. Quan, J. Moon, A.W. Wetzel, et al., 2017, Whole-brain serial-section electron microscopy in larval zebrafish, *Nature* 545:345, copyright 2017.

in theory and in experiment, and these are continuing. The number of neurons that we can monitor simultaneously, and the quality of these recordings, continues to grow. Complete maps of the connections among tens of thousands of neurons have been achieved in the fly brain, and there is intense effort in other systems. Different organisms, from hydra to octopus, are emerging rapidly as model systems in which to make coordinated explorations of neural networks and behavior. There remains, however, a gap between theory and experiment. New, larger data sets need new tools for analysis, but these analysis methods should be grounded in deeper theoretical ideas. Many quantities that theory points to as being crucial still are not so easy to measure. The hope for the coming decade is that there will be not only continued, parallel progress in theory and experiment, but new ideas about how to build bridges between the two.

COLLECTIVE BEHAVIOR

Collective behaviors in animal groups provide some of the most familiar examples of emergent phenomena in living systems. Most of us have seen the ordered patterns of geese flying in formation, the more fluid flocking behaviors of other species (see Figure I.3), the analogous schooling behaviors of fish, and the apparently chaotic motions in swarms of insects. But animals can do more than just move together; they can also organize themselves to accomplish large construction projects, such as the termite nest in Figure 3.15. All of these phenomena are emergent: There is no blueprint for the nest, no commander broadcasting a common movement direction to all the individuals in a flock, school, or swarm; instead, order arises out of the interactions among individuals in the group. More subtly, the very existence of ecology is an emergent phenomenon, since it is not obvious why a large number of different species can coexist, stably, in a single environment (Chapter 5). The focus of this section is on the collective behavior of multicellular organisms; collective behaviors in unicellular organisms are discussed in relation to communication (Chapter 2) and active matter (Chapter 5), although one can hope for unifying principles. The theme of our discussion is the interplay between theory and experiment: Qualitative observations inspire theories, new and more quantitative observations test these theories and show how living systems have found unexpected regimes of order and fluctuation, and theorists are sent back to search for new mechanisms that can generate the observed behaviors. It now is inescapable that collective behavior provides examples of new physics, beyond the examples from emergent phenomena in the inanimate world.

Flocks and Swarms

The emergence of a common movement direction in flocks, schools, and swarms is tantalizingly close to familiar ordering phenomena in the inanimate world. Could it be that all the birds in a flock agreeing to fly in the same direction is like all the spins in a ferromagnet agreeing to point in the same direction? In the 1990s, physicists began to explore models that embody this intuition. These models could be expressed as rules by which individuals in the group adjust their movements in relation to their neighbors, or equations of motion for animals acted upon by what had been called social forces in the earlier biological literature.

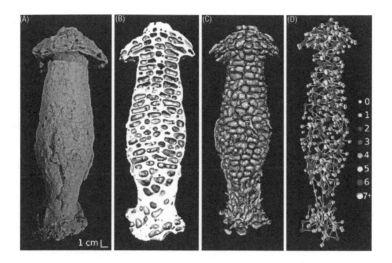

FIGURE 3.15 Collective behaviors in animal groups, such as the large construction projects of termite nests, provide examples of emergent phenomena in living systems. Three-dimensional structure, reconstructed via X-ray tomography of a *Cubitermes* nest, from a set collected in equatorial forest regions of the Central African Republic and Cameroon. (A) A conventional photograph of the nest. (B) A virtual slice through the middle of the nest, constructed from X-ray tomography, showing the different chambers. (C) A virtual "cast" of the nest, illustrating the three-dimensional chambers and galleries. (D) A representation of chambers and galleries as a network, where each node corresponds to a chamber and each edge to a corridor. The color of the nodes reflects their degree, i.e. the number of corridors connected to that chamber, as shown in the legend. SOURCE: Reprinted by permission from Springer: A. Perna, C. Jost, E. Couturier, S. Valverde, A. Douady, and G. Theraulaz, 2008, The structure of gallery networks in the nests of termite *Cubitermes* spp. revealed by X-ray tomography, *Naturwissenschaften* 95:877, copyright 2008.

These equations could be simulated, or coarse-grained to derive the analog of fluid mechanics for a flock, using renormalization group ideas. Both simulation and analysis agreed on a first striking result, that these systems could break symmetry and produce ordered motion in a particular direction even for a hypothetical flock or swarm confined to two dimensions. Such symmetry breaking in two dimensions is not possible for a system in thermal equilibrium with local interactions, and thus is a harbinger of the qualitatively new physics that is possible in living systems.

The early work on theories of flocks and swarms became a foundation for what is now the lively field of active matter physics, as described in Chapter 5. This theoretical work also prompted efforts to collect more compelling quantitative data on collective behavior. One approach is to bring the behavior into the laboratory, studying schooling fish in a tank or swarming insects in a box. The other approach is to bring the laboratory into the field, setting up multi-camera video to reconstruct the trajectories of all the organisms in the group as they move through their natural environment. In both approaches, a central role is played by analyzing the observed fluctuations in movement velocities around the mean, and statistical physics gives us a natural language of correlation functions to use in this analysis.

Analysis of correlation functions in flocks of European starlings showed, surprisingly, that correlations among velocity fluctuations are independent of scale (see Figure I.3). Scale invariance for fluctuations in flight direction can be understood by realizing that this is a system with local interactions, and the system breaks a continuous symmetry, so there will be "massless" modes. But there is no generic expectation for scale invariance of correlations in speed fluctuations. Although swarms of midges exhibit no overall velocity ordering, they too show scale invariant fluctuations in velocity, and analysis of correlations in both space and time reveals dynamic scaling, with an exponent closer to ballistic propagation rather than diffusion of information through the swarm (see Figure 3.16). Ballistic propagation is seen also in flocks, during events where the entire flock turns. All of these correlation structures are outside the predictions of the generic

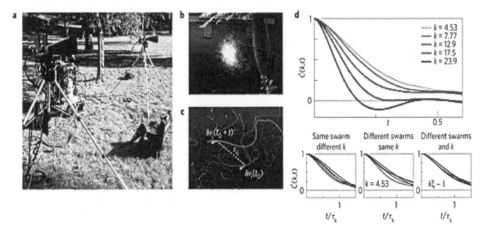

FIGURE 3.16 Swarms of midges exhibit scale invariant fluctuations in velocity. (A) A system of three synchronized high-speed cameras is used to collect video sequences of midge swarms in their natural environment. (B) A swarm of approximately 300 midges. (C) Two trajectories within the swarm. (D) Spatiotemporal correlation functions of the velocity, as a function of time t (seconds) and the Fourier variable k conjugate to distance. Upper panel: normalized correlation function in one natural swarm at various values of k. Bottom panels: correlation as a function of the rescaled time , t/τ_k, in various attempts to rescale the data. (*left*) Rescaling by a k-dependent time for all k in one swarm. (*center*) Comparing many swarms at the same k. (*right*) Measuring the static correlation length ξ for each swarm, and choosing $k\xi \sim 1$, then rescaling time. This is evidence for dynamic scaling, $\hat{C}(k,t) = \hat{C}(t/\tau_k, k\xi)$, with $\tau_c = k^{-z}g(k\xi)$; further analysis shows that $z = 1.12 \pm 0.16$. SOURCE: Reprinted by permission from Springer: A. Cavagna, D. Conti, C. Creato, L. Del Castello, I. Giardina, T.S. Grigera, S. Melillo, L. Parisi, and M. Viale, 2017, Dynamic scaling in natural swarms, *Nature Physics* 13:914, copyright 2017.

active matter models, and there are continuing efforts to find a theory that captures these behaviors. Methods range from the renormalization group analysis of a wider range of microscopic theories to direct inference of statistical descriptions from the data using maximum entropy methods. This last approach connects the study of collective animal behaviors to the study of sequence variation in protein families (Chapter 3) and patterns of activity in networks of neurons (Chapter 3).

Flocks of birds, swarms of insects, and schools of fish function in unconfined environments. In contrast, communities of ants and termites that construct tunnels and structures in soft materials such as soil must move collectively in their confined and crowded nests. Such densely packed and disordered conditions in non-living systems lead to a breakdown of flow, through glassiness and jamming (Chapter 5). Physicists studying the traffic of confined fire ant colonies have revealed that they routinely move through foraging tunnels that are comparable in dimension to the individual ants. Movement is hindered not only by the spatial restrictions, but by social interactions when ants moving in opposite directions encounter one another and pause to touch antennae ("attenation"), presumably to exchange information. Models that incorporate these interactions exhibit a phase transition as a function of the attenation time, similar to the fragile glass or jamming transition. Longer attenation times likely allow for more effective information flow through the colony, but this is useless if the colony is jammed. Real colonies appear to function close to the transition.

Social Insects and Superorganisms

Flocks of birds, swarms of insects, and schools of fish are undeniably emergent phenomena. These also are social behaviors, but evidently this term covers a much wider range of possibilities. In some cases, the collective behavior is so compelling that what emerges is a "superorganism," as with social insects such as termites, ants, and social wasps. These superorganisms breathe, feed, grow, breed, and modify their environments, as with the termite nests in Figure 3.15.

To appreciate the analogy between a superorganismal insect colony and individual organism, consider a colony of fire ants composed of hundreds of thousands of individuals. The superorganism is composed of individual colony members in the same way that an organism is built of cells. Instead of specialized organs, superorganisms consist of specialized castes responsible for different functions, all of which contribute to the survival and reproduction of the whole group. Thus, superorganisms display tremendous cooperation and integration in roughly the same way that cells of a single organism work together to help the individual succeed. Importantly, from a physicist's perspective, the processes which define the behavior of the superorganism are emergent, such that the rules by which the higher levels function are often unknown and might not be easily predictable from the behavior of a single organism.

Ants and other invertebrates have external skeletons (exoskeletons) that give the individual organisms shape, structure, and strength. The superorganism also has a nest that provides protection and a location where food is returned and where young animals are reared. Nests promote division of labor among individuals, modulate communication and information distribution, and regulate the physical environment. Indeed, social insect nests are often referred to as part of the "extended phenotype" of the colony. Importantly, no single insect has a conception of how the nest should be built or what it should look like when it is complete; insects do not have managers directing the construction process. Instead, social insects use micro-scale rules that lead to the formation of the complex colony exoskeleton. One important aspect of colony self-organization is the concept of stigmergy. Stigmergy is a process of indirect coordination and activation of behaviors through environmental signals. For example, rapidly growing structures within a nest may act as strong stimuli for additional construction and grow quickly until a positive stimulus plateau is reached, triggering negative feedback leading to the reduction in construction. Stigmergic models are often "agent based" and consider insects (agents) to be engaged in sets of limited behaviors when encountering particular environmental stimuli.

The biological physics community's understanding of flocks and swarms began with somewhat complicated agent-based models from the biological literature and went through phases of simplification and deeper theoretical analysis, followed by dramatic improvements in quantitative measurement that exposed new statistical physics problems. The understanding of social insects seems somewhere near the beginning of this process, and it is encouraging to see new experiments probing the collective behaviors of honeybees, ants, and others using modern physics-based approaches. There are significant technical challenges in tracking individuals through much denser groups, and in some cases having to work in opaque environments, and it is reasonable to expect progress on these experiments over the next decade.

Perspective

In a flock, as in a fluid, natural macroscopic variables are spatial averages over the velocities of the component parts, be they birds or molecules. Still, finding the correct effective theory for these coarse-grained variables is an unfinished project—these collective behaviors belong to a universality class beyond what we have understood from conventional physics problems. In contrast, it does not seem likely that spatial averages over the behavior of individual termites will capture their contribution to nest building. A statistical physics of social insects will require development of approaches in which coarse-grained variables change their character as we change the scale of our observations, leading to more new physics. As in neural networks (Chapter 3), the distinction between systems that can search for a quasistatic equilibrium and systems that generate spontaneous and self-sustaining dynamics is blurring. In flocks and swarms, and with social insects, the search for theories proceeds in parallel with dramatic improvements in experimental observations, and there are opportunities for substantial leaps forward in the coming decade. The world of collective behaviors is much larger than described in this section, and it is possible that the deepest insights will come from taming an example that currently is only barely explored.

4

How Do Living Systems Navigate Parameter Space?

Any attempt at a "realistic" description of biological systems leads immediately to a forest of details. Making quantitative predictions about the behavior of a system seems to require knowing many, many numerical facts: how many kinds of each relevant molecule are inside a cell; how strongly these molecules interact with one another; how cells interact with one another, whether through synapses in the brain or mechanical contacts in a tissue; and more. The enormous number of these parameters encountered in describing living systems is quite unlike what happens in the rest of physics. It is not only that as scientists we find the enormous number of parameters frustrating, but the organism itself must "set" these numbers in order to function effectively. Many different problems in the physics of living systems, from bacteria to brains, revolve around how organisms navigate this parameter space through the processes of adaptation, learning, and evolution.

As will be familiar from examples in previous chapters, the biological physics community has made progress on understanding adaptation, learning, and evolution by engaging with the myriad details of particular examples. But standing behind these details is an approach to living systems more generally. The physicists' approach to describing any particular functional mechanism invites us to see that mechanism as one possibility drawn from a large space of alternatives. In this view we think not about one system but about a distribution or ensemble of systems, much as in the statistical physics of disordered systems. Crucially, the relevant ensembles of biological mechanisms are neither random nor truly disordered, but sculpted by the constraints of functionality. The work surveyed in this chapter includes different approaches to characterizing these functional ensembles, and the dynamics through which these ensembles are selected. Table 4.1 provides an overview of these problems.

ADAPTATION

When we step from a dark room into the bright sunlight, we are momentarily blind, but then the cells in our eye "adapt." More generally, sustained, constant sensory inputs tend to fade away. During adaptation, the concentrations of various molecules in the individual cells of our eye are changing, in effect shifting the parameters that describe the response of these cells to light. Related forms of adaptation happen in all cells as they respond to signals in their environment.

TABLE 4.1 Navigating Parameter Space

Discovery Area	Page Number	Broad Description of Area	Frontier of New Physics in the "Physics of Life"	Potential Application Areas
Adaptation	159	Understanding how living systems deal with signal spread across an enormous dynamic range, with changing statistical structure.	Near-perfect adaptation without fine tuning; optimal coding and adaptation to input statistics; from control of gene expression to sloppy models.	Adaptive sensors.
Learning	166	Formalizing and quantifying colloquial ideas of learning in animals, humans, and machines.	Statistical physics of inference; phase transitions and "aha;" connecting molecular complexity to learning algorithms.	Neural networks in artificial intelligence; principles of machine learning; natural language processing.
Evolution	170	Taming the complexity of evolutionary dynamics through targeted experiments and simple models; bringing this understanding back into the real world.	Evolutionary dynamics are driven by the tails of distributions; finding regimes where evolution is predictable; new experiments to track thousands of coexisting strains of microbes; the immune system as evolution in microcosm.	Vaccine design; tracking epidemics.

Sensory Adaptation

Some of the earliest experiments on the responses of single sensory neurons showed that the perceptual fading of constant inputs has a corollary in a slow decay of the neural response to the same constant inputs. But adaptation is more than subtracting a constant. In many photoreceptor cells, for example, constant background light results in a reduction in the amplitude of the response to each additional photon along with a speeding up of this response. These effects are linked through the molecular cascade that amplifies the single photon response (see Figure 1.11), since the gain of the system depends on the time that individual molecules spend in their active states. This picture of scaling amplitudes and time scales in the adaptation of photoreceptor responses was understood and connected quantitatively to experiment even before the identity of the components in the cascade were known, testimony to the power of phenomenological theories.

Neurons that respond to sensory inputs also adapt to statistical features of these inputs beyond their mean. Single cells at the output of the retina, for example, adapt to the dynamic range over which light intensities are varying, and to the spatial and temporal correlations in these variations. In some cases, these adaptive dynamics can be traced down to the dynamics of particular ion channels, which have long time scales of inactivation after opening during an action potential. Similarly, at different stages of the auditory system neurons adapt to the dynamic range and temporal statistics of incoming sounds. Adaptation to input statistics is predicted by theories of neural coding that maximize the information captured in a limited number of action potentials given that the natural sensory world has intermittent dynamics (Chapter 2), and in some cases it has been possible to show that the form of adaptation really does optimize information transmission, quantitatively.

Although adaptation often is described as a strategy for dealing with slowly changing signals, these changes in response to changing input statistics can be so fast that they are difficult to resolve; reliable changes in the form of the response can occur essentially as soon as the system can reliably infer that the distribution of inputs has changed. More generally, efficiency arguments suggest that the time scales of adaptation can be related to the time scales on which the statistical features of the environment vary. This happens, even in the responses of a single neuron to injected current (see Figure 4.1). In this case, at least, more detailed analyses show that the dynamics are best described not by a single adaptation time scale that changes in response to the inputs, but instead by a near continuum of time scales in parallel. The result is that adaptation discards constants by taking a fractional derivative, or equivalently by comparing signals with a memory of the past that decays only as a power-law.

Adaptation can be seen not only in neural signal processing but also in the sensory responses of single celled organisms, such as bacterial chemotaxis (Chapter 1). In chemotaxis, adaptation makes the cell's probability of running or tumbling

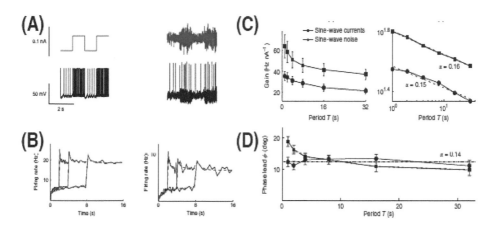

FIGURE 4.1 Single neurons exhibit multiple time scales of adaptation, resulting in nearly scale-invariant behaviors. (A) Raw voltage responses of a cortical neuron (*bottom*) to injected currents (*top*). At left, the current changes periodically, as a square wave. At right, the mean current is constant but the variance of the current changes periodically. (B) Counting the action potentials in the voltage traces from (A), and averaging over many periods to give the rate versus time. An increase in current (*left*) or current variance (*right*) is associated with a sudden increase in rate, which then relaxes, a sign of adaptation. The time scale of the relaxation gets slower as the period of the changes gets longer. (C) Modulation of the rate of action potentials in response to sinusoidal variations of current or current variance. Response declines as a (small) power of the sine wave period. (D) With sinusoidal inputs, rate is phase shifted by an angle that is almost independent of period. This phase shift agrees with the exponent from (C), consistent with the response being a fractional derivative of the input. SOURCE: Reprinted by permission from Springer: B.N. Lundstrom, M.H. Higgs, W.J. Spain, and A.L. Fairhall, 2008, Fractional differentiation by neocortical pyramidal neurons, *Nature Neuroscience* 11:1335, copyright 2008.

sensitive to the derivative of the concentration of attractive or repellent molecules, ignoring the absolute value. Theory makes clear that this is an essential part of the cell's strategy for advancing up the concentration gradient, and careful experiments show that the adaptation is nearly exact, so that, for example, after a step increase in concentration the behavior returns precisely to its baseline. Early models for adaptation envisioned two parallel pathways of response to input, one fast and one slow, which are combined with opposite signs to generate the final output. In this broad class of mechanisms, perfect adaptation requires fine-tuning of the responses in the two pathways so that they cancel, which seems implausible. An alternative is adaptation through feedback, which in some limits can ensure that any steady output is at a fixed level independent of inputs. Detailed analysis of the molecular events in Figure 1.12 shows that this in fact is what happens, providing a concrete example of how complex biochemical mechanisms are sculpted by the cell's need to solve the physics problems involved in climbing the gradient.

Adjusting Gene Expression and Protein Copy Numbers

In chemotaxis the adaptation mechanism involves biochemical reactions, crucially the attachment and detachment of methyl groups on the receptor by the enzymes CheR and CheB. These reactions occur on the time scale of seconds. Cells can also adapt to their environment more slowly by changing the levels of gene expression, or equivalently the number of copies of different proteins. These crucial regulatory processes were discovered by studying the way in which bacteria respond to changes in the available nutrients. For a generation, the focus of these explorations was on the regulation of single genes in bacteria, for example the gene for one crucial enzyme in the consumption of a particular sugar. This work provided the paradigms for how information encoded in regulatory sequences of DNA (Chapter 2) and information encoded in the concentration of transcription (Chapter 2) combine to control the expression of one gene.

In many cases, it is interesting to ask how the expression levels of many genes are coordinated to accomplish functions. Part of the problem is to understand the landscape of performance as a function of the many expression levels. One important example of this is the simple conversion of nutrients into growth or biomass, where the relation between the number of enzyme molecules and the conversion rate is determined by the classical chart of biochemical reactions in core metabolism. A more subtle example is the electrical activity of neurons, which depends on the numbers of different ion channel proteins through equations that are known very precisely.

The human genome, like that of many animals, encodes more than 100 types of channels, and a single neuron chooses perhaps seven different kinds of channels from this large set of possibilities. Each channel type itself is described by many parameters, including the rates at which they open and close, controlling the flow of electric current into the cell, and the dependence of these rates on the voltage across the cell membrane. The result is that just one neuron is described by 40 or even 50 numbers, which can be different in every cell, and there are billions of cells in the brain.

One can measure the properties of individual channel molecules, but it is more difficult to make an independent measurement of the number of channels of each type that are in the cell membrane. Thus, at a minimum, describing the electrical dynamics of neurons requires us to fit these numbers to the overall behavior of the cell. By the 1980s, building these sorts of models was a major activity in the neuroscience community. It was known, but not widely discussed, that inferring the number of each type of channel was a challenging problem.

Our mathematical description of ion channel dynamics rests on a firm foundation, one of the classical chapters in the interaction of physics and biology, as described in Chapter 2. The result of all these developments is that, in contrast to the typical situation in describing networks of interacting proteins in a cell, the equations that describe the dynamics of ion channels and their interaction with the voltage across the cell membrane are known quite precisely. In the 1990s, theoretical physicists suggested that relying on this knowledge allows a new approach: rather than thinking of these as models tuned to describe the behavior of particular cells, one can take the family of models seriously as a theory of possible cells.

If cells are committed to making particular types of channels, then the universe of possible cells is defined by the number of each channel type that is synthesized and inserted into the membrane to become functional. This theoretical space of possibilities is the same space that real cells explore as they control the expression of ion channel genes; see the discussion of these control mechanisms in Chapter 2. Figure 4.2 shows the behaviors in different slices through the space of possible cells. Importantly the calculations start from combinations of ion channels that occur in particular, well-studied neurons.

Relatively small changes in the number of channels can lead to qualitative changes in the pattern of electrical activity, and this is more obvious in the high-dimensional space of possibilities facing the cell. On the other hand, there are directions in this high-dimensional space where variations in behavior are modest. Each time the cell generates an action potential, calcium ions enter the cell, and these ions are pumped back out (or into internal storage spaces) more slowly, so that the calcium concentration provides a record of electrical activity; this is what makes possible the use of calcium-sensitive fluorescent proteins to monitor the electrical activity of neurons, as in Figure 3.13. Of all the ions that contribute to electrical current across the cell membrane, calcium is special because it both carries electrical current and serves as a signaling molecule for various biochemical processes inside the cell. This immediately suggests that cells could tune their pattern of electrical activity by increasing or decreasing the synthesis of particular ion channels in response to changing internal calcium concentration. Real molecular mechanisms will be sensitive to concentration averaged over some limited time, and by using mechanisms that have different averaging windows the cell could achieve even more fine-grained control.

The theoretical proposal that neurons regulate the number of channels in response to their own patterns of electrical activity was confirmed almost immediately by experiments in a variety of systems, from the small networks that generate digestive rhythms in the crab gut to cells in the mammalian cerebral cortex, responsible for our thoughts and actions. The exploration of this phenomenon, and its underlying mechanisms, has become a substantial effort in the mainstream of neurobiology, spreading far from its origins as a theoretical physics problem. There are many potentially general lessons.

First, if there is pressure for the organism to achieve certain dynamical behaviors, but no explicit preference for one molecular implementation over another, then

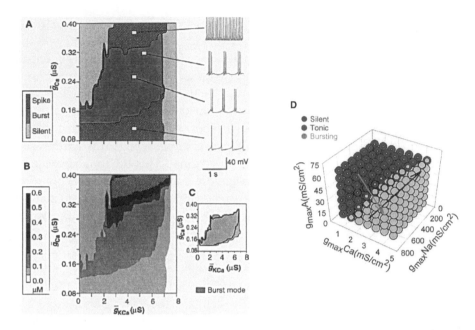

FIGURE 4.2 Possible electrical dynamics of a neuron as a function of the number of ion channels in the membrane. At left, the neuron uses seven different types of channels, and behavior is mapped versus two of these, a channel that is selective for calcium ions (Ca) and one that is selective for potassium (K) but dependent on calcium; other channel copy numbers are held fixed. By convention, the number of channels is measured by their maximal contribution to the electrical conductance across the membrane. (A) Patterns of membrane voltage versus time include silence, the repeated generation of single action potentials or spikes, and bursting with two or three spike per burst. (B) The average concentration of calcium inside the cell. (C) Superposition of (A) and (B), showing that bursting corresponds to a well-defined range of calcium concentrations. (D) A simpler cell with five types of channels, but the five-dimensional space is explored fully and projected into three dimensions corresponding to the Ca channel as before, a sodium (Na) channel, and the "A current" channel; conductances measured per unit area of the cell membrane. Green (black) arrow denotes the direction of highest (lowest) sensitivity. The size of the gold ball inside the blue ball, for example, indicates the probability that variations in the other two parameters will lead to bursting as opposed to silence. SOURCES: (A–C): From G. LeMasson, E. Marder, and L.F. Abbott, 1993, Activity-dependent regulation of conductances in model neurons, *Science* 259:1915, reprinted with permission from AAAS. (D): M.S. Goldman, J. Golowasch, E. Marder, and L.F. Abbott, 2001, Global structure, robustness, and modulation of neuronal models, *Journal of Neuroscience* 21:5229, https://doi.org/10.1523/JNEURO-SCI.21-14-05229.2001, copyright 2001 Society for Neuroscience.

there is no single channel protein whose number needs to be controlled precisely. This is not because cells are incapable of precise control (see, e.g., Chapter 2), but rather because the mapping from molecular mechanisms to macroscopic functions is many-to-one, echoing the ideas of Chapter 3. Second, what needs to be controlled are not the numbers of individual channels, but rather combinations. This predicts that while protein copy numbers are variable, correlations in these fluctuations carry the signature of functional constraints. Finally, different combinations of protein copy numbers have vastly different impacts on functional behavior.

Sloppy Models and Reduced Dynamics

A different approach asks not about particular settings for parameters of the system, but more explicitly about the distribution or ensemble of parameters that are consistent with the observed behavior. This corresponds to constructing a statistical mechanics in parameter space, and predicts the distributions, for example, of protein copy number variations that are seen in real systems. This distribution has a geometry, being compact along directions that correspond to combinations of parameters whose variation generates big effects, and broad along directions that have small effects (see Figure 4.3). The surprise is that these distances in parameter space are almost uniformly distributed on a logarithmic scale: There is one combination of parameters that is most tightly constrained, another which is allowed to vary twice as much, another four times as much, and so on. It has been suggested that systems with this sort of behavior form a well-defined class of "sloppy models," neither robust nor finely tuned but with a full spectrum of parameter sensitivities. There is a substantial effort underway to understand the origins of this behavior, its connections to other ideas in statistical physics, and its implications for the dynamics of adaptation. It would be exciting to connect these ideas with other examples in which many-to-one mapping arise, such as the sequence/structure mapping for proteins (Chapter 3).

The theory of dynamical systems provides us with settings in which behaviors become universal, and thus explicitly independent of most underlying parameters. If we think, for example, about models for genetic networks that can describe a developing cell making choices among alternative fates, then in the neighborhood of the decision point the dynamics takes a stereotyped form. Building outward from these decision points allows construction of a geometrical model for the dynamics more globally, in which coordinates are abstract combinations of gene expression levels. As it becomes possible to follow gene expression levels through the steps of cellular differentiation, this approach makes it possible to search for the simplified collective coordinates and to classify the impacts of perturbations, with almost no free parameters.

Perspective

Some form of adaptation occurs in almost every living system, matching its behavior to relatively short-term variations in the demands of the environment. The phenomena range from the gradual fading of constant sensory inputs to the intricate control of gene expression, and more. Even seemingly simple examples

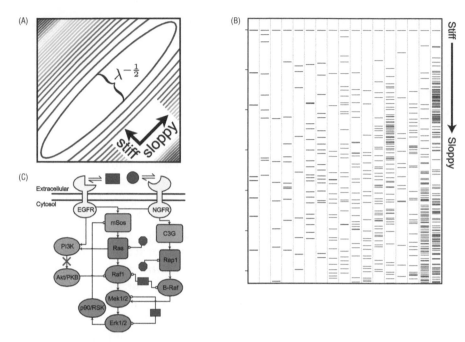

FIGURE 4.3 Sloppy models. (A) Contours show the (mean-square) difference in the behavior of a model as a function of two parameters θ_1 and θ_2. There are "stiff" and "sloppy" directions, combinations of the original parameters. The typical distance in parameter space needed to cause a small change in behavior is determined by the eigenvalues λ of an appropriate matrix. (B) The spectrum of eigenvalues in a wide range of models for biochemical reaction networks, ranging from embryonic development to hormonal signaling to circadian rhythms. (C) A model for growth factor signaling, corresponding to column (i) in (B); this model has 48 parameters. SOURCE: R.N. Gutenkunst, J.J. Waterfall, F.P. Casey, K.S. Brown, C.R. Myers, and J.P. Sethna, 2007, Universally sloppy parameter sensitivities in systems biology models, *PLoS Computational Biology* 3:e189, copyright 2007.

are deeper than they first appear, and new forms of adaptation—often suggested by theory—continue to be discovered. This circle of ideas provides some of the most concrete examples of the idea that real living systems should be seen as examples drawn from a larger set of possible systems. There is a vigorous theoretical effort to understand how and when the high-dimensional parameter spaces of these systems can be collapsed to lower dimensionality, a kind of emergent simplicity. In the coming decade we can hope to understand whether this simplicity is generic, or whether it is itself selected by evolution.

LEARNING

The word learning is used, colloquially, to describe many things—learning a language, learning a rule, learning to play a musical instrument, learning physics.

There is a long history of both psychologists and computer scientists formalizing these colloquial ideas, in the hopes of describing human behavior or learning machines. The biological physics community entered the subject through models for networks of neurons, as described in Chapter 3.

Statistical Physics Approaches to Learning

The functions that are accomplished by a neural network are determined by the strengths of connections or "synaptic weights," among all the neurons in the network. An important idea, which echoes ideas from many different sections of this report, is that one should think not about particular settings of the synaptic weights, but about a probability distribution over these weights. Given only a limited number of examples of what the network should be doing, such as assigning names to images of faces, then there are many combinations of synaptic weights that are consistent with these examples. There may also be some tolerance for error. If one makes the analogy between error and energy, so that low energy states are close to the correct answer, then a tolerance for error is analogous to temperature.

For real materials, it is natural to plot a phase diagram (e.g., mapping gas, liquid, and solid to different parts of the plane defined by temperature and pressure). Statistical mechanics teaches us that these sharply defined phases emerge from a probability distribution over all the microscopic states of the material in the limit that number of atoms or molecules becomes large. For neural networks, there thus will be sharp phases when the number of neurons or connections becomes large, which surely is relevant for real brains. Natural coordinates for the phase diagram are the tolerance for error and the number of examples that the learner has seen. In the simplest perceptron model—where a single neuron takes many inputs and classifies them into two groups—not only is there a phase transition into the state where the network has learned the correct classification, but in this case the transition is discontinuous (first order) so that the fraction of errors drops abruptly as the network is exposed to more and more examples, as if it experienced an "aha!" moment.

Learning can be thought of as the inference of some underlying parameters (e.g., the synaptic weights) from given data (e.g., examples of correct input/output pairs). Parallel to the statistical physics of learning is the statistical physics of inference. This is interesting as an approach to inference problems solved by the brain, but also as a way of thinking about data analysis and data acquisition strategies such as compressed sensing. Closing the circle are ideas about how olfactory signaling, for example, may instantiate compressed sensing. Statistical physics approaches to the original problem of learning in model neural networks have had a resurgence in response to the deep network revolution in artificial intelligence (Chapter 7), but it is too soon to say how these ideas will influence thinking about the brain itself.

Perhaps the most dramatic example of learning by the human brain is the learning of language. Instead of thinking about a probability distribution over synaptic weights in a network of neurons, we could think of a probability distribution over the parameters of model languages. In particular, if we have grammatical structures that are defined by rules (e.g., replacing nouns by noun phrases), there are probabilities that these rules will be applied in constructing long grammatical statements. Recent work shows that ensembles constructed in this way have phase transitions as a function of the natural parameters in the underlying rules, and that the different phases can be distinguished by their order or correlation. In this approach, then, there is a phase transition between a kind of incoherent babbling and the construction of potentially meaningful sentences. These are first steps in a long and ambitious program.

Connecting with Real Neurons

The introduction of probabilistic ideas into learning also has had implications for the design and analysis of experiments. Songbirds learn their songs, and continue listening to their own songs to stay in tune. If this auditory feedback is disrupted, songs will drift. More systematically, experimentalists can play noise to the bird whenever he sings a note below a certain pitch, and the bird will learn to compensate, driving the pitch upward over a period of hours or days (see Figure 4.4). Under steady conditions, the distribution of pitch across multiple examples of a single note gives a measure of the bird's own tolerance for errors, in the language of statistical physics models for learning in networks. Experiments show that this distribution has long tails, which means that small deviations from the correct pitch are heavily penalized, but this becomes less steep at larger deviations. Placed in the larger theoretical context of learning, this implies that the bird will have difficulty following cues that would drive larger, immediate changes, but could easily follow repeated small changes over the same total excursion. This sort of behavior is seen in many learning problems, and the songbird experiments agree quantitatively with experimental predictions.

Rather than using theory to understand macroscopic learning behaviors, other groups in the biological physics community have tried to push down to the molecular events at real synapses to understand how learning rules are implemented. Part of what was so exciting about the very first symmetric models for neural networks is that they could learn a new memory by changing the synaptic weight between two neurons in proportion to the correlation between their activities. This is a mathematically precise version of an old idea in the biological and even psychological literature that neurons that "fire together wire together."

A deeper theoretical examination of synaptic dynamics, however, shows that limits on the number of distinguishable states of individual synapses quickly leads

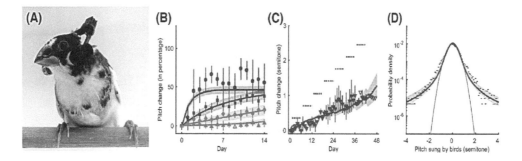

FIGURE 4.4 Songbirds learning to shift pitch in response to altered auditory feedback. (A) A Bengalese finch with headphones, so that experimentalists can control what the bird hears while it is singing. (B) The pitch of a single note is "pushed" by altering the sound of the bird's song as heard through miniature headphones. Pitch shifts are 0.5 (brown), 1 (blue), 1.5 (green), or 2.0 (cyan) semitones. Small shifts are ~50 percent compensated, while the largest shifts are hardly compensated at all. (C) In contrast to (A), large shifts can be compensated if done in a series of smaller steps (dotted lines). (D) Distribution of pitch variations before altered feedback (brown), with much longer tails than a Gaussian distribution (gray). Theory links the difference between (B) and (C) through (D), quantitatively, as shown by the smooth curves. SOURCES: (A) Reprinted by permission from Springer: S.J. Sober and M.S. Brainard, 2009, Adult birdsong is actively maintained by error correction, *Nature Neuroscience* 12:927, copyright 2009. (B–D) B. Zhou, D. Hofmann, I. Pinkovievsky, S.J. Sober, and I. Nemenman, 2018, Chance, long tails, and inference in a non-Gaussian, Bayesian theory of vocal learning in songbirds, *Proceedings of the National Academy of Sciences U.S.A.* 115:E8538.

to new memories overwriting the old if the network continues learning over an organism's long lifetime. This problem can be solved if the synapses themselves have dynamics with multiple time scales, as in Figure 4.5. It is known that molecular mechanisms involved in changing the strengths of synapses have many, many molecular components, and so more complex dynamics are to be expected. What is important is these dynamics are not just more complicated but need to be selected to have properties that solve a physics problem faced by the network. There is an important challenge in connecting more detailed experiments on the microscopic mechanisms of synaptic plasticity to a larger theoretical framework for learning in networks. Experimental studies that have tracked synaptic dynamics in the live mammalian brain suggest that the adult neocortex, which is thought to store some memories for the adult lifetime, has different subsets of synapses with different lifetimes, in the spirit of these theoretical considerations.

Perspective

The study of learning has had many independent lives: in psychology and animal behavior; in mathematics and computer science; and in neurobiology, genetics,

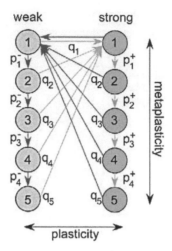

FIGURE 4.5 Persistence of memory and the internal dynamics of synapses. A model synapse can be strong (blue) or weak (gold), but hidden behind the synaptic strength are multiple internal states. When conditions are such as to trigger strengthening of the synapse, system transitions into blue state 1; conversely weakening of the synapse causes a transition into gold state 1. Transitions among the internal states occur spontaneously, with transitions to deeper states occurring more slowly, thus generating a cascade of time scales. If we measure the signal-to-noise ratio for stored memories as a function of time in storage, this cascade model achieves a gentle decay that is not possible with simpler dynamics. SOURCE: Reprinted from R.S. Fusi, P.J. Drew, and L.F. Abbott, 2005, Cascade models of synaptically stored memories, *Neuron* 45:599, copyright 2005, with permission from Elsevier.

and pharmacology. All these groups have touched different aspects of the problem. The biological physics community is unique because it has engaged with learning at all levels, from the molecular events at synapses to animal behavior, through both theory and experiment. This is important, because while there has been great progress in each of the many directions, major open questions exist about how these different directions are connected: How is molecular complexity at the synapse related to the efficacy of learning? How do networks learn effectively when the number of synaptic connections is much larger than the number of examples that the animal—or the artificial neural network—has seen? The biological physics community is playing a key role in sharpening these questions, and it is reasonable to expect substantial progress over the coming decade.

EVOLUTION

Over longer time scales, evolution can change almost anything in biology, from the rules of the genetic code to the structure of proteins, the logic of gene regulatory networks, and the ways in which organisms learn. The fact that all living systems have arisen through such an evolutionary process imposes a general constraint that tunes parameter values throughout the rest of biology. This makes evolutionary optimization a key simplifying principle throughout biology: Biological systems have purpose and function to the extent that evolution can measure and select for those properties, and they navigate to highly constrained regions of parameter space to do so. The biological physics community began to engage more deeply with these issues at the start of the 21st century, starting by developing theories for evolutionary dynamics in the simplest possible contexts. Even these simple examples had surprises, and natural formulations as statistical physics problems in which the mean behavior of a population is controlled by the extreme tails of the distribution over individuals. From this has grown a vigorous program of both theory and quantitative experiment, connecting abstract ideas from physics to the detailed behavior of real organisms.

Statistical Dynamics in Fitness Landscapes

In the extreme, it is possible that evolution selects for functional performance close to the relevant physical limits. Examples of near optimal performance include the diffraction-limited optics of insect eyes, photon counting in vision and molecule counting in chemotaxis (see Figures 1.11 and 1.12), and more. There are many efforts in the biological physics community to turn these observations into theoretical principles from which aspects of system behavior and mechanism can be derived, as described above. It is even possible to imagine optimization principles for evolutionary dynamics itself. The mutation rate in copying DNA from one generation to the next is reduced by proofreading mechanisms, analogous to Maxwell's demons (Chapter 2), but this comes at an energetic cost. The rate of evolution itself is subject to evolutionary change, and this leads to the regular appearance of "mutator" strains of bacteria that can adapt more rapidly to novel and stressful environments.

A corollary of ideas about optimization is that the parameters of living systems should not be more constrained than is required to reach some criterion level of function. In large families of proteins, for example, patterns of conservation and diversity allow us to identify regulatory elements and highlight proteins and protein domains that have particular functional roles. Correlated patterns of amino acid substitutions within a protein family can be a signature of natural selection to restore physical interactions between residues that are in contact in the three-dimensional structure of the protein. As discussed in Chapter 3, researchers from the biological physics community have used methods from statistical physics to describe the distribution of sequences as being as variable as possible consistent with the observed correlations. This leads first to the possibility of drawing new sequences from the distribution, circumventing the sequential nature of normal evolution and synthesizing new proteins; a strong confirmation of this theoretical framework is that a large fraction of these new proteins fold and function. Analysis of the models allows a disentangling of direct physical interactions from indirect correlations, and this information can be used to predict the three-dimensional structure of the corresponding proteins.

The conditions under which evolutionary dynamics allow for optimization are much less clear: Evolution does not itself have a direction or purpose. Instead, macroevolutionary processes are the collective outcome of an enormous number of individual cell divisions, each of which can introduce errors due to random mutations, combined with the effects of natural selection and genetic drift that act on this novel variation. The nature of available genetic variation does not necessarily allow for full exploration of the relevant parameter space, and even in cases where such variation exists, evolution cannot always favor it. For example, in small populations there are important limits to the efficiency of natural selection, and evolution can actually in some circumstances lead to a degradation in function. The biological physics community has played a key role in recent efforts to understand these dynamics of stochastic evolutionary processes and the limits they place on optimization arguments in living systems.

An important contribution from the biological physics community has been the calibration of ideas about evolutionary dynamics through the study of simplified models. Perhaps the simplest evolutionary problem is a finite population of organisms that can always mutate to achieve slightly higher fitness. But as mutations arise and compete with one another, beneficial mutations can go extinct before reaching sufficient population size to take over the population as a whole. At the start of the 21st century it was realized that in these problems the evolution of the population as a whole is driven by the tail of the distribution of high-fitness individuals, connecting to other statistical physics problems; even the overall rate of evolutionary progress toward higher growth rates has a subtle dependence on population size, mutation rates, and the selective advantage of each mutation. More recently it has been possible to bring similar rigor to the analysis of much more complex evolutionary scenarios; examples include spatial structure, fluctuating environments, and perhaps most importantly the interactions between evolutionary and ecological dynamics that create and maintain diverse communities of organisms.

The statistical physics approach to evolutionary dynamics identifies multiple regimes for these dynamics, two examples of which are illustrated in Figure 4.6. Although the engine of evolution is random mutation, there are regimes in which the future trajectory of evolutionary change can be predicted. There is evidence that viruses which infect humans, such as those that cause COVID-19 and the seasonal flu, are in this regime. In an unexpected turn, the biological physics community's approach to evolution thus has clear practical implications, in particular for the design of vaccines, as discussed in Chapter 7.

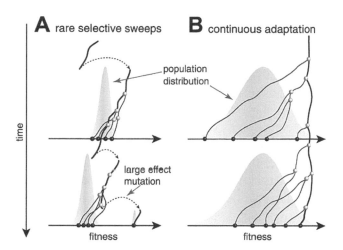

FIGURE 4.6 Statistical dynamics of evolution in two different regimes. (A) Evolution proceeds via rare large effect mutations (dashed arrows) that occur in a population with little fitness variance. All individuals are roughly equally likely to pick up the large effect mutation, rendering evolution unpredictable from sequence data alone. (B) If evolution proceeds by the accumulation of many mutations, each having a small effect, the successful lineage (thick) is always among the most fit individuals. Being able to predict relative fitness therefore enables researchers to pick a progenitor of the future population. SOURCE: R.A. Neher, C.A. Russell, and B.I. Shraiman, 2014, Predicting evolution from the shape of genealogical trees, *eLife* 3:e03568. Creative Commons License Attribution 4.0 International (CC BY 4.0).

Evolution as an Experimental Science

The explosion in genome sequencing has turned observations of evolution in the wild into a data rich, quantitative enterprise. It is now relatively straightforward to sequence near-complete genomes from thousands of individuals from essentially any species, including humans, and use the observed patterns of genetic variation within the population to infer key aspects of evolutionary history. This enterprise has helped us understand the demographic history of humans, leading to productive exchange between evolutionary biologists and archaeologists, and has shed light on adaptation in response to pathogens such as malaria and the plague. In other species, genome-based investigations of evolutionary history have led to insights on population structure, local adaptation, and speciation.

In microbial and viral populations, we can track evolutionary changes by sequencing populations as they adapt to the human gut or as viral epidemics spread across the world. This allows us to go beyond inferences of evolutionary history from sequence data in the present, and to observe evolution acting in real time. Technology for analyzing ancient DNA offers the promise to directly observe change through time in populations where this would otherwise be impossible, such as humans.

However, all of these studies of natural populations are inherently observational. While we sometimes have the opportunity to track multiple populations—observing intra-patient evolution of HIV across many infected individuals, or tracking the evolution of *Pseudomonas* infections across multiple cystic fibrosis patients—it is impossible to precisely control parameters or conduct replicate studies. Thus to make evolution an experimental science, there have been parallel efforts to bring evolution into the laboratory.

Laboratory evolution experiments have been conducted in a wide range of organisms, including obligately sexual eukaryotic organisms such as the fruit fly *Drosophila melanogaster* and the nematode *Caenorhabditis elegans*. However, microbial and viral populations are in many ways the ideal model systems for these studies. In these organisms, we can conduct hundreds or thousands of evolution experiments in parallel and we can freeze clones or whole-population samples to create a "fossil record" for future study. Unlike the actual fossil record, these frozen samples can actually be resurrected later and compared directly to their descendants. Microbes and viruses have manageable genome sizes that make it possible to sequence many individuals from many replicate populations through time at a reasonable cost, and in many of these organisms, genetic tools allow manipulations such as inserting fluorescent markers or elevating mutation rates. Finally, we can leverage our deep understanding of cell biology and genetics in some of these organisms to interpret results in a functional context.

The most well-known and deeply studied evolution experiment is the long-term evolution experiment (LTEE). As discussed in Chapter 6, this study has propagated 12 initially identical independent replicate populations of the bacterium *Escherichia coli* in the same environment since 1988, a total of nearly 75,000 generations to date. This experiment originated in the evolutionary biology community, and has been used as the basis for important insights into the evolution of novelty, the interactions between ecological and evolutionary dynamics, the evolution of mutation rates, and many other questions. It has also generated a rich data set (including a fossil record of samples frozen every 500 generations) for the broader community, and biological physicists have played a key role in analyzing these data and making connections to theory.

While the LTEE is a remarkably rich resource, it involves only 12 replicate populations of a single organism evolving in response to a single selective pressure. Because evolution is an inherently stochastic process, numerous groups have made efforts to expand the scale of replication in order to quantify the probabilistic outcomes of adaptation, and have explored the generality of conclusions in other organisms and environments. The physics of living systems community has played a central role in many of these efforts, both in developing theoretical predictions and in conducting more highly replicated experiments that allow for closer quantitative comparison between theory and experiment.

Guided by theoretical studies showing that microbial populations are often in a regime where small fractions of the population can play a critical role in driving evolution of the population as a whole, and that large changes can result from the accumulation of many small advantages, one important experimental effort has been to push for higher resolution and more precise phenotypic measurements. This has included developing new methods to detect strains of single celled organisms that constitute only one ten-thousandth of a large population, and measuring the relative fitness of mutants to within one part in a thousand. As in other areas of physics, profound tests of our understanding come from precision experiments, but not so long ago, it would have been difficult to imagine such precision measurements on the dynamics of evolution itself.

Theorists from the physics of living systems community have also worked to extend models of evolutionary dynamics to include important additional features, such as the role of recombination within linear genomes, the emergence of ecological interactions, and the effects of spatial structure. These same theorists have established connections with experimental groups to test these theoretical ideas (or, in several cases, conducted experiments themselves). For example, this has led to major advances in our understanding of adaptation in spatially expanding populations, with implications both for microbial populations and for evolution more generally (see Figure 4.7). This combination of theory and experiment in the study of evolution is an example of how the intellectual scope of biological physics has expanded in the 21st century.

The Immune System

Many of the conceptual issues in evolutionary dynamics are illustrated in microcosm by the adaptive immune system. To respond to the enormous range of challenges from our environment, the immune system synthesizes billions of distinct types of antibody molecules. In contrast to other protein molecules, the genome does not contain the precise instructions for making these antibodies. Instead, the genome has multiple sequences for separate V, D, and J segments of an antibody, and individual cells in the immune system edit their own DNA to combine one of each segment into a full sequence. This recombination process is an important example of chromosomal dynamics (Chapter 3), and in this process random lengths can be deleted from the ends of the V, D, and J segments, and random nucleotides can be inserted. Further steps in the development of the antibody repertoire include removing those sequences that encode antibodies against the body's own molecules, as well as hypermutation in some cells to further diversify the population. The many different antibodies can bind to different kinds of molecules and molecular fragments from invading viruses or bacteria, and those cells making antibodies that engage in these binding events generally reproduce more

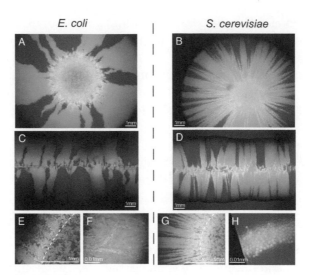

FIGURE 4.7 The combination of theory and experiment in the study of evolution, as shown here for adaptation in spatially expanding populations, is an example of the widening intellectual scope of biological physics. Evolution in spatially expanding populations is illustrated here by the competition between fluorescently labeled lineages in colonies of the bacterium *Escherichia coli* (*left*) and the yeast *Saccharomyces cerevisiae* (*right*). (A, B) Spatial gene segregation emerges as populations expand in both cases, but differences in the population dynamics at the front lead to different patterns of diversity. (C, D) The influence of geometry; linear expansions lead to different patterns of gene segregation. (E, G) Continuous patches of boundary regions at a magnification of 51× for bacteria (E) and yeast (G). (F, H) Images at single cell resolution (100×). (F) Tip of a bacterial sector dies out. (H) Section boundary at the frontier in yeast. SOURCE: O. Hallatschek, P. Hersen, S. Ramanathan, and D.R. Nelson, 2007, Genetic drift at expanding frontiers promotes gene segregation, *Proceedings of the National Academy of Sciences U.S.A.* 104:19926.

quickly. Thus, the antibody repertoire evolves over the lifetime of the organism, with diversity being generated at random and subject to both positive and negative selection pressures.

In the late 2010s, the exploration of antibody diversity was revolutionized by the possibility of using DNA sequencing to make deep surveys of antibody diversity in single organisms, first in model systems such as zebrafish and then in humans. The biological physics community pioneered these experiments, along with the theoretical analysis that followed. Over the course of a decade, precise probabilistic models were developed for the generation of diversity, providing a framework for inferring the parameters of recombination, deletion, and insertion events (see Figure 4.8). This theoretical work shows that the total entropy of antibody sequences is large (nearly 50 bits in humans), so that the actual repertoire at any moment is only a small sampling from the set of possible antibody molecules. Furthermore, only ~20 percent of this entropy arises from the combinatorial choices of V, D, and

J regions from the genome; the overwhelming majority of the diversity comes from the random insertion and deletion events. This is important because the enzymes that catalyze insertion and deletion can be controlled, and indeed there are clear connections between these controls and changing patterns of antibody diversity from embryonic to adult life.

Quantitative understanding of how diversity is generated provides a foundation for measuring selection in response to infections and leads to the surprising conclusion that the mechanisms for generating diversity already are biased strongly toward sequences that turn out to be functionally relevant in fighting infections. There are interesting theoretical questions about what it means for the antibody repertoire to be well matched to the distribution of possible challenges from the environment, with connections to other living systems that must represent information about the world with limited physical resources (Chapter 2). The enormous entropy of sequences means that most will be unique to individual members of a species even if they share the same genome. But the distribution is predicted to have anomalously large fluctuations in the probability itself, so that some sequences are overwhelmingly more likely than others. A crucial test of this prediction is to predict the probability that different individuals will share the same sequences. As shown in Figure 4.8B, in a group of hundreds of people, most sequences will be unique to each individual, but hundreds of sequences are shared among a large fraction of the group. This pattern is in excellent agreement with theoretical predictions. This is an especially interesting example because it makes explicit how the biological physics community has been able to tame the variability across organisms—in this case, humans—even to the point of showing how parameter-free predictions for the distribution of this variability arise from more fundamental theoretical considerations.

The examples of evolution in the immune system discussed so far involve the analysis of snapshots of the system at single moments in time. It is more challenging to get at the dynamics. An important test case for these ideas is also an important human health problem, HIV infection. As discussed in Chapter 7, input from the biological physics community played an important role in realizing that HIV evolves rapidly, even over the lifetime of a single patient, and that effective treatments should take this into account. More recently, it has become possible to sequence samples from the population of viruses over time in many individual patients, sometimes in conjunction with surveys of the patient's antibody repertoire. These experiments in the clinic provide an unusual opportunity to observe fundamental dynamics of co-evolution between the virus and the immune system. Ideas from statistical physics have been used to make clear that this process is far from equilibrium, and to disentangle the flow of information from the virus into the immune system versus the reverse process where the immune system is driving evolution of the virus.

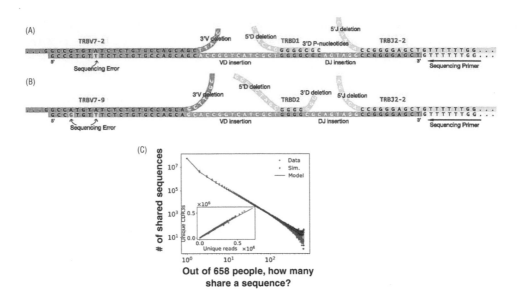

FIGURE 4.8 The biological physics community pioneered DNA sequencing experiments that revolutionized the exploration of antibody diversity. Statistical physics models provide a framework for inferring the parameters of recombination, deletion, and insertion events. (A) Schematic showing how particular V (pink), D (blue), and J (green) segments chosen from the genome are spliced together, with insertions and deletions, to generate the observed antibody sequence (gray). To be fully realistic, the observed sequence includes measurement (sequencing) errors. (B) As in (A), but showing that the same sequence could be explained as having been constructed from a different set of V and D regions. A rigorous probabilistic model expresses the probability of seeing any particular sequence as a sum over all these possibilities. (C) In sampling T-cell receptor sequences from 658 humans, the number of sequences that are shared among exactly K individuals. Theory and experiment agree with no free parameters. SOURCES: (A–B) A. Murugan, T. Mora, A.M. Walczak, and C.G. Callan, 2012, Statistical inference of the generation probability of T-cell receptors from sequence repertoires, *Proceedings of the National Academy of Sciences U.S.A.* 109:16161. (C) Y. Elhanati, Z. Sethna, C.G. Callan, T. Mora, and A.M. Walczak, 2018, Predicting the spectrum of TCR repertoire sharing with a data-driven model of recombination, *Immunological Reviews* 284:167. Creative Commons License CC BY-NC-ND 4.0.

Perspective

The evolutionary relatedness of all living systems is a unifying theoretical principle on which physicists and biologists can agree. The biological physics community has engaged with evolution, both theoretically and experimentally, in several ways. Ideas of optimization, discussed in many sections of this report, represent an attempt to predict the outcome of evolution while circumventing its dynamics. In the opposite direction, many theoretical and experimental observations are concerned with the imprint that evolutionary history leaves on the diverse collection of current organisms and individuals. Biological physicists also made important contributions to understanding the evolutionary process itself, particularly in recent years. There is a growing body of both theoretical and experimental work that aims to predict and measure, in quantitative detail, how populations create and maintain genetic variation, how efficiently natural selection can operate, and how predictable and repeatable adaptation will be. These advances are now contributing to understanding the somatic evolution of cancers, the operation of adaptive immune systems, the spread of viral epidemics, and more (Chapters 6 and 7).

PART II

CONNECTIONS

5

Relation to Other Fields of Physics

For decades, biological physics was seen as sitting at the interface between physics and biology. As discussed in Part I, this location on the map of science does not do justice to the field: The physics of living systems now stands on its own as a branch of physics, rather than being a temporary alliance at the border between disciplines. The phenomena of life provide a continual supply of challenges to our fundamental understanding of physics. Responding to these challenges requires new concepts, new principles, new theoretical and experimental techniques, and new instruments. While biological physics has an independent existence, many of these new developments are deeply connected to progress in other fields, and this is part of the beauty of physics. This chapter surveys, briefly, some of the many points of contact between biological physics and the broader physics community, summarized even more briefly in Table 5.1.

In this chapter, as elsewhere in the report, there are subtleties in drawing boundaries. Part I of this report has defined biological physics as bringing the physicist's style of inquiry to bear on the phenomena of life, asking the kinds of questions and searching for the kinds of answers that characterize our understanding of the inanimate world. But in this search for a physics of life, the community often has found questions and answers that have resonance with other areas of physics, or perhaps define a new subfield that overlaps with biological physics. Living systems also may provide a clear, even pedagogical example of concepts from other parts of physics, or crucial applications of new experimental methods. This chapter is devoted to this rich exchange between biological physics and the rest of physics. Part of the emergence of biological physics as a branch of physics is that this exchange goes in both

TABLE 5.1 Connections to Other Fields of Physics

Other Physics Fields	Relationship to Biological Physics
Elementary particle physics	Source of experimental tools and theoretical ideas now used in biological physics; recipient of data analysis methods based on neural network ideas from the biological physics community.
Nuclear physics	One of the sources for post–World War II funding for interactions between physics and biology.
Atomic, molecular, and optical physics	Source for spectroscopic techniques and the evolution of microscopy; interaction led to Nobel Prizes in super-resolution images (Chemistry, 2014) and optical tweezers (Physics, 2018).
Statistical physics	Biological physics as a source of general statistical physics problems; living systems as decisive testing grounds for new theoretical ideas.
Soft condensed matter physics	Polymers, membranes, self-assembly, and other soft matter problems inspired by phenomena in the living world.

directions, with the physics of living systems being both consumer and producer of ideas and methods whose primary impact is outside the field.

It might seem far-fetched to think that biological physics is connected to elementary particle physics. But the detectors that are used in today's X-ray diffraction experiments on protein structure, for example, have their roots in particle physics experiments. Conversely, the machine learning methods that are used today in the analysis of raw data from particle physics have their roots in models of brain function that began in the biological physics community. On the theoretical side, ideas from quantum field theory are central to thinking about the collective behavior of flocks and swarms, for example, but this reflects the merging of field theory and statistical physics that occurred in the 1970s, in the wake of the renormalization group. The fact that one can understand aspects of a flock's behavior, quantitatively, by thinking about broken symmetry and massless excitations (Chapter 3) is testimony to the unifying power of theoretical physics.

For many years after World War II, significant interactions between physics and biology were supported by the U.S. government as an adjunct to nuclear physics. This work included the production of rare isotopes that were used as tracers in explorations of many different living systems, and the study of the effects of radiation on organisms, which led to the discovery of basic DNA repair mechanisms. For the biological physics community this is largely a closed chapter, but it is fertile ground for historians of science.[1]

The emergence of biological physics in the modern form described in this report is approximately contemporary with the renaissance of atomic, molecular,

---

[1] See, for example, A.N.H. Creager, 2014, *Life Atomic: A History of Radioisotopes in Science and Medicine,* University of Chicago Press, Chicago, IL.

and optical (AMO) physics, including observation of the first Bose-Einstein condensates. By then there had already been two or three decades of dramatic impacts of laser physics on the exploration of the living world. Notably, as fast pulse lasers were being developed, one of the first experimental targets for these instruments was the photosynthetic reaction center (Chapter 1). More generally, lasers made possible an enormous expansion of spectroscopic techniques, including many that are widely used in the study of living systems. Even more profound was the impact on microscopy. Lasers are essential to the widespread use of confocal microscopy, and high-powered pulsed lasers are essential for two-photon fluorescence microscopy, which allows high-resolution imaging deep into highly scattering media, such as the brain. Total internal reflection microscopy, nonlinear methods such as coherent anti-Stokes Raman microscopy and stimulated Raman scattering microscopy, fluorescence speckle microscopy, light sheet microscopy, structured illumination microscopy, and fluorescence recovery after photobleaching (FRAP) all have had substantial impact on biological physics, and the continued development of these methods by the biological physics community has had an impact on biology more broadly (Chapter 6).

Some of the most profound interactions between AMO and biological physics involved developing a deeper understanding of basic physical principles. Although it was known, in principle, that the "diffraction limit" was not really a limit on our ability to reconstruct details in images, it was a revolution to develop methods whereby such super-resolution images become routine (see Figure I.2). These methods, based on stochastic localization or stimulated emission depletion, were recognized by a Nobel Prize in 2014. Similarly, although it was known in the 19th century that beams of light can apply forces to objects, it would take until the late 20th century to understand the conditions under which these forces could trap neutral particles. Crucially, optical traps or tweezers can produce forces on the scale of those generated by single biological molecules, which opened a whole generation of new experiments in the biological physics community (Chapters 1 and 2); this development was recognized by a Nobel Prize in Physics in 2018 (see Box 1.3).

Single molecule manipulation, both with optical and with magnetic traps, made it possible to connect one of the classic problems of statistical mechanics with controlled, quantitative experiments. All textbooks on the subject consider the random chain polymer, which is a compelling illustration of entropic forces. The calculation typically imagines taking a single long molecule and pulling on it, and the theory generates predictions for the force or stiffness. With single molecule manipulation, one can do exactly this experiment, and one can use DNA synthesis to be sure that the polymer is precisely N units long. As shown in Figure 5.1, experiments are sufficiently accurate that one can resolve the difference between a full elastic theory of the polymer and the simplifying freely jointed chain model.

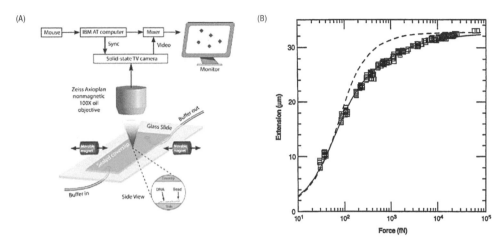

FIGURE 5.1 DNA as an example of polymer elasticity. (A) Schematic of apparatus for applying controlled foces to a single DNA molecule using a magnetic bead, and measuring the resulting length or extension of the molecule. The molecule is 97,000 base pairs long. (B) Results for extension versus force. The solid line shows predictions of the full elastic theory for a polymer of length L = 32.8 ±0.1 μm (corresponding to 97,000 base pairs) and a persistence length of A = 53.4 ±2.3 nm. Dashed line shows predictions for the simplified model of a freely jointed chain. SOURCES: (A) S.B. Smith, L. Finzi, and C. Bustamante, 1992, Direct mechanical measurements of the elasticity of single DNA molecules by using magnetic beads, *Science* 258:1122, reprinted with permission from AAAS. (B) C. Bustamante, J.F. Marko, E.D. Siggia, and S. Smith, 1994, Entropic elasticity of λ-phage DNA, *Science* 265:1599, reprinted with permission from AAAS.

Polymers with $N \sim 100,000$ units, as in Figure 5.1, are at the border between the molecules of AMO physics and the matter of condensed matter physics. Condensed matter physics is divided into "hard" and "soft," where hard condensed matter is focused on problems where quantum mechanics is crucial, notably the behavior of electrons in solids. Soft condensed matter today encompasses many-body problems in the classical (non-quantum) limit. Many problems involving polymers, membranes, fluids, glasses, and other systems with complex landscapes are classical statistical physics problems, as with the polymer example in Figure 5.1. The soft matter community has substantial overlap with the biological physics community, and many soft matter problems are interesting in part because they capture some aspects of living systems in simpler and more easily controlled contexts.

As noted in Chapter 3, ideas from polymer physics also have been important in thinking about the structure and dynamics of chromosomes. These polymers are dense, and their configurations are shaped in part by active, non-equilibrium mechanisms. The nuclear pore complexes, which control transport in and out of cell nuclei, provide examples of polymer brushes, and it now is thought that different states of the brush can be identified with open and closed states of the pore, but again it is essential that these systems are driven away from equilibrium, allowing active sorting of molecules between the cell nucleus and cytoplasm.

Early interest in another soft matter system, surfactant solutions, was motivated partly by oil recovery techniques and partly by cellular membranes composed of lipid bilayers. These systems often exhibit structure at mesoscopic scales, driven by competition between different forms of interactions, such as the hydrophilic and hydrophobic interactions in the case of surfactants or lipids. The existence of that structure and its sensitivity to thermal fluctuations or other perturbations lies at the core of soft matter research conducted in the 1980s and 1990s. There are echoes of this work in the current excitement surrounding phase transitions in real cell membranes and liquid-liquid phase separation in the cytoplasm (Chapter 3).

The complex phases that emerge in lipids and surfactants are examples of self-assembly. Although the structures are intricate, there is no blueprint. Self-assembly is an important meeting ground for the biological physics and soft matter physics communities. Living systems offer an astonishing array of structures that self-assemble, and using the term more loosely one can even think about structures that are actively constructed by communities of organisms, as in Figure 3.15. From the soft matter perspective, it is attractive to start with the very simplest examples and try to build up.

An example of self-assembly is provided by the lenses in the eyes of many aquatic animals, including cephalopods. Here the functional goal of self-assembly is to build a structure with desirable optical properties. Light passing through a curved lens is refracted more at the edges, distorting the image (spherical aberration). To avoid this, these eye lenses have an index of refraction that depends on the curvature. The squid lens is made of a single type of protein, but with subtle variations in amino acid sequence. These proteins have largely repulsive interactions, and the variations are in linker regions that bring the proteins into contact, allowing the formation of a gel. Variations in linker density drive gradients in the density of the gel and hence refractive index, avoiding aberration but also avoiding condensation into phases that would be opaque. Living systems make use of an astonishing variety of physical principles for building optical devices, in part because there are no intrinsically reflective parts such as metal surfaces. In insects, single photoreceptor cells can act as optical waveguides, pigment granules are transported through evanescent waves to act as attenuators (as with our pupil), stacks of cells and extracellular structures serve as interference-based reflectors, and more. In other examples, there is self-assembly of populations of organisms to perform crucial optical functions, as with the symbiosis between clams and algae in Figure 5.2, which also self-assemble.

Phases and phase transitions are central to classical statistical mechanics, and a major theme of soft matter physics is to push these concepts beyond their original context of thermal equilibrium. A paradigmatic example is the jamming transition

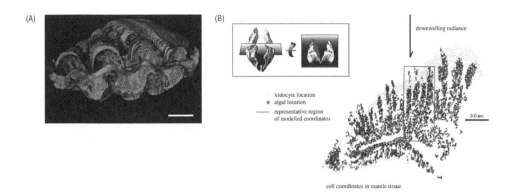

FIGURE 5.2 Many aquatic animals have self-assembling optical devices. Algae is an interesting example in that it uses another organism, rather than proteins, for this assembly. Giant clams living on intensely irradiated shallow tropical reefs host photosynthetic algae cells within their mantle tissue. An array of light-scattering skin cells on the clam is evolved to deliver optimal doses of light to these internal algae, resulting in very efficient, damage-resistant photosynthesis. (A) Photograph of *Tridacna crocea* with prominent green iridocytes. Scale bar: 2 cm. (B) Locations of algae (gray dots) and iridocytes (yellow dots) in a single histological section. The approximate direction of downwelling radiance in the ocean relative to this epithelial tissue section is shown with a black arrow. *Inset:* cartoon showing orientation of the clam shells visible in (A) relative to tissue section in this figure. SOURCE: A.L. Holt, S. Vahidinia, Y.L. Gagnon, D.E. Morse, and A.M. Sweeney, 2014, *Journal of the Royal Society Interface* 11:20140678.

in collections of randomly packed particles, which has the literal feel of a liquid/solid transition, but in the absence of a temperature or thermal motion. Epithelial tissues also undergo something like a liquid/solid transition, and recent work emphasizes that this rigidity transition, like jamming, is from a fluid to a rigid disordered state rather than an ordered one (see Figure 5.3).

The classical statistical physics of disordered systems connects with biological physics through many problems beyond cell movement. Understanding of glassiness in random heteropolymers provides a benchmark for thinking about protein folding, ultimately leading to the view that amino acid sequences must be selected by evolution to avoid the glass transition (Chapter 3). This example encourages thinking more generally about disordered systems in which the underlying randomness is shaped by some non-trivial physical constraints. Proteins do not just fold, they also exhibit allostery, in which the binding of a regulatory molecule to one site on a protein can influence the structure at very distant sites, for example changing the binding energy for a second type of molecule there. Theoretical approaches used to explore complex energy landscapes in disordered systems can be exploited to tune mechanical networks to display analogous allosteric behavior.

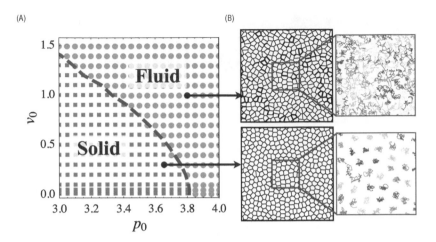

FIGURE 5.3 A major theme of soft matter physics is to push the concepts of phase and phase transitions beyond their original context of thermal equilibrium. An example of this is seen in epithelial tissues, which undergo a transition from a fluid to a rigid disordered state. (A) Phase diagram for a model of cell movements in a dense tissue. The fluid-solid transition is shown as a function of p_0, which characterizes the cell's preferred shape, and v_0, which characterizes the propulsion velocity of individual cells. (B) Instantaneous tissue snapshots (*left*) show the differences in cell shape across the transition. Cell tracks (*right*) show dynamical arrest due to caging in the solid phase and diffusion in the fluid phase. SOURCE: D. Bi, X. Yang, M.C. Marchetti, and M.L. Manning, 2016, Motility-driven glass and jamming transitions in biological tissues, *Physical Review X* 6:2021011. Creative Commons License Attribution 3.0 International (CC BY 3.0).

Theories of neural networks are a rich source of problems in statistical physics, so much so that one of the categories on the electronic archive of physics papers, arXiv.org, is "neural networks and disordered systems." The relation of these statistical physics problems to real brains is discussed in Chapter 3, and application to the deep networks that are powering the current revolution in artificial intelligence is discussed in Chapter 7. Recent attempts to build analytic theories of structural glasses in high spatial dimension—models that might provide a mean-field approximation to real glasses—make strong connections to transitions that occur in neural network models, closing the circle.

Living systems are always away from equilibrium, and many crucial functions occur on a scale far from the thermodynamic limit, where the impact of single molecules can be felt clearly. As discussed in Part I of this report, the identification of these functions as physics problems, and understanding how they emerge, are central to biological physics. But the past two decades also have seen an explosion of activity in the statistical physics of small, non-equilibrium systems, more generally, often taking inspiration from the phenomena of life. The study of fluctuations in these systems has led to a number of remarkable results. These "fluctuation theorems" show, for example, that the distribution of microscopic response of far-from-equilibrium systems contains equilibrium information, and this can be extracted by comparing processes and their time-reversed versions. Some of the first and still clearest tests of these ideas are in experiments with single biological molecules, adapting methods that were developed in the biological physics community, as for example in Figure 5.4. This stream of work also has emphasized that the equivalence of entropy as a measure of heat flow and entropy as a measure of available information is more than formal, and has physical consequences. Subsequent theoretical work has derived thermodynamic uncertainty relations, which for example connect the precision of molecular scale clocks to the rate of energy dissipation. There is a widespread sense in the community that all these ideas should feed back into our understanding of living systems, for example providing a new class of physical limits that can be compared to the performance of real molecular

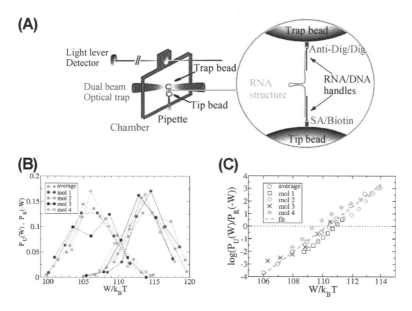

FIGURE 5.4 Testing the principles of non-equilibrium statistical mechanics with single biological molecules. (A) A single RNA molecule that can fold into a "hairpin" structure is held between two beads, which in turn are held by a dual beam optical trap (see also Figure 2.1). The distribution of work done in unfolding the molecule as the traps are moved apart (B) and the negative work done as the molecule refolds (*left*). Results are shown for four different molecules with the same sequence, each taken through 50–100 folding/unfolding transitions. (C) Although these experiments are done under non-equilibrium conditions, the (log) ratio of the distributions in (B) has a simple dependence on the work done, as predicted by general "fluctuation theorems." SOURCE: Reprinted by permission from Springer: D. Collin, F. Ritort, C. Jarzynski, S.B. Smith, I. Tinoco, Jr., and C. Bustamante, 2005, Verification of the Crooks fluctuation theorem and recovery of RNA folding free energies, *Nature* 437:231, copyright 2005.

clocks such as the cell cycle or circadian rhythms, and more generally providing insight into the energetics of biochemical control.

On a larger scale, physicists long wondered whether the agreement of birds in a flock to fly in the same direction could be seen as a transition to order, perhaps analogous to the alignment of spins in a magnet, but different because the "spins" (velocity vectors for the individual birds) are moving. As explained in Chapter 3, in the 1990s physicists began to study models that embody this intuition, and this launched the now lively field of active matter. The theory of active matter focuses on the derivation of effective theories for the generic behavior of these systems on length scales long compared with their component parts or the range of interactions. The renormalization group tells us that these behaviors should be independent of the short distance details. Importantly this means that one can observe related phenomena in flocks of birds and in schools of fish, in networks of protein filaments and motor molecules in single cells, and in populations of cells organizing themselves on a surface, as in Figure 5.5. In some cases, these predictions of universal behaviors are confirmed, while in other cases living systems have managed to circumvent generic expectations. This is generating a productive dialogue between the biological physics and active/soft matter communities, even driving the development of synthetic active systems.

Understanding animal movement depends on understanding the environment in which this movement happens. For organisms that swim or fly, this leads to potentially challenging problems in fluid mechanics, but these problems rest on a firm foundation. Many organisms, however, move on or through granular environments such as sand and soil. These materials flow, shift, and crack under stress, prototypes of the complex rheology that is a major theme in soft matter physics. Granular materials are collections of particles that interact via repulsive frictional contact forces; they remain solid below a yield stress, and flow frictionally above this stress. Diverse organisms, from ants to camels, contend with and utilize the properties of these soft materials. A newly hatched juvenile sea turtle, using its flippers to make its way to the ocean, must not drive the sand beyond its yield stress, and deadly sidewinder rattlesnakes climbing sandy desert dunes face the same problem. In contrast, slithering snakes and sand-swimming lizards create localized "fluids" and propel themselves within these. In granular systems, dissipation from frictional interactions dominates inertia, so that these quite macroscopic creatures face some of the same problems as bacteria, living at low Reynolds number (Chapter 1). In a very different regime of soft material behavior, a snail takes advantage of shear thinning in viscoelastic fluid as waves of muscle activity propagate along its foot. In all these examples, the physics of soft materials and the physics of animal behavior in relation to these materials are advancing together.

As emphasized above, a distinguishing feature of soft matter physics is that it is essentially classical rather than quantum mechanical. A related but distinct set

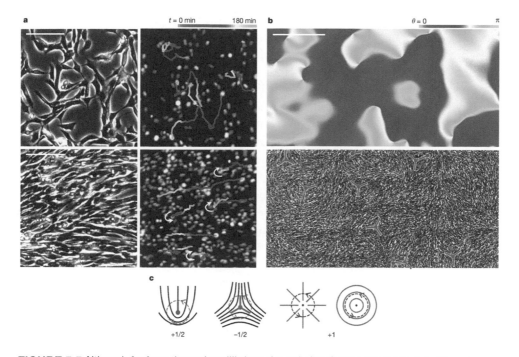

FIGURE 5.5 Although far from thermal equilibrium, the ordering that we see in active matter can support the same kinds of topological defects that arise in equilibrium systems. Here we see a population of neural progenitor cells organizing themselves on a surface. (A) Individual cells in a two-dimensional culture. Top panels show cells at low density (1,100 cells/mm^2), observed in phase contrast (*left*) and fluorescent channel (*right*, nuclei are marked in pseudocolor); lines trace the trajectories of single cells. Bottom panels show cells at high density (3,000 cells/mm^2); lines again trace the trajectories of single cells, now ordered. White arrows (right bottom panel) mark reversals of the tracked cells. Scale bar: 100 μm. (B) Color (*top*) indicates the angle of the local alignment of cells, calculated from the phase contrast image (*bottom*). Singular points where all colors meet in the top figure are the topological defects, as shown explicitly in the bottom image with the winding numbers +1/2 (red) and –1/2 (blue) indicated. Scale bar: 1 mm. (c) Types of topological defects characterized by their winding numbers. SOURCE: Reprinted by permission from Springer: K. Kawaguchi, R. Kageyama, and M. Sano, 2017, Topological defects control collective dynamics in neural progenitor cell cultures, *Nature* 545:327, copyright 2017.

of questions is addressed by the nonlinear dynamics community. The problems of nonlinear dynamics have a venerable history, reaching back to the three-body problem in celestial mechanics. Many of the issues of stability and ergodicity, despite obvious physical consequences, became the province of pure mathematics for much of the 20th century, while fluid turbulence drifted out of physics into engineering. But there was a resurgence of interest from the physics community in the 1970s and 1980s, driven by a combination of theoretical and experimental developments. There has been a productive interplay between nonlinear dynamics and biological physics ever since.

Perhaps the most dramatic development in the resurgence of nonlinear dynamics as part of physics was the demonstration that there are a limited number of ways in which systems with small numbers of variables could make the transition from regular to irregular or chaotic motion. This work resulted in a complete analysis of these transitions using the renormalization group, and the verification of the theory in experiments on fluids and other systems. While the motivation for this work clearly came from the search for a more manageable version of the transition to turbulence, the low dimensional dynamical systems that were studied first came equally from fluid mechanics and from the dynamics of population growth. The tools that were built up to analyze low dimensional dynamical systems—bifurcations; fixed points and attractors; periodic, quasiperiodic, and chaotic behavior; entrainment and synchronization—have found applications in the analysis of many different living systems, from gene regulatory networks to neurons to ecosystems. While ecology was the traditional source of nonlinear dynamics problems, genetic networks have formed a productive modern nexus, connecting the biological physics community to nonlinear dynamics, but also to the systems and synthetic biology communities (Chapters 6 and 7).

While stability and instability are familiar concepts, nonlinear dynamics identified an intermediate class of excitable systems. The prototypical example is the dynamics of neurons: In response to a brief injection of current, the voltage across the membrane of a neuron typically returns to its initial value, but if the current is large enough there will be a large amplitude transient that takes on a nearly stereotyped form—the action potential. Similar behaviors are seen in very different systems, such as the signaling systems that slime molds use to communicate as they make the transition from living as isolated cells to forming an aggregate. Such spatially extended excitable media support novel excitations, such as spiral waves (see Figure 5.6). In some regimes the heart also acts an excitable medium, and spiral waves can be triggers of tachycardia and fibrillation. More generally, models of the heart provide problems for nonlinear dynamics, and this interaction has progressed to the point where it is not unreasonable to think of certain heart conditions as disorders of dynamics.

The classical nonlinear dynamical problems of ecology are low dimensional, from oscillations in predator/prey systems to chaos in seasonal population growth. These problems remain important, not least because these models are actually used to manage fisheries and other human food sources. But in many ways the frontier has shifted to high-dimensional systems, ecologies in which many different kinds of organisms are present simultaneously. This is what is seen in many environments, but there are questions about how this is possible: Why doesn't the single fastest-growing species win out? There are many ideas about how ecological diversity is maintained: There can be multiple resources, with different species becoming more or less specialized, there can be spatial structures that inhibit competition, or it

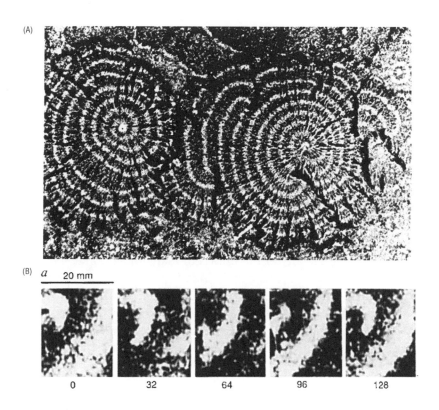

FIGURE 5.6 Spiral waves are supported by spatially extended excitable media, like the signaling systems that slime molds use to communicate. (A) The slime mold *Dictyostelium discoideum*, at an intermediate stage in its aggregation. Note the spiral waves, and the onset of a streaming instability at the edges of the pattern. (B) Electrical activity in canine heart muscle, monitored by a voltage-sensitive dye and imaged with a charge-coupled device (CCD) camera. Black is the resting voltage, white the maximum depolarization. Images are labeled by time in milliseconds. SOURCE: (A) H. Levine and W. Reynolds, 1991, Streaming instability of aggregating slime mold amoebae, *Physical Review Letters* 66:2400. (B) Reprinted by permission from Springer: J.M. Davidenko, A.V. Pertsov, R. Solomonsz, W. Baxter, and J. Jalife, 1992, Stationary and drifting spiral waves of excitation in isolated cardiac muscle, *Nature* 355:349, copyright 1992.

could be that persistent diversity is not a steady state but just very slow dynamics. All these ideas, and more, are naturally expressed in dynamical systems language. To make progress often requires some mixing with concepts of statistical mechanics, passing to the limit of very large systems and approximating interactions among species as drawn from ensembles of random matrices. This is a very active field, with theory now being supplemented by large scale, controlled experiments in microbial ecologies.

Finally, what about the connection of biological physics to the physics of the universe as a whole? There are at least two questions. First, in the spirit of cosmology, how far back is it possible to trace the history of life on Earth? Second, what would life look like on another planet?

Regarding the deep history of life on Earth, there are classical experiments showing that early atmospheric conditions allowed for the synthesis of moderately complex organic molecules, including amino acids. But there is a huge gap from this to something one might think of as alive. In life today, information passed from one generation to the next is stored in one kind of polymer (DNA), but much of the business of life is carried out by another kind of polymer (proteins). The discovery that some RNA molecules act as catalysts—a function largely filled by proteins today—suggested that there was an early "RNA world." The first living organisms might have been self-replicating polymers, but there are active efforts to understand how protocells might form, allowing for compartmentalization and encapsulation of an RNA-based genome. While these ideas remain in many ways very speculative, there are serious efforts to map, for example, the landscape of catalytic function versus sequence in RNAs of modest size. This has much in common with ideas about sequence/structure mapping in families of proteins (Chapter 3), and would provide some foundation for more rigorous discussion. In a different direction, several groups are trying to construct artificial cells with limited numbers of components, perhaps providing models for the origin of life but certainly creating a path to understanding by building.

Taking the analogy to cosmology seriously, one might say that we have many ideas about the Big Bang, but what is missing is the cosmic microwave background: What can be measured that carries the fingerprints of those earliest moments? By sequencing the genomes of more and more of the different organisms alive today, it seems possible to construct an evolutionary tree that points to common ancestors further and further back in time. But this assumes that all relationships fit on a tree, and this is not quite true. Bacteria can exchange genetic material without reproducing, and it has been conjectured that this was more prevalent in the distant past. If this is correct, there was a murky ground of information exchange out of which the modern tree of life grew. At present, there is no way to get an image of this ground.

The question of life on other planets of course is given greater urgency by the 1995 discovery that there *are* other planets orbiting stars similar to the sun. As more and more such planets are discovered, the range of conditions comes closer and closer to those which seem hospitable for life. What should we expect to see? Measurements will be observations from afar, so what is needed are essentially spectroscopic signatures. It is unreasonable to think that life on Earth is so special that all possible life forms will use the same molecules. Recent work suggests focusing instead on the complexity of molecular ensembles: To be alive is not to have one crucial molecule, but rather to have many species of molecules connected by a reaction network. In this direction, it is important to understand whether the molecules of life on Earth form a network with special properties, distinguished from a random jumble of molecules of similar size and elemental composition.

Perhaps the most important result of thinking more concretely about how to search for life on extra-solar planets is that we are not so sure what features of life on Earth constitute its defining physical characteristics.

6

Biology and Chemistry

Phenomena in nature do not come labeled as belonging to biology, chemistry, or physics. But scientists from these different disciplines ask different kinds of questions, and seek different kinds of answers. Part I of this report is focused on how the phenomena of life generate questions for the physics community. In that process, knowledge gathered in the biology community provides a foundation for asking new physics questions about the phenomena of life, and ultimately for discovery of new physics. This chapter is focused on the flow of knowledge back from the physics community into the biology community, and the way in which this knowledge provides a foundation for new biology. These are not crisp distinctions, but are intended to capture the spirit of interaction between the disciplines, which has been so extraordinarily productive over more than a century. Along the way are ideas and methods that could be equally well categorized as chemistry, further enriching the disciplinary mix.

TOOLS FOR DISCOVERY

Optical physicists have greatly influenced the study of living systems through their many innovations in light microscopy. Innovations in electron microscopy have been similarly influential. Notably, it was the invention of the light microscope that enabled the discovery, in 1665, of cells as the basic units of living organisms. The creation of the electron microscope was equally pivotal for the discovery of cellular organelles. Advances in microscopy have continued to propel biological discovery to the present day.

Whereas the first light microscopes used optical absorption to create image contrast, by the 1800s, dark field microscopy had emerged as a means of using light scattering to view cells. This was an important advance, as many cells are poor absorbers of light, but they can be imaged at greater contrast using photophysical processes, such as scattering, that cells enact well. Most cells also shift the phase of light to a substantial extent, and the invention of phase contrast microscopy—recognized by a Nobel Prize in 1953—enabled routine inspections of the micron-scale morphologies of many cell types that are poorly revealed by bright-field microscopy and light absorption. To this day, phase contrast imaging remains a mainstay technique in research and clinical contexts worldwide. For teaching, a phase contrast microscope equipped with a time-lapse camera also allows a compelling demonstration of how a single photon can interfere with itself. In some contexts, differential interference contrast (DIC) microscopy has replaced phase contrast imaging due to its superior sensitivity to refractive index gradients inside the cell. By comparison, polarized light microscopy remains a niche technique for viewing biological specimens that are birefringent.

Fluorescence Microscopy Becomes Dominant

Fluorescence microscopy techniques have had especially great influence on biological physics investigations, and on biology more broadly, particularly in recent decades. This impact is due in large part to the ability to mark and thereby identify specific, chosen components of biological cells with a wide range of different fluorescent dyes, genetically encoded fluorescent proteins, fluorescent nanoparticles, fluorescently labeled nucleic acids, or fluorescent markers of other kinds. The resulting ability to image select components or molecular constituents of cells and tissues, combined with the many available fluorescent labels with different targets and emission colors, has spurred innovation in a wide set of fluorescence microscopy methods.

One of the most important advances in fluorescence microscopy for biological discovery came from the isolation, cloning, and sequencing of naturally occurring fluorescent proteins such as green fluorescent protein (GFP) from the jellyfish *Aqueorea victoria*. Methods of molecular biology were harnessed to fuse the coding sequence of GFP to the coding sequence of almost any gene in any organism. Understanding the physics and chemistry of photon absorption and fluorescent re-emission enabled the engineering of GFP mutants in a rainbow of colors and with exceptional photophysical properties (see Figure 6.1). These advances allowed specific fluorescent "tagging" of any protein in its natural context of the living cell, enabling elucidation of that protein's dynamics and function in situ. This revolutionized the use of fluorescence microscopy as a research tool, and was recognized by a Nobel Prize in 2008.

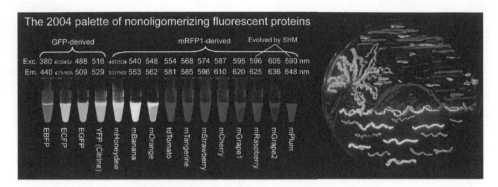

FIGURE 6.1 Genetic engineering of naturally occurring fluorescent proteins has led to a broad palette of colors, shown here as of 2004. At right, a playful example of painting with bacteria expressing different proteins. SOURCE: R.Y. Tsien, 2009, Constructing and exploiting the fluorescent protein paintbox (Nobel lecture), *Angewandte Chemie International Edition* 48:5612.

In addition to genetic encoding of fluorescent tags, the chemistry community has made important innovations in organic labels, starting with the very bright and photostable Alexa and Cy fluorophores introduced in the late 1990s and early 2000s, and the more recent development of the Janelia Fluor(R). There is an increased demand for dyes that are cell permeable and conjugate genetically encodable tags like HaloTag and SnapTag. These engineered dyes are often brighter and more photostable than fluorescent proteins, thereby providing higher signal-to-background ratios, higher localization precision, and longer trajectories when used to track individual biomolecules. These organic dyes are also engineered to expand the spectrum of available genetically encodable labels, facilitating multi-color imaging of labeled endogenous proteins which would have been difficult otherwise. These advancements in specific fluorescent labeling techniques, enabled by molecular biology, physics and chemistry, have thus propelled the tandem development of ever more sensitive, higher resolution, and faster fluorescence microscopes as biological discovery tools.

Wide-field epi-fluorescence microscopy is the simplest of these methods to implement and was the earliest to emerge, but it does not provide three-dimensional optical sectioning. One important advance from physics that revolutionized the ability of biologists to visualize the behavior of single fluorescently labeled molecules in a very thin optical section was total internal reflection fluorescence microscopy (TIRFM). By exciting fluorescence via the evanescent wave that is produced when light reflects off of a low refractive index surface at the water/microscope coverslip interface, TIRFM provides a simple form of optical sectioning, albeit limited to within a hundred or so nanometers from the surface at which the evanescent wave emerges. This very thin plane of excitation has been exploited to excite only the small number of fluorescently labeled molecules within the evanescent wave in a dilute bulk solution of fluorescently labeled molecules, providing exquisitely high signal-to-background single molecule imaging. Together with the realization that one could literally see beyond the nominal diffraction limit (below), this enabled the dynamic tracking of motions of single molecules at the nanoscale. This has allowed the characterization of the stepping of single motor proteins along cytoskeletal filaments or DNA, statistics of single protein-protein binding interactions, and microrheological measurements of material properties, to name a few.

By achieving a more versatile form of optical sectioning and enabling three-dimensional imaging at the sub-micron scale, confocal fluorescence microscopy revealed many new facets of cells and tissues. But this approach is limited in its ability to image deep into optically opaque tissue.

Pulsed Lasers Lead to New Microscopies

The advent of ultrashort-pulsed lasers led to the development of multiphoton fluorescence microscopy, which can penetrate far more deeply into opaque tissue, including in live animals and humans. Specifically, two-photon fluorescence microscopy has become widespread as a way to image cellular properties and dynamics in the nervous system, lymph nodes, muscles, and other scattering tissues that were previously difficult to inspect at cellular scale in intact form. In the past few years, three-photon fluorescence microscopy has begun to emerge as a means of imaging even deeper (more than 1 millimeter) into thick tissue, due to the even longer attenuation length of the illumination wavelengths used as compared to those used for two-photon imaging.

Other microscopy modalities based on nonlinear optical effects include second-harmonic generation microscopy, which reveals ordered polar polymeric structures such as those in striated muscle and collagen, and third-harmonic generation microscopy, which is suited to inspecting cell interfaces. These optical harmonic methods are not used as widely as fluorescence imaging, as they are less versatile for inspecting a broad range of tissue and cell types, but they have the key virtue of sensing intrinsic optical signatures of tissue and thus not requiring any exogenous labels. Other nonlinear optical imaging methods probe biomolecular vibrations and include Raman scattering microscopy, coherent anti-Stokes Raman microscopy, and stimulated Raman scattering microscopy. The molecular specificity of these approaches is a strong suit, although isolating particular vibrational modes within the complex biomolecular environment of a cell is often challenging.

Breaking the Optical Diffraction Limit

One of the most important microscopy advances in recent decades is the invention of super-resolution optical imaging, which achieves imaging resolution finer than the limits set by conventional diffractive considerations (see Figure I.2). Nearly all super-resolution methods rely on fluorescence contrast. The first such approach, stimulated emission depletion (STED) microscopy, emerged in the early 1990s and introduced the strategy of using an optical nonlinearity as a means of shrinking the spatial support of the optical point spread function.

In this same time period, the capacity emerged to image and probe the photophysical properties of single fluorescent molecules, which provided a powerful new approach to observe biological processes in action. An especially potent version of this relies on fluorescence resonance energy transfer (FRET) between a pair of single fluorophores, as the efficiency of energy transfer depends on the distance between the two molecules and hence provides a "molecular ruler" with the sensitivity to detect changes in macromolecular conformations of less than 10 nanometers.

From this field of single molecule biophysics emerged the stochastic localization microscopy techniques as an alternative to STED microscopy for acquiring nanoscopic information. Stochastic localization imaging methods use particular classes of fluorophores that are either photoactivatable or photoswitchable. Activating a small fraction of the flurophors converts a dense collection of molecules into a sparse array of point fluorescent sources, and these points can be localized with a precision of tens of nanometers, much better than the usual diffraction limit to image resolution. Successive rounds of activation and deactivation sample different subsets of molecules, gradually piecing together the entire image, all at a resolution of tens of nanometers, a kind of pointillism on a molecular scale.

Taken together, the super-resolution optical methods have opened entirely new avenues of research into nanoscopic properties of cells and their macromolecules. Notable examples include the discoveries of fine ultrastructure within actin filaments of the cellular cytoskeleton, recent evidence for phase condensation of RNA polymerase molecules during transcription, unprecedented insights into chromatin structure and its role in gene regulation, and the ability to characterize individual cells based on their RNA expression patterns with near single molecule sensitivity.

Microfluidics

The field of microfluidics refers to the science and technology of fluidic dynamics at the scales of microliters to femtoliters, at which surface tension and capillary effects have predominant roles. Microfluidic devices can channel, transport, sort, and mix fluids (and the specimens they may contain) from distinct sources, often using only pneumatic control mechanisms, but sometimes via other methods such as those based on electrowetting. Microfabrication of microfluidics devices has led to integrated systems for processing, testing, combining and even performing logical operations on fluidic specimens with minute volumes. Such capabilities have had broad-ranging impact on biological physics and related fields, particularly when microfluidic devices are combined with additional approaches to characterize or manipulate specimens, such as via optical measurements or biochemical reactions. Example applications involving microfluidic control include flow microcytometry analyses of the properties of individual cells, screening of small model organisms (e.g., nematodes, fly embryos, zebrafish larvae) within microfluidic chambers, identification of optimal conditions for protein crystallization, assessments of ligand/receptor binding affinities, and amplification of genetic material.

Notably, microfluidics technology has played a crucial role in the development of high-throughput, single cell assays. For example, in systems biology, microfluidic approaches have enabled researchers to move beyond population-level characterizations of cells in bulk and instead to directly analyze large numbers of individual cells, thereby capturing the variability in cellular properties across the population. By characterizing this variability at the level of individual cells, systems biologists and biological physicists have gained key insights into cellular signaling and decision-making processes. Furthermore, such single cell measurements, particularly in the context of transcriptomic or RNA sequencing analyses, have also become central to developmental biology, where they have made it possible to disentangle biochemical signaling networks that drive specific developmental transitions.

In biotechnology, microfluidic devices have greatly reduced costs for biochemical procedures, owing to the use of extremely small sample volumes and their associated fast mixing times, which accelerate biochemical reactions and enable low-cost, sophisticated biotechnological applications. For example, a low-cost rapid antigen test based on microfluidics detects SARS-CoV-2 viral particles in one second. Microfluidic technology has also helped usher in a new era of so-called single cell "omics," which allow genome-wide studies to be performed at the resolution of individual cells. For instance, biotechnology companies use microfluidics within single cell transcriptomic assays to identify rare populations of diseased cells, such as metastatic cancer cells circulating in the blood; for non-invasive prenatal tests using blood samples from a pregnant mother; and for immune profiling studies, such as to probe a patient's immune system regarding its ability to neutralize viral particles of a new SARS-CoV-2 strain. Looking ahead, the miniaturized, versatile designs of microfluidic devices are enabling a new era of personalized medicine. Overall, the use of microfluidic devices and approaches has become an integral mainstay in biological physics and the biotechnology industry.

Increasing Data Rates

Experimentalists in the biological physics community are well positioned to create basic technologies that enable rapid biological exploration. Such technologies are necessary to increase the rate at which theoretical predictions are tested, to make new discoveries, and to contribute applied technologies to society.

The speed of collecting data imposes limits on our exploration of life, and many advances have been enabled by the ability to gather data more quickly. In neuroscience, the biological physics community developed a microtome capable of cutting extremely thin slices of tissue (tens of nanometers thick) in an electron microscope. This advance made it possible to automate the collection of three-dimensional, nanoscale imaging to map out the network of synaptic connections between neurons (the connectome, Chapter 3), pushing the field forward. The entire field of genomics now relies on the ability to sequence single DNA molecules rapidly. The first methods of single DNA molecule sequencing came out of the biological physics community, which has continued to play a key role in the development of nanolithography methods and nanopore technologies that significantly speed up the process.

Such tools are needed to carry out high-quality experiments capable of generating large amounts of high-quality data. Similar new technology is needed in other areas, such as protein and organelle purification, cell and animal care, and the development of transgenic organisms. These frontiers of measurement often are explored by the biological physics community in response to physicist's questions about the phenomena of life, but the resulting methods are transferred to the larger biology community at ever increasing rates.

MOLECULAR AND STRUCTURAL BIOLOGY

One of the most important results of the interactions between physics and biology in the 20th century is the understanding of life's basic mechanisms as being the results of specific interactions among a set of identifiable molecules. This view of life as molecular machinery is the defining feature of modern molecular biology. Structural biology is the study of the molecular structure and dynamics of these molecules, typically proteins and nucleic acids, which provide the working machinery of the cell. The underlying goal of determining these structures is to understand their function and interactions in the complex environment inside the cell. While much of this effort is in support of basic scientific inquiry, structural biology is also a critical contributor in the discovery and development of new pharmaceutical therapeutics, which are almost always targeted to a specific macromolecule.

The dominant techniques used to determine these structures have been X-ray crystallography, nuclear magnetic resonance (NMR) spectroscopy, and more recently, cryogenic electron microscopy (cryoEM). Macromolecular crystallography (MX) is a mature technique with very high throughput (capable of characterizing many thousands of structures per day) but requires crystallization of samples. NMR is a particularly useful technique for determining the dynamics of macromolecules in solution as well as membrane proteins in a solid state. Following a dramatic expansion in the early 2010s dubbed the "resolution revolution," cryoEM has now been firmly established alongside MX and NMR as an essential structural biology technique. These efforts were recognized by a Nobel Prize in 2017.

In particular, cryoEM can provide structures of proteins that were previously intractable by other methods, including large protein complexes like the ribosome, integral membrane proteins, and highly heterogeneous or conformationally dynamic systems; an example in Figure 6.2 is the very recent structure of the flagellar motor (Chapter 1). These three principal approaches are supported by a wealth of other methods such as small angle scattering (SAS), hydrogen deuterium exchange mass spectrometry (HDX-MS), electron paramagnetic resonance (EPR), FRET, and

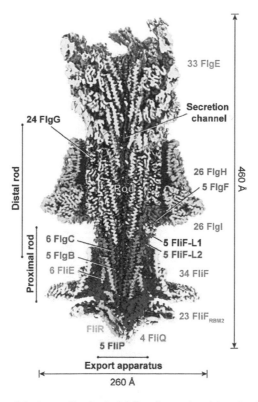

FIGURE 6.2 An atomic model of an entire bacterial flagellar motor determined using cryoEM. This supramolecular protein machine provides motility and determines pathogenicity. The overall structure includes 175 subunits with a combined molecular mass of more than 6,000,000 Daltons, and represents a tour de force of structural biology. Shown here is a cross-section through the structure. SOURCE: J. Tan, X. Zhang, X. Wang, C. Xu, S. Chang, H. Wu, T. Wang, H. Liang, H. Gao, Y. Zhou, and Y. Zhu, 2021, Structural basis of assembly and torque transmission of the bacterial flagella motor, *Cell* 184:1.

others. Increasingly, an integrated approach is required for a complete understanding of the structure and function of these complex macromolecular machines.

One way to measure the success of structural biology is through the growth of the Protein Data Bank (PDB), an open-access, public archive of the atomic coordinates of molecular structures. Launched in 1971 with just seven entries, the archive now holds roughly 170,000 structures and serves as a critical resource for both academic researchers and drug developers. The number of structures in the PDB continues to grow rapidly. At the same time, protein folding prediction algorithms, accelerated by machine learning, hold promise for a purely computational approach to providing structures that would be much faster and cheaper than traditional experimental methods.

Almost all of the protein structures in the PDB were determined from proteins in a purified form, separated out from their natural cellular environment. A major frontier in structural biology concerns the precise interactions of these macromolecules in the context of the cell—in particular, how and where macromolecules come together to form short-lived, transient but functional complexes. A complete understanding of the cell at atomic resolution will not be possible in the absence of this understanding. Achieving this will require a combination of techniques including cryo light microscopy (cryoLM), milling cells and tissues using focused ion beams (FIBs), and cryo electron tomography (cryoET). These three methods, currently undergoing rapid development, allow molecular complexes of interest to be located in the context of the cell (cryoLM), reduce bulk cells to thin lamella (FIB) suitable for high-resolution imaging by electron microscopy, and enable collection of a tilted series of images of the thin lamella in the transmission electron microscope (cryoET) that can then be converted to a three-dimensional volume using mathematical methods similar to those used in X-ray computed tomography or magnetic resonance imaging. A remaining challenge is then to identify individual molecules in the three-dimensional volumes, which have a very low signal-to-noise ratio and are also closely packed and crowded within the complex machinery of the cell.

While all these methods have their origins in physics, especially in the biological physics community, there has been a substantial effort to export these methods to a wider range of biologists. This has sped up, enormously, the exploration of the molecular structures relevant to the mechanisms of life. While this export was a slow process for X-ray crystallography, it was a bit faster for NMR and faster still for cryoEM. The result is that quite advanced methods of structural biology are broadly accessible. Nonetheless, from the standpoint of biology one can identify challenges that require solutions from the physics community.

Macromolecular crystallography. Understanding biological dynamics across many time scales is an exciting new frontier. These dynamics will be explored in existing and next generation synchrotrons as well as X-ray Free Electron Lasers (XFEL), which provide capabilities for serial femtosecond X-ray crystallography. These approaches will generate important new structures that will help to understand chemical reaction mechanisms catalyzed by enzymes and the molecular motions underpinning biochemical phenomena in living systems. Physics will undoubtedly continue to play an important role in moving the fundamental technology forward as well as in interpreting the results. New sources create opportunities that can be realized only with new detectors. There also is much to be done to support sample preparation for both XFEL and synchrotrons that will require fundamental physics insights, especially for data collection and dynamics at physiological temperatures.

Nuclear magnetic resonance. Biological molecules are not static, rigid objects. Flexibility is essential for function, and few interactions are truly lock and key. The unique analytic powers of NMR have been applied in structural biology and biological physics to clarify these dynamical mechanisms, including identifying allosteric effects in molecular recognition, and especially elucidating the role of conformational exchange in protein function. As such, NMR can provide the crucial missing link between static structures and functional insights for drug discovery. The major challenge for NMR achieving this impact is its low throughput, high required amount of expert intervention, and limited (though growing) ability to analyze very large molecules. A number of broad technical efforts will make NMR a more routine tool for complex biological systems. These include rendering NMR analysis tools suitable for large complex biopolymers (e.g., proteins of 100 kDa and larger) through development of higher magnetic fields and associated hardware at a price and volume that can serve numerous investigators; and rendering the tool conveniently accessible to non-specialists through improvements in the efficiency and automation of the crucial project steps (sample preparation, data collection, and data analysis).

Cryogenic electron microscopy. To further mature cryoEM techniques, advances are needed in sample preparation, instrumentation, and analysis. A deeper understanding of how macromolecules interact with substrates and the air-water interface would inform improvements to sample preparation. In terms of instrumentation, a large gap persists between the physical estimates of the number of macromolecules required to reconstruct a high-resolution, three-dimensional map, as well as the size of the macromolecules for which such reconstructions are possible, and what can currently be achieved in practice. Improvements may come from the development of laser-based phase plates, aberration corrections for electro-magnetic lenses, improvements to the electron detectors (cameras), and reductions in signal loss due to radiation damage and specimen movement during imaging. In terms of analysis, many macromolecules may exist in a continuum of conformational states; recent approaches using manifold embedding have the potential to uncover the work-cycle of a molecular machine as it passes through a continuum of states and map out its free-energy landscape.

CryoLM/FIB/CryoET. Major challenges remain in the use of these technologies, some of which will be addressed by further engineering, technical, and computational developments. There are also fundamental physical obstacles, for example the need to precisely target macromolecules of interest in three dimensions within the cell. Currently cryoLM is used for this targeting, but it lacks the resolution required to ensure that the milled lamella will contain the region of interest.

An important complement to methods of structure determination is the ability to simulate the dynamics of these molecules. Conceptually simple, simulation is both subtle and computationally demanding. Motions of atoms in these large molecules are largely classical, but the forces are determined by quantum mechanics of the electrons. It is not feasible to solve the full quantum mechanics problem, so molecular dynamics depends on semi-empirical models for the forces between atoms in a protein, and between these atoms and the surrounding water molecules. These models have improved over decades, as have simulation methods themselves, so that one can now expect reasonably accurate estimates of equilibrium structures, binding free energies, and other quantities of functional importance. The long path to this level of precision was recognized by a Nobel Prize in 2013.

But brute force simulation is not enough. A typical protein molecule has thousands of atoms, and the output of a simulation is a sample trajectory through this enormously high-dimensional space. Theoretical ideas from biological physics play a central role in analyzing these simulations by extracting simplified descriptions of the dynamics.

Solving the structures of many different proteins has provided a scaffolding on which to build a more precise understanding of life's basic mechanisms. But the emergence of these structures from interactions among many amino acids is itself a profound problem. Proteins are unusual in that many of them fold into compact and nearly unique structures. This raises deep questions about how the amino acid sequences of real proteins avoid the competing interactions that would frustrate a typical random sequence's search for a well-defined equilibrium conformation. Put another way, what are the physical principles that distinguish functional proteins from all possible polymers of amino acids? As explained in Chapter 3, this problem has been a focus of interest in the biological physics community. Out of this work has come ideas about the funnel-like structure of the energy landscape for folding (see Figure 3.4), the many-to-one nature of the mapping from sequences to structure, and a statistical mechanics in sequence space describing the evolution of protein families. All of this has resulted in a coherent theory of protein folding.

A more practical formulation of the protein problem is concerned not with the general question of why proteins fold, but with the prediction of the folded structure for particular amino acid sequences. This prediction remains out of reach for direct molecular dynamics simulations, except in very special cases. On the other hand, the energy landscape picture suggests that the folding process has some hidden simplicity. There have long been efforts to learn the mapping from short sequences to local structures, such as helices and sheets, through a combination of simulation and generalization from known structures. Most recently, there has been an extraordinarily successful effort to learn the sequence/structure mapping more globally, using deep networks to generalize from the large number of examples now available. While this is an exciting step forward, there is still a long way to go in understanding the full repertoire of protein structures.

Most of the work on protein folding has been focused on the ability of proteins to self-assemble while not interacting with any other molecules—that is, in the dilute limit. In the cell, however, proteins are not free polymers. The situation is much more complex, and protein folding often begins as the protein is being synthesized on the ribosome. There are chaperone molecules that help to guarantee successful folding, and in such a dense environment there is a serious danger of aggregation being thermodynamically competitive with folding. Misfolding and aggregation are triggers for diseases, notably the prion diseases, and cells have significant machinery devoted to avoiding these errors. Understanding all these problems is part of a more general transition from thinking of biological molecules in isolation, as has been traditional in structural biology, to thinking of them as an interacting system, forming the underpinnings of molecular biology.

Molecular biology examines the molecular basis of biological processes in and between cells. The view of the biological cell has come a long way from the picture of the cell as a featureless vessel filled with chemical species undergoing reactions through random collision. Approaches such as cryoEM, fluorescence correlation spectroscopy, and dynamical rheology have indicated that the actual picture is much more interesting, involving cytoskeletal networks, membrane-bound and membrane-less organelles, ribosomes, proteins, nucleic acids, and small molecules—all of which are in a state of constant flux that can be dynamically modulated by the cell in response to changing conditions. While not so long ago characterization of a single protein was viewed as a major breakthrough, it now is appreciated that each protein is generally linked together with many others into complex "molecular machines" that carry out specialized and coordinated tasks. Approaches from the biological physics community have been helpful, and often necessary, in advancing understanding of the behavior of this molecular machinery and establishing the principles of its function. By abstracting and simplifying the highly complex inner workings of living systems, these approaches enable the application of simple models that have predictive value.

The capacity of biological macromolecules to act as sophisticated and highly efficient cellular machines—switches, assembly factors, pumps, or motors—is realized through their conformational transitions, that is, their folding into distinct shapes and selective binding to other molecules. As the number of molecular structures in the protein database grew, there was a clear need for methods to characterize the conformational dynamics of these structures. For decades, scientists could only investigate biochemical processes on a bulk level. The forces that molecules exert on each other and the displacements that they undergo in the course of reactions were not directly measurable. Two different approaches have been developed to follow the time histories of individual molecules. In single-molecule manipulation methods (as in Chapter 1), the dynamics of the individual molecules are measured by attaching them to an external probe, which is used to exert defined forces on the molecules in order to characterize their mechanical properties or induce conformational changes. In single-molecule detection methods, the molecule is tagged with a fluorescence label, or a pair of labels that can undergo FRET, and the dynamics of the molecule are followed in real time from a change in the intensity of the fluorescence of the single probe or from the change in the FRET signal. These approaches have made it possible to follow, for the first time, the dynamics of individual molecules as they interact and undergo transformations, and to control and even alter the fate of these transformations.

Single-molecule methods elucidate details that are typically lost to ensemble averaging or asynchrony when studied by traditional "bulk" methods of biochemistry. The resulting time trajectories contain a wealth of unaveraged information that is directly amenable to mechanistic interpretation. However, the analysis and interpretation of these trajectories to reveal the kinetic barriers and time scales of the associated biological process required reformulation of many of the traditional concepts of thermodynamics and kinetics. Approaches rooted in non-equilibrium statistical mechanics have produced general theories of force-induced conformational dynamics, which enable the extraction of intrinsic kinetic rates and activation free energy barriers from single-molecule data. Single-molecule manipulation methods also present a natural approach to test some of the fundamental relationships in statistical mechanics (Chapter 5).

GENES, GENOMES, AND EVOLUTION

Genetics and genomics are at their heart the study of heredity, and describe how genetic variation leads to phenotypic differences between individuals. The tools of genetics also provide powerful methods to understand cellular and molecular biology, by exploiting genetic perturbations and patterns of inheritance to dissect the physiological basis of important traits. Much of this work has traditionally viewed genetic networks as isolated systems that affect specific traits according to essentially digital rules of logic. However, in recent years the field of systems biology has recognized that the functioning of a cell depends on the quantitative details of physical interactions between biological components across space and time. This has led to efforts to describe genetic networks not only as logical structures but as stochastic dynamical systems. In the biological physics community, this perspective has been pursued in many directions—to examine the formation of patterns of gene expression in space and time (Chapter 1), the flow of information through genetic networks (Chapter 2), and the possibility of collective or emergent behaviors in these networks (Chapter 3).

In exploring genetic networks, biological physicists have developed methods that now are being adopted more widely by the biological community. An important example is provided by tools to interrogate the expression levels of many genes simultaneously, in single cells. There are two broad strategies. In the first, single cells are routed through microfluidic devices in which their mRNA molecules are extracted, encapsulated in droplets, and then amplified and sequenced (see above). In the second, cells are fixed and treated so that fluorescently tagged sequences can diffuse in and bind to complementary sequences of mRNA; these molecules can then be counted, one at a time, and successive rounds of washing and labeling with different sequences allows surveying of many different genes (see Figure 2.6). These methods have been used to provide objective definitions of cell types based on clusters in the high-dimensional space of expression levels, to identify very rare populations of cells, and more.

Our understanding of genetics has also been revolutionized in recent years by advances in molecular biology and in sequencing technology that makes it possible to analyze and manipulate genetic and phenotypic variation on a genome-wide scale. For example, systematic large-scale screens of CRISPR mutant libraries, transposon mutagenesis libraries, and knockout collections allow us to comprehensively survey the effects of every gene in model organisms or human cell lines, and the interactions between them. Many exciting experiments today bring these core biological methods together with physics-based methods for measuring gene expression, chromatin dynamics (Chapter 3), and other aspects of cell state.

Our understanding of genetics has also been enhanced by viewing sequence variation and corresponding phenotypic changes in their evolutionary context. This broad field of evolutionary and quantitative genetics aims to draw inferences from variation observed in natural populations. In quantitative genetics, the goal is to find statistical associations between genetic and phenotypic changes among closely related organisms, and to use these signatures to map the genetic basis of diseases, behaviors, and other important traits. Essentially, this work attempts to infer the structure of the genotype-phenotype map. Research in biological physics has played a role in developing methods for inference and for analyzing the structure of the resulting genotype–phenotype maps. For example, methods from statistical physics have helped to categorize types of epistasis and analyze how evolution across genotype–phenotype maps with a given statistical structure will tend to lead to regions of the landscape with specific properties. For more than a century, evolution was an observational science. Early efforts to bring evolution into the laboratory in the 1950s were led by physicists and chemists. A huge step forward came with the launch of the long-term evolution experiment in 1988. This experiment now has followed a dozen replicate populations of the bacterium

Escherichia coli over more than 75,000 generations, keeping a "fossil record" along the way. As described in Chapter 4, the biological physics community has contributed to understanding the data that emerge from this experiment, and to taking experimental evolution in new directions.

CELL AND DEVELOPMENTAL BIOLOGY

Cell biology is concerned with elucidating the structure and function of the cell, the "basic unit of life." This field aims to determine how biomolecules self-assemble into functioning organelles, or subcellular compartments, that perform specific functions necessary for energy production, waste removal, or self-propagation; how systems of organelles mediate whole-cell functions such as motility or phagocytosis; and how cells interact with each other and their microenvironment to mediate tissue-scale physiological functions such as muscle contraction or glandular secretion. The biological physics community has been interested in all these phenomena, and has produced ideas and methods that have spread into the larger cell biology community.

Developmental biology is the study of how a single cell becomes a complex organism, and in particular how cells make decisions that determine their fates and identities during this process. It has long been known that signals driving these decisions are carried by the concentrations of particular molecules, called morphogens. Now classical biological studies have identified many of these morphogens, and the biological physics community has pushed to understand how much information these molecules carry, how that information flows through the genetic networks for which they provide input (Chapter 2), and how these interactions lead to spatial patterns (Chapter 1). Beyond morphogens, the field has come to appreciate that physical forces influence all stages of development, from the early embryo, to gastrulation and establishment of the body plan, to organogenesis. The field now seeks to understand how developmental morphogenesis is influenced by intrinsic force generation and tuning of tissue stiffness and fluidity as well as extrinsic factors including microenvironmental mechanics, pressure, and fluid flow. In the past decade, advances in animal models, live-imaging approaches, and biophysical measurements have shed light on how the physics of cells and tissues mediate morphogenesis.

The folding of proteins and nucleic acids has been a major focus of the biological physics community, as described in Chapter 3, and this line of inquiry continues in the exploration of self-assembly of cellular structures such as the cytoskeleton and cell membranes. As an example, studies of lipid–lipid and lipid–protein interactions in vitro and later applied to studies of lipid and protein dynamics in living cells have led to fundamental and well-accepted concepts about the structure and function of cellular lipid bilayer membranes such as the fluid mosaic model, the notion of "lipid rafts" as cell signaling platforms, and the fission and fusion of membranes that mediate processes such as endo and exocytosis and cell division. There continue to be exciting questions at the frontier of this subject.

Another important contribution of physics to cell biology is in the areas of cell motility and mechanobiology. These subfields are aimed at elucidating how cells generate and react to forces and material properties; thus, by their very definition they require physics. The biological physics community has contributed to understanding across a range of scales. At the smallest scale, research is aimed at understanding how individual proteins generate and respond to force. The study of force- and motion-generating motor proteins that use ATP and the cytoskeleton or DNA as substrates grew out of the 1950s studies of muscle contraction and is an expansive subfield of biological physics. This field was aided early on by biophysicists applying methods of diffraction to give rise to models such as the sliding filament theory of muscle contraction. Investigations of motor proteins at the atomic level are aimed at elucidating the intramolecular rearrangements that take place in response to substrate binding interactions and energy released by ATP hydrolysis to drive conformational changes that generate force and motion or changes in protein-protein binding affinity. Insights into the structural basis of motor protein force generation or cell adhesion molecule bond strength have been aided by crystallography methods with roots in the physics of diffraction. More recently, as in Figure 1.3, single-molecule approaches utilizing optical and magnetic traps and microfluidics systems that were directly developed by physicists and adapted by biologists have allowed quantitative characterization of force generation by single motor proteins, catch bond behavior by individual cell adhesion molecules, and force-induced unmasking of protein-protein interaction sites.

At the scale of cells, soft matter and polymer physicists are advancing a quantitative and predictive understanding of how ensembles of mechanoactive proteins and biopolymers give rise to emergent properties of organelles or whole cells (Chapter 5). For example, this approach has been applied to understand the formation of the mitotic spindle or the waveforms of flagellar motility, wildly different processes that are both generated by the same basic elements of microtubules, motor proteins, and microtubule crosslinkers. These subfields of physics have also made important contributions to understanding the theoretical basis for how systems of cytoskeletal filaments give rise to cell and tissue material properties of stiffness and viscoelasticity. This has had an important impact, for example, on understanding the mechanoactive behavior of lung tissue that must expand and contract on time scales that are orders of magnitude longer than the motion generated by a motor protein.

In stem cell biology, the seminal discovery that fate determination could be controlled by the physical cue of tissue stiffness, led by a physicist, has revolutionized the field. The cellular response to these cues can be seen in macroscopic morphology, as in Figure 6.3, and also in the patterns of gene expression inside the cells. This discovery, and the cell biological mechanisms that have been pursued as a result, now form the basis for applications in tissue engineering and regenerative medicine in which physical properties of tissue scaffolds are carefully constructed to control cell fate for engineered tissue implants.

The notion that cells sense, respond to, and modulate tissue stiffness and viscoelasticity has also had a major impact on cancer cell biology and cancer research more generally. The vast majority of the research effort in cancer has historically been devoted to uncovering the biochemical mechanisms and genomics underly-

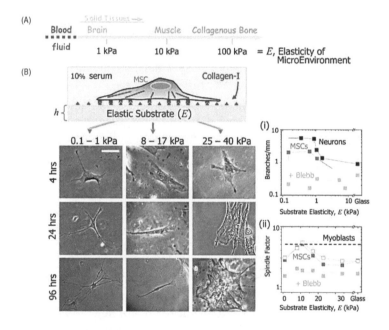

FIGURE 6.3 The discovery that fate determination could be controlled by the physical cue of tissue stiffness has revolutionized stem cell biology. Tissue elasticity and differentiation of mesenchymal stem cells (MSCs). (A) Solid tissues exhibit a range of stiffness, as measured by the elastic modulus, E. (B) Substrates with varying stiffness can be synthesized by manipulating the level of crosslinking, controlling of cell adhesion by covalent attachment of collagen-I, and controlling the thickness h. Stem cells of a standard type are initially small and round but develop increasingly branched, spindle, or polygonal shapes when grown on matrices respectively in the range typical of brain ($E \sim 0.1$–1 kPa), muscle ($E \sim 8$–17 kPa), or stiff crosslinked-collagen matrices ($E \sim 25$–40 kPa). Scale bar: 20 µm. Inset graphs quantify the morphological changes versus stiffness: (i) cell branching per length of primary mouse neurons, MSCs, and blebbistatin-treated MSCs and (ii) spindle morphology of MSCs, blebbistatin-treated MSCs, and mitomycin-C treated MSCs (open squares) compared to C2C12 myoblasts (dashed line). SOURCE: A.J. Engler, S. Sen, H.L. Sweeney, and D.E. Discher, 2006, Matrix elasticity directs stem cell lineage specification, *Cell* 126:677.

ing cancer. That situation has changed over the past decade with an explosion of research on the physics of cancer.

Physical mechanisms and properties can strongly influence tumor progression. The field of mechanobiology, which considers mechanical effects on biological processes such as mechanical changes of nuclei, cells, and tissues, encompasses an important class of physical mechanisms. Tissue stiffness is an example of an important mechanobiological property. Dysmorphic nuclei are common in cancer, as are changes in chromosome number, and further changes in nuclear properties are emerging. Tumor architecture is likewise abnormal compared to normal tissue, and it is clear that some solid tumors such as breast tumors are palpably stiffer than surrounding normal tissue. It has also been noted that stiffer tissues tend to exhibit more mutations. Changes in tissue stiffness can occur long before cancer is detectable; for example, liver stiffness is an excellent predictor of liver cancer. A proposed mechanism for how tissue stiffness might promote DNA damage is illustrative because it involves coupling between the mechanical process of cell migration and the biochemical process of DNA repair: Cell migration through pores that are more constrictive in stiffer tissue can promote segregation of fluid, including DNA repair factors, from the chromosome in the nucleus, thereby suppressing the repair of double-stranded DNA breaks. The coupling between mechanics and biochemistry in cancer is also illustrated in gene expression, which depends on the spatial structure of chromatin and how it varies over time. As a third example, the epithelial-to-mesenchymal transition marks the transition from a state of epithelial tissue in which cells do not change their relative positions (what in physics would be called a solid) to one in which the cells migrate and invade normal tissue (a fluid state).

More generally, the coupling of physical and biochemical processes is likely important to many aspects of cancer as well as many other biological processes. The local environment around a tumor—its microenvironment—interacts constantly with a tumor and affects its progression. Physical aspects of the tumor microenvironment are known hallmarks of cancers, and there is evidence that they are functionally linked to metabolism, the immune cell interactions with tumors, drug transport into tumors, the proliferation of cells, the ability of cancer cells to develop drug resistance, the ability of cells or clusters of cells to migrate out of the tumor to cause metastasis, and the plasticity of cancer stem cells. The heterogeneous physical properties of the tumor microenvironment can change significantly in space and time due to changes in local blood flow and oxygenation, or due to chemotherapy or other treatments. Thus, the tumor and its microenvironment can be viewed as a system of many components that interact and evolve via coupled physical and biochemical processes.

From a very broad perspective, understanding how microscopic changes lead to collective behavior in highly complex, adaptive many-body systems lies at the core of the mission of biological physics. Cancer is a disease of altered genes and gene expression. This genetic information is transferred within cells and between cells and across a range of scales in space and time, eventually leading to collective behavior—symptoms of disease—at the system level. More specifically, a fitness landscape links genotypes and phenotypes to reproductive success and depends on local environmental conditions. Mutations and changes in gene expression thus evolve within the fitness landscape. The question of how subtle changes at the microscopic level disrupt processes in many-component systems to cascade into the collective phenomenon of cancer is a fascinating and impactful area of focus for biological physics.

Another recent and major breakthrough in cell biology that came from the biological physics community was based on understanding of the principles of phase separation (Chapter 3). For nearly 150 years, cell biologists had thought the only way to spatially sequester biochemical reactions to subcellular compartments is to encapsulate them inside a membrane or immobilize them on a scaffold. The notion that specific biomolecules could coalesce into discrete, non-membrane-bounded droplets within the cell cytoplasm or nucleoplasm by phase separation amounted to the first discovery of a novel principle of subcellular protein organization in more than a century. This has revolutionized thinking in cell biology about the organization and function of signal transduction, the genome, and protein processing. The therapeutic potential for this discovery is being explored as evidence gathers that liquid-liquid phase separation or its misregulation may be at the root of protein aggregation diseases such as Alzheimer's disease, Parkinson's disease, or amyotrophic lateral sclerosis.

A major thrust in the application of physics to developmental biology centers on how systems of motor proteins and cytoskeletal filaments in cells couple to cell-cell and cell-extracellular matrix (ECM) adhesion to generate the forces required for cells to pull on their neighbors or the ECM scaffold to drive tissue movement and morophogenesis. Actomyosin contractions can locally constrict parts of the cell, as in apical constriction, in order to change cell shape or allow a cell to exert traction in order to move over a surface, as in cell migration. These active forces at the level of single cells can be translated into dramatic changes in shape at the tissue level. For example, forces generated by actomyosin networks drive cell shape changes required for transforming early vertebrate embryos from a spherical mass of cells (the morula) into a fluid-filled blastocyst with an inner cell mass and an outer layer of cells called the trophoblast. Researchers are actively working to elucidate how morphogenesis in this and other tissues and organs is mediated by the spatial and temporal dynamics and magnitude of force generation at the single cell level.

In addition to generating forces, developing tissues are subject to a variety of extrinsic physical constraints, including compression by the ECM, attachment to neighboring tissue, and sculpting by surrounding contractile tissues. Physical constraints can prevent or direct tissue growth, as well as lead to mechanical instabilities or tissue buckling. For example, growing tissues subject to compressive forces from the ECM, contraction, or cell growth in neighboring tissues can fold to relieve the resultant strains, as observed in the developing eye and intestine of the chick embryo. Cells also can sense physical cues and transform them into biochemical signals that inform cell behavior and fate decisions in a process termed "mechanosensation." For example, airway epithelial cells cultured in wide tubes to simulate proximal airways or narrow tubes to simulate distal airways experience curvature-dependent tension that dictates the expression of either proximal or distal fate markers. The mechanism of such cellular mechanosensation from the atomic to the tissue scale is a major thrust in the fields of cell and developmental biology.

Physicists have applied the concepts of fluid dynamics to the study of developmental cell migration and tissue movement during morphogenesis, similarly to how fluid dynamics has been applied to the epithelial-to-mesenchymal transition (EMT) in cancer cell biology. This has provided accurate predictors of whether cells migrate independently or collectively (or somewhere in between). In fluid-like tissues, cells elongate, rearrange, and move past their neighbors, whereas in solid-like tissues, neighbors are maintained, cell shape remains essentially the same, and rearrangements are minimal. The transition between these two states, the jamming transition, is a concept from soft condensed matter physics (Chapter 5) and can theoretically be caused by changes in cell contractility or adhesion. In epithelial cell sheets, the jamming transition is predicted to occur when a single cell shape parameter reaches a specific value; hexagonal cells correspond to solid tissue and deviations from hexagonal are associated with fluidity. An example of a solid-to-fluid transition during development is the onset of neural crest cell migration, which requires a reduction in cell-cell adhesion mediated by internalization of cadherin cell-cell adhesion molecules. This allows neural crest cells to fluidize and thus migrate collectively to establish the neural tube and eventual spinal cord.

Fluid-filled lumens play diverse mechanical and biochemical roles during development, and several experimental techniques have been devised to quantify pressure within them. Lumens form when cells release solutes and generate an osmotic gradient that directs fluid flow. The fluid in the lumen is contained by cell-cell adhesions in the tissue surrounding them, and its expansion generates hydrostatic pressure. Pressure in lumens has recently been shown to play a role early on in blastocyst development, as well as later in development, such as when the neural tube becomes filled with cerebrospinal fluid that exerts pressure on the neuroepithelium and during lung development where pressure drives formation of complex tissue architecture and influences cell differentiation. Cells that line fluid-filled lumens can be subjected not only to pressure, but also to fluid flow. Flow generates shear forces along the walls of the lumen. A wide range of fluid velocities are present throughout the embryo, from weak flows generated by beating cilia to fast flows generated by the beating heart. It is well established that embryonic left-right patterning is established by fluid flow generated by motile cilia, while midway through embryogenesis, once the heart has formed and started beating, blood flow shapes the development of the vascular network.

The development and application of methods to measure and manipulate force, pressure, stiffness, and flow in developing organisms will be key to uncovering their full role in developmental biology. On the flip side, the developing organism provides a rich platform for the discovery of new principles of physics and material science in dynamic living tissues.

FROM NEUROSCIENCE TO PSYCHOLOGY

Our modern understanding of voltage and charge has its origins in the discovery of "animal electricity" in the late 1700s. All living cells have a voltage difference across their membrane, and the dynamics of voltage changes are crucial to the functioning of muscles, the nervous system, and the heart. As explained in Chapter 2, the elucidation of the basic mechanisms of this electrical signaling, through a combination of theory and quantitative experiment, forms a classical chapter in the emergence of biological physics. The path from quantitative phenomenology to the identification and structure of the ion channels responsible for the electrical activity of cells has been recognized by three Nobel Prizes.

The understanding of ion channels has had broad implications for biology. Many organisms, including humans, have genes for more than 100 different kinds of channels. Individual neurons in the brain use a handful of these, and the decision processes through which cells decide on the correct mix of channels are an important part of brain development. As with many genes in complex organisms, the genes encoding ion channel proteins also come in separate pieces along the chromosome, and the splicing together of these pieces is an important example of how this mechanism can fine-tune cellular functions. The different channels have a wide range of structures, selectivities, kinetics, and sensitivities to drugs. As a result, ion channels are central to modern drug discovery in cardiology, nephrology, neurology, and psychiatry. More subtly, it has gradually been appreciated that ion channels play key roles even in non-neuronal cells.

Throughout the brain, neurons communicate through synapses, where electrical signals in neurons are transduced into a chemical form and then back into electrical signals. The same dynamics occur at the junctions between nerves and muscles. This report has explored the dynamics of synaptic transmission as part of the problem of processing single photon signals in the retina (Chapter 1), and as the substrate for connections in neural networks (Chapter 3). As with ion channels, the modern picture of synaptic transmission has its roots in work from the biological physics community, has been recognized by multiple Nobel Prizes, and has implications far beyond the examples presented here. As an example, molecular components responsible for synaptic transmission are shared by all cells that release vesicles, which contain everything from hormones to waste products. In a very different direction, the processes of synaptic transmission are the targets of the most widely prescribed drugs for mood disorders that affect the lives of tens of millions of people in the United States alone. There is a continuous path from work in the biological physics community, through the broader application of these results in neurobiology, to an impact on mental health.

Neuroscience as a field was actively constructed from multiple more well-established biological disciplines—physiology and pharmacology, anatomy and cell biology, biochemistry, and more. Perhaps because of this history, there has been a relatively rapid absorption of ideas from the biological physics community into the mainstream of neuroscience. This has happened with new experimental methods, with new theoretical ideas, and with new approaches to data analysis, resulting in a continuum of activity from theoretical physics to experimental neurobiology. It is a remarkable feature of this larger community that some of the most sophisticated physics-based methods are driving developments in our understanding of the human brain, reaching from neuroscience to psychology and neurology.

Functional magnetic resonance imaging is the most widely used method for visualizing human brain activity. Much of what is known about the dynamics of information processing in the human brain now has come from fMRI studies. Today, virtually all leading academic psychology departments have researchers who conduct fMRI studies in human subjects, using behavioral paradigms that probe sensory perception, diverse modes of cognition and decision-making, language processing, social interactions, motor control, and more.

Our ability to image the human brain in action is a direct outgrowth of work in the biological physics community. The electrical activity of neurons is supported by their metabolism, and oxygen is essential. Oxygen is carried throughout the body by hemoglobin in the blood, which is a paradigmatic protein—one of the first to have its structure solved by X-ray diffraction. Binding and unbinding of oxygen to hemoglobin changes the optical absorption spectrum of the protein, so that blood in our veins is a different color from blood flowing through our arteries, and the fact that oxygen binds directly to an iron atom at the active site of the protein provides opportunities for even more spectroscopic tools. NMR in particular was used intensively to probe protein structural changes in response to oxygen binding. At some point it was realized that the oxygen binding, and the changing spin state of the iron atom, affects not only the NMR spectrum of the protein, but the relaxation dynamics of proton spins in the surrounding water. Thus, measurements of proton spin relaxation can generate a map of blood oxygenation, and indirectly neural activity, in humans, in real time; an example of these experiments is shown in Figure 6.4. It is hard to overstate the impact that these developments have had on cognitive science and psychology.

In many ways, the use of physics-based techniques to explore the human brain closes a circle. In the 19th century, scientists routinely crossed boundaries among subjects now distinguished as physics, chemistry, biology, and psychology. The physics community's interest in human perception was especially strong, and productive. It is not only that physicists were interested in the mechanics of the ear or the optics of the eye, they also were interested in the inferences that the brain draws from these raw sense data. Some went so far as to wonder if someday we might be able to unify not just physics and biology, but the natural sciences and aesthetics—if we really understand our perceptions, we should understand why we find some things to be beautiful.

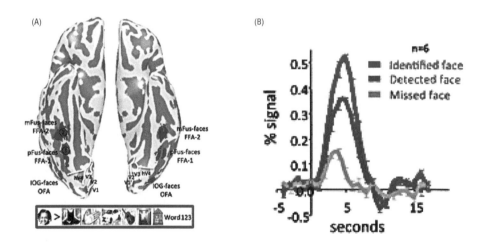

FIGURE 6.4 Measurements of proton spin relaxation can generate a map of blood oxygenation, indi-rectly providing us with a real-time look at human neural activity. Functional magnetic resonance imag-ing (fMRI) of the human brain during face recognition. (A) Inflated view of the human brain, showing the cortical surface. Red regions are those in which the fMRI signal is reliably larger in response to seeing human faces than seeing other body parts, objects, places, or characters. (B) Time course of fMRI signal averaged over the regions identified in (A). Signal is largest for faces that are identified by the subject, smaller when recognized as a face but not identified, and smaller still when missed. This suggests that the fMRI signals in these brain regions are correlated with perceptual decisions beyond the visual input. SOURCE: K. Grill-Spector, K.S. Weiner, K. Kay, and J. Gomez, 2017, The functional neuroanatomy of human face perception, *Annual Review of Vision Science* 3:167.

7

Health, Medicine, and Technology

The project of writing this report has overlapped almost completely with the worldwide COVID-19 pandemic. Millions of people around the world have died, and the isolation required to restrict the spread of the disease has taken a severe toll on society. A bright spot in this dark chapter of human history is the extraordinary response of the scientific community. As always with the solution of major technological problems, what success the world has had in combating COVID-19 has been woven from many threads, but many results, ideas, and methods that have emerged from the biological physics community proved foundational for this success. The society has effective responses to the pandemic because many scientists focused their extraordinary skills on the problem, but also because they could build on an extensive intellectual infrastructure. This infrastructure supports much more than our response to the pandemic, reaching into almost every aspect of health and medicine, and many areas of technology more broadly. This chapter explores some of this infrastructure, how parts of it emerged from the biological physics community, and how it has influenced the progress of human health, medicine, and technology more generally; an overview is given in Table 7.1.

The practice of medicine has been revolutionized by our ability to see what is happening inside the body and in isolated cells, on scales from single molecules to whole organs. Our understanding of disease, and our ability to respond to it, have been profoundly affected by our ability to resolve the structures of the relevant molecules down to the positions of individual atoms, and to design new molecules that emulate, extend, or block the functions of those that occur naturally. Our understanding of the basic mechanisms of information flow and control in living cells

TABLE 7.1 Broad Societal Impact Through "Physics of Life" Intellectual Infrastructure

"Physics of Life" Topics: Intellectual Infrastructure Supporting Health, Medicine, and Technology	Capabilities or Promise of Biological Physics	Connected Topic, Field, or Application
Imaging, diagnostics, and treatment	See what is happening inside the body and in isolated cells, on scales from single molecules to whole organs.	Medical physics; medical diagnoses.
Molecular design	View structures of the relevant molecules down to the positions of individual atoms, and design new molecules that emulate, extend, or block the functions of those that occur naturally.	Drug design.
Synthetic biology	Engineering cells that can be harnessed to address problems.	Personalized medicine; synthesis of clean fuels.
Predicting and controlling evolution	Creating a vaccine by predicting the future trajectory of the virus's evolutionary change.	COVID-19; seasonal flu.
Biomechanics and robotics	Enables building truly new things, and predicting how these things will behave; engineering and design guided by the physical principles at work in living systems.	Robots.
Neural networks and artificial intelligence	Emulate human and animal performance at challenging tasks, ranging from walking on rough terrain to understanding language.	Ongoing revolution in artificial intelligence.

created the opportunity for a new kind of engineering, using these mechanisms as building blocks in much the same way that computer chips use transistors, capacitors, and resistors; these engineered cells can be harnessed to address problems as diverse as personalized medicine and the synthesis of clean fuels. While popular accounts of evolution often emphasize that it is driven by random mutations, creating a vaccine against next year's seasonal flu requires predicting the future trajectory of the virus' evolutionary change, and there has been remarkable progress on this fundamental problem.

Many of these different themes, and more, came together as the scientific community responded to COVID-19. As soon as SARS-CoV-2 was isolated there were rapid efforts to determine the structures of crucial molecular components, including the infamous spike protein that enables viral entry into cells. Methods for single molecule counting and visualization provided extraordinarily sensitive probes for the presence of the virus, and sequencing methods focused attention on parts of the antibody repertoire that were central to an effective immune response. Methods for tracking and predicting the evolution of influenza virus, which had come directly out of the biological physics community's engagement with evolutionary dynamics, were immediately adapted to the new virus and became the worldwide standard. In particular, these methods led to the first detection of community spread of COVID-19 in the Seattle area as early as January 2020, months before the scale of the public health crisis facing the United States (and the world) was evident. A generation of work that connects the biological physics community with ecology and epidemiology resulted in the rapid development of effective models for spread of the virus, within which the impact of different public health measures could be assessed. This provided direct input to public health policy in the United States and around the world, including the difficult decisions to shut down institutions and whole states. Acting locally, many members of the biological physics community have been central to their institutions' efforts to expand testing capabilities, helping to provide infrastructure to support a safe return to in-person teaching.

The biological physics community studied the spread of droplets experimentally, using particle image velocimetry and other physics techniques, and theoretically, using fluid dynamics simulations. Such studies provided evidence that aerosols are extremely important in transmission of the virus, at a moment when this was controversial. This technical work was adapted into accessible videos to demonstrate the efficacy of masks in blocking droplets to a larger audience. Finally, the ingenuity of biological physicists has been directed toward developing substitutes for medical supplies in short supply during peaks of the pandemic, such as N95-like filter materials made using repurposed cotton candy machines—inexpensive workarounds that may be useful in future pandemics. The community also developed and circulated designs for cheap and easy-to-build ventilators and oxygen concentrators that can be constructed from commonly available parts.

The pandemic inspired many members of the biological physics community, as with the scientific community more broadly, to engage with a problem of immediate concern to the society at large. The evidence of success, and the intellectual excitement that surrounds such an urgent problem, will have lasting impact. In addition, biological physicists have directed efforts toward many other public health problems beyond the COVID-19 pandemic. One example is the recent development of microscopes made largely out of folded paper that cost under a dollar and have a resolution of 2 microns. Similarly, centrifuges can be made from paper, string, and plastic for 20 cents that can spin medical fluid samples at up to 125,000 rotations per minute. Although currently used primarily as tools for teaching, one can imagine that these inexpensive instruments will have broader impact in the developing world.

One profound consequence of understanding physics is that it enables building truly new things, and predicting how these things will behave. While the lightbulb could be invented by trial and error, building a working smartphone requires design, and depends on our understanding of how electrical currents flow in the millions of devices that are packed into each computer chip. Progress in biological physics holds the promise of engineering and design guided by the physical principles at work in living systems, beyond the emulation or harnessing of particular mechanisms. Examples include robots and computers that emulate human and animal performance at challenging tasks, ranging from walking on rough terrain to understanding language. Although there has been revolutionary progress, it is reasonable to expect that addressing the more basic scientific challenges outlined in Part I of this report will lead to even more powerful and life-changing discoveries and inventions in the future.

IMAGING, DIAGNOSTICS, AND TREATMENT

Since the first microscopes were built in the 1600s, our ability to visualize or image the phenomena of life has been central to scientific progress. As is clear from many examples in Part I of this report, progress in imaging also is central to today's biological physics. Importantly, these same imaging methods also have revolutionized the practice of medicine. These developments have spawned major industries and shaped societal expectations regarding medical care. The impact of biological physics on human health can be discerned even before a baby is born, when a fetal heartbeat is detected by Doppler sonography and the fetus is imaged by ultrasound tomography. If the parents' fertility has been examined, then sperm motility will likely have been assessed with technologies such as microfluidic methods. Later, when a child injures a limb or requires a dental cavity to be filled, X-ray imaging will be used to inspect for broken bones or identify tooth decay. As individuals age, routine clinical screening may involve optical endoscopy, flow cytometry, and electrocardiography—all physical methods. Optical coherence tomography and the fundus camera provide assessments of retinal health. In patients with treatment-resistant depression or chronic pain, transcranial magnetic stimulation can often provide relief. In those with suspected neurological disorders and diseases, physical methods such as magnetic resonance imaging (MRI), positron emission tomography (PET), electroencephalography (EEG), or electromyography (EMG) may be critical for diagnoses. When cancer is suspected, flow cytometry and X-ray computed tomography or PET imaging may be used for diagnosis, while light microscopy and fluorescent histopathological stains may be used for detailed characterizations of a cancer and its level of invasiveness. Cancer treatment may involve the targeted delivery of X rays, gamma rays, or even protons to solid tumors. Microfluidic and gene sequencing assays may be used to detect circulating tumor cells. Emerging, personalized cancer therapies involve gene sequencing and genome editing of immune cells, guided by increasingly sophisticated theoretical approaches.

Medical physics largely emerged after the discovery of X rays and their preferential absorption by the body's hard tissues, chiefly in the musculoskeletal system, in contrast to the far greater X-ray transparency of soft tissues. Within only a few years after this discovery, medical researchers began investigating the use of X-radiography for non-invasive imaging of the body. Further development and refinement of X-ray–based medical imaging continues to the present day. The invention of tomographic three-dimensional X-ray imaging, now widely known as X-ray computed tomography (CT), is used in hospitals worldwide and has an essential role in a wide range of image-based diagnostics; an example is in Figure 7.1. Ongoing research in this area remains vibrant and includes pursuits such as the innovation of new solid-state X-ray detectors, image processing techniques to empower new medical diagnostics, and machine-learning based methods for automated detection of tissue abnormalities.[1]

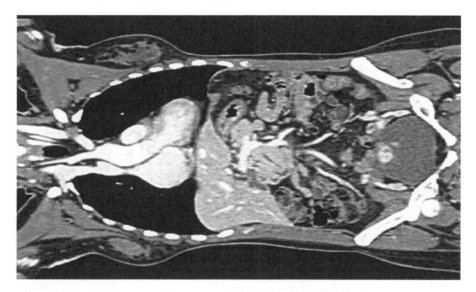

FIGURE 7.1 Modern X-ray computed tomography (CT) is used in hospitals worldwide for a wide range of image-based diagnostics, showing here a patient's full body after an operation, illustrating the level of detail now available. SOURCE: F. Paglia, L. D'Angelo, D. Armocida, L. Samprisi, F. Giangaspero, L. De Vincentiis, and A. Santoro, 2021, A rare case of spinal epidural sarcoidosis: Case report and review of the literature, *Acta Neurologica Belgica* 121:415.

---

[1] Anecdotally, if you are reading this report soon after its release, you probably have experienced some of this revolution. As a child, your X rays were likely taken with a large instrument that projected an image onto photographic film. The film was developed and stored at your doctor's office for future reference. Today, if you need a chest X ray, it is likely that you stand with a compact X-ray source in front of you and an "area detector" behind you. The detector, adapted from technologies used in physics laboratories, registers an image electronically and transmits it directly to a computer. No more film, no more file cabinets. This reflects progress in computers and information technology more broadly but would be impossible without the new X-ray sources and detectors.

Other portions of the electromagnetic spectrum also have important roles in medical imaging. For instance, gamma ray cameras are scintillation cameras with key roles in nuclear medicine (see Figure 7.2). By injecting a patient with a radiopharmaceutical—a pharmacologically active compound that will bind to a specific molecular target and emit gamma rays—radiologists can use the gamma ray camera to determine where in the body the radiopharmaceutical has accumulated. This approach has played important roles in cancer diagnosis and assessment, as well as in the diagnosis of abnormalities of the thyroid, heart, and lungs. Notably, the speed of imaging can be sufficiently fast to allow tracking of a radiotracer through the beating heart, enabling inspection of the contractile pattern. As with X rays, gamma-ray imaging can also be done in a tomographic manner, in this case called single photon emission computed tomography (SPECT).

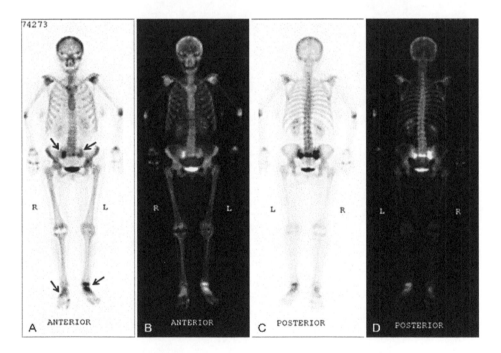

FIGURE 7.2 X rays are not the only portion of the electromagnetic spectrum with an important role in medical imaging; gamma ray cameras are also used for imaging and diagnosis. A patient with back pain is given a molecule (methylene diphosphate) that accumulates in metabolically active bones. This molecule is tagged with technetium-99, and images are formed from gamma rays emitted by this isotope. Front (anterior, A and B) and rear (posterior, C and D) views of the skeleton, as indicated, shown in both positive and negative contrast. Arrows in (A) indicate accumulation of the tracer molecule in the sacral ala, left ankle, and right midfoot; follow-up tests confirmed bilateral sacral ala fractures. SOURCE: R.S. Adler, 2019, pp. 1678–1685 in *Goldman-Cecil Medicine, 26th Edition* (L. Goldman and A. Schafer, eds.), Elsevier, Amsterdam, 2019.

Gamma rays also play a critical role in PET, another widespread form of non-invasive medical imaging. The patient is administered a beta-emitting radiotracer, which localizes to tissues or cells with selective characteristics. Typically, this compound has a specific biological activity or receptor, allowing physicians to target a particular biological process (such as in metabolism) or dynamic within the body. This stands in contrast to X-ray imaging, which is usually better suited for revealing tissue morphology, although exogenous X-ray contrast agents can be used to track physiological processes. Due to their typically brief half-lives, PET radiotracers are often produced at cyclotrons at or near medical facilities where PET imaging is done. Upon emission within the body, positrons generally travel only a short distance within tissue before annihilating with an electron to produce a pair of gamma ray photons traveling in opposite directions. The coincident detection of this photon pair enables a determination of a line along which the emitter must have resided, and by rotation of the detector array, a three-dimensional tomographic image can be computed.

Today, PET imaging is a workhorse technique for cancer diagnosis. Moreover, thanks to its ability to target specific biological processes of interest—often limited only by the ability of radiochemists to produce a suitable tracer—PET also has a forefront role in the growing radiological subfield of molecular imaging. For instance, neuropsychiatry has used PET to examine differences in the distributions of dopamine receptor subtypes in the brains of normal versus depressed human subjects. Related methods for PET imaging of the dopamine system allow the diagnosis of Parkinson's disease. Ongoing research seeks new radiotracers, new detectors, and new methods of image analysis.

Like X-ray CT and PET, MRI is also a computed tomography. However, it does not require the use of ionizing radiation and can be used to inspect soft tissue details with an exogenous contrast agent, leading to unique diagnostic roles in neurology, cardiology, orthopedics and sports medicine, and hepatology. MRI is based on the principles of nuclear magnetic resonance: Individual atomic nuclei, when placed in a strong magnetic field, can absorb and emit radio waves. The frequency of the waves depends on the field, and this means that by applying a magnetic field that varies in space one can map the positions of the atoms, today with a resolution of roughly 1.5 mm. The simplest choice, and the one most widely used in medical applications, is to probe the hydrogen nuclei in the water that comprises a large part of all living material; other nuclei can also be chosen, such as for NMR spectroscopy studies of metabolism, but these are far less common in a diagnostic context. By choosing different sequences of radio wave pulses, MRI signals become sensitive to different aspects of energy exchange or relaxation (T_1 and T_2) between individual hydrogen atoms and their surroundings, highlighting specific tissue features. MRI contrast agents, which typically involve chelates of gadolinium, have further expanded the range of medical applications beyond those in which the tissue of interest naturally exhibits suitable MRI contrast. Magnetic resonance images can also be sensitive to the functional activity of tissues, leading to opportunities in visualizing the human brain in action, as noted in Chapters 3 and 6, as well as to measuring concentrations of specific molecules down to millimolar concentrations and millimeter resolution.

Acoustic waves underlie ultrasound imaging, a widely used means of imaging the body's soft tissues in a non-invasive manner. A modern ultrasound probe typically comprises both acoustic emitters and an array of detectors. An acoustic impedance matching gel is applied between the probe and the tissue to minimize unwanted loss of acoustic energy at the tissue surface. Ultrasound waves enter the body, reflect off body tissues, and can be detected upon exiting the body. The round-trip "time of flight" for this process enables a determination of the depth at which reflection occurred in the tissue. Image contrast comes from the variability in the acoustic reflectivity of different tissue types; phased arrays enable tomographic imaging. The imaging resolution is limited by the acoustic wavelength and typically is sub-millimeter; unfortunately, higher-frequency ultrasound waves enabling finer resolution attenuate more readily in biological tissue and cannot penetrate as deeply. New forms of ultrasound imaging continue to emerge, enabled by progress in the design of emitter-detector arrays, new forms of detectors, and new computational engines for real-time image determination. For example, an emerging form of ultrasound imaging is photoacoustic imaging, which relies on the faint ultrasound emission that occurs during non-radiative decay of an optical excitation. Although the signals involved are weak, photoacoustic imaging has generated excitement because it can combine the molecular specificity of an optical probe with the imaging depth of ultrasound. Of course, optical emissions also have an enormous role in medical imaging. Nearly all of modern pathology rests on the ability to examine excised tissue specimens under a light microscope. Classical histopathological stains, such as hematoxylin and eosin (H&E), bind to specific tissues, alter their optical absorption spectra, and thereby enable a majority of pathological diagnostics that remain in common use. More advanced pathological techniques involve the use of fluorescence stains, such as immuno-fluorescent labels or fluorescent in situ hybridization (FISH) probes. These are the methods that have been pushed to the point of counting single molecules and surveying the expression of many genes, as in Figure 2.6. Unlike H&E, these fluorescence-based methods in pathology have the advantage of revealing specific proteins or RNA sequences in the tissue specimen, which can enable more precise diagnoses. While not strictly an imaging technique, fluorescence-activated cell-sorting (FACS) combines immunofluorescence and flow cytometry and has also enabled major advances in medical diagnostics, particularly in hematology, immunology, and cancer biology.

Beyond the benchtop microscope, visible light underlies nearly all of endoscopy, which is widely used to image the gastrointestinal system, lungs, bladder, throat, ears, brain, and other body tissues. The invention of fiber optics was a crucial development in physics that made such diverse diagnostics possible. Today, endoscopes are an essential component of not just diagnosis but also minimally invasive surgery, as they can be equipped with a working channel that allows in situ specimen acquisition and the introduction of surgical tools through the endoscope itself. Equally important is the surgical microscope, which comes in many forms and is used in diverse surgical procedures. Today, leading-edge research aims to create new endoscopes capable of advanced forms of optical imaging, for example to achieve cellular resolution or to use nonlinear optical mechanisms, as well as surgical microscopes with robotic manipulation capabilities.

Ophthalmology owes a tremendous amount to the field of optics, without which it would be impossible to provide proper eyeglasses or to perform vision correction surgeries by using an excimer laser to modify the shape of the cornea (LASIK) so as to ameliorate the eye's optical aberrations.

Ideas from biological physics have also been pivotal to a wide range of prosthetic devices that can be used to facilitate body function or ameliorate dysfunction. For those who have suffered the loss of a limb, biomechanics has led to prosthetic arms and legs with increasingly impressive mechanical properties. Cardiac pacemakers facilitate proper timing of the heartbeat by providing electrical synchronization when the heart's electrical conduction system is impaired. For Parkinson's disease, an electrical neurostimulator may be implanted into the brain to reduce tremor. Hearing loss may be counteracted with a digital hearing aid; if deafness is severe, a cortical prosthetic may be implanted into the inner ear, allowing direct electrical stimulation of auditory afferents in response to sound. Emerging methods for retinal prostheses promise to restore sight to the blind. There are even emerging brain-computer interfaces, which allow direct communication between the human brain and external electronics for people who are fully paralyzed. These developments depend both on advances in technology and on theoretical ideas about the representation of information in the brain.

Some sense for the liveliness of this enterprise comes from the continued emergence of startup companies based on technologies that have emerged from the biological physics community, as described in Box 7.1.

MOLECULAR DESIGN

Biological molecules perform an astonishing array of functions. It is an old dream to harness and adapt these functions for engineering purposes. One approach is to use an evolutionary strategy, generating large numbers of variants, selecting for molecules closer to the desired function, and repeating. This approach has been made practical, and was recognized by a Nobel Prize in 2018. A different approach is to actually design molecules, in the same way that humans design other engineered objects.

BOX 7.1
From Science to Startup

In the past few decades, biological physics research has directly spawned multiple successful startup companies that have had a substantial impact on the biotechnology and life science industries. In the early 1990s, companies emerged with gene chip technologies, and evolved precursors to the next-generation sequencing companies that largely supplanted the gene chip approach to genomic and RNA expression analyses. By the early 2000s, single molecule approaches to DNA sequencing, which had emerged from the biological physics community, were becoming industrialized. Technologies based on optical methods for single molecule fluorescence detection, zero-mode optical wells, and intellectual property licensed from biological physics groups led to companies that now are publicly traded, and competing commercial approaches based on nanopores had a similar origin.

Microscopy modalities developed in biological physics have also given rise to substantial economic activity, both in confocal and two-photon imaging systems. Moreover, the market for ultrashort-pulsed lasers has been propelled by the predominance of the Ti:sapphire laser as the most common illumination source for two-photon microscopes and ultrafast spectroscopy. These ultrafast laser systems make up roughly 10 percent of sales for the world's largest laser manufacturers.

Advances in imaging methods coming from the biological physics community have generated startup companies targeting optical brain imaging, nonlinear optical approaches to in vivo cancer detection, and the application of optical voltage imaging to drug screening. In just the past few years, roughly 20 companies have begun developing multiplexed image-based approaches to analyzing the RNA and protein content of tissue specimens for pathology.

A natural target for molecular design are protein molecules. Protein functions are determined to a large extent by their three-dimensional folded structures, and structures are determined by amino acid sequences. So the targets of design are particular structures, while the space of possibilities is the set of amino acid sequences. Protein design requires us to understand the sequence/structure mapping, a problem that has appeared in previous sections of this report.

Proteins are unusual because many of them are polymers that fold into well-defined compact structures. A fundamental problem in modern biological physics is to understand how this is possible (Chapter 3): What is it about the amino acid sequences of real proteins that allows for more or less unique structures, avoiding the competing interactions that lead to glassy behavior in typical random sequences? There is also a practical version of the problem (Chapter 6): Given a sequence that is known to fold, what is the resulting folded structure? Protein design is yet a third version of the problem, inverse to the practical folding problem.

The exploration of principles underlying the emergence of folded proteins provides several important hints about design. First, in a successful design, the energy landscape for the protein will have a form that assists rather than frustrates the search for the folded structure, funneling the molecule toward its final configuration as in Figure 3.4. Second, real structures are highly designable, so that many sequences map to nearly the same structure. Third, within this large set of sequences, there is a kind of smoothness so that it is possible to capture the structural constraints by keeping track of only pairwise correlations among amino acid substitutions.

The success of pairwise models means that one can generate new sequences that will fold into known structures, as discussed in Chapter 3. It is important that these new sequences can be very far from the original sequences; as such, this approach allows us to probe the sequence space globally rather than just locally. The funneled structure of the energy landscape suggests that, although high dimensional, the folding of a protein is simpler than it might have been without the competing local minima that characterize computationally hard optimization problems in computer science and physical systems such as glasses that fail to find their minimum energy configuration in a reasonable time. In a startling development, the problem of predicting protein structure from the amino acid sequence has inspired the development of a direct machine learning approach, "AlphaFold," which achieves a precision close to that of experimental structure determination by X-ray crystallography.

A corollary of the ideas mentioned above is that the number of protein sequences seen in nature today is much larger than the number of protein folds, or folding motifs, that have been found by studying protein structures. It is not clear just how strong this limitation really is. Has evolution discovered all the possible folding motifs, or can we synthesize new ones? Successful efforts aimed at discovering new motifs have been very limited. One of the first successes toward this goal has been the protein called Top7 (see Figure 7.3), which is able to fold into a unique structure but does not have many of the properties of real proteins. Although much effort has been dedicated to de novo protein engineering, genuinely new folds remain rare, and searches for new functionality have focused on combining known motifs. Basic questions about the universe of possible protein structures are closely tied to our hopes for further progress in the practical problem of protein design.

Beyond protein design, structure-based drug design (SBDD) is an integral element of most pharmaceutical industries developing therapeutics or vaccines. SBDD refers to the design and optimization of chemical or protein entities that interact with the molecular machines in the cell in order to select those most suitable for further optimization as potential clinical therapeutics (i.e., drug candidates). The basic approach is to develop a detailed understanding of the three-dimensional structure of the protein target, its interaction with the potential drug, and how the shapes and charge of the protein and drug affect their interactions. These fundamental atomic relationships correlate to the efficacy and safety of the drug in clinical trials.

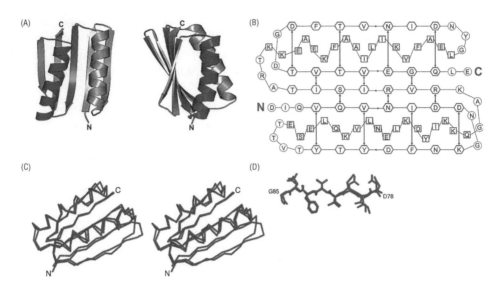

FIGURE 7.3 The protein Top 7 was the first designed protein with a fold that does not occur in nature. (A) A ribbon diagram of the protein. (B) Two-dimensional schematic of the target fold. Amino acids in strands are represented by hexagons, those in helices by squares, and those in the other local structures by circles. Purple arrows are hydrogen bonds. The amino acids shown are the designed sequence. (C) Comparisons between the predicted structure of the designed protein (blue) and the experimental structure (red). These structures are very similar, with a root-mean-square (rms) differ- ence of 1.4 Å among backbone atoms. (D) Closer look at a small region of the protein (from ASP 78 to GLY 85) where the rms difference is only 0.54 Å. Although a successful folded structure is obtained, folding of Top 7 is complex and does not show the typical cooperative, two-state mechanism observed in natural proteins of comparable size. SOURCE: B. Kuhlman, G. Dantas, G.C. Ireton, G. Varani, B.L. Stoddard, and D. Baker, 2003, Design of a novel globular protein fold with atomic-level accuracy, *Science* 302:1364, reprinted with permission from AAAS.

The basic science underlying this process is supported by proteomics, structural genomics, information technology, cloning, expression, and purification of protein targets, as well as by high-throughput crystallography, NMR, and most recently cryo electron microscopy (cryoEM). The revolution in computer technology has also been integral to all of these efforts while also ushering in the prospect of in silico identification and optimization of drug targets, as with protein design above.

The year 2020 has especially highlighted the value of SBDD. The COVID-19 pandemic put a brief pause on the activities of most labs focused on protein structure and SBDD as the pandemic shut down normal activities, but this was very rapidly followed by a massive retooling to devote the full force of these efforts toward understanding the structure and function of SARS-CoV-2. Research groups focused on determining structures using X-ray crystallography, NMR, cryoEM, or computational methods turned their full attention to the machinery of the novel coronavirus; see, for example, Figure I.7. Efforts have been focused on understanding the "trimeric spike," the invasion mechanism of the virus and a target for vaccines and antibody therapies, as well as the basic viral replication mechanism, a target for small molecule drug therapeutics. In addition, it was fundamental work to understand the structure of the earlier coronavirus spikes that allowed for the possibility of rapid design of vaccines within days of the viral sequence being published. Current efforts are contributing to the development of second-generation vaccine candidates and potentially also a universal coronavirus vaccine that will safeguard us from future pandemics.

Can one go beyond SBDD and utilize the protein's full energy landscape when designing new drugs? SBDD is driven by the idea that binders with higher binding affinity to a target should be more efficacious than those with lower binding affinity to the same target. This approach is incomplete and therefore the ability to understand the kinetics of drug binding together with protein motions is now becoming a new challenge. To elucidate these relationships, energy landscape theory has been generalized to create a computational framework that is able to construct a complete ligand-target binding free energy landscape. In another example, this strategy has been generalized to develop therapeutic drugs that regulate multiprotein complexes. The challenge here is to identify molecular compounds that can impact the protein-protein binding interface. Success requires identifying these binding interfaces and designing compounds that compete with these interface interactions. The open question is when these expanding ideas from physics will allow for the design of drugs that could not be discovered by the current strategies.

SYNTHETIC BIOLOGY

Synthetic biology emerged in the early 2000s as a number of researchers, many with physics and engineering backgrounds, took an interest in designing synthetic living systems that either mimicked real ones or behaved in completely new ways. The goals of this type of research are both practical, leading to useful technologies, and fundamental, to better understand what is life by attempting to build it. Research in synthetic biology is powered by technological advances in genetic engineering, live cell imaging, sequencing, protein engineering, and so on, and often pursued in interdisciplinary teams that bring together physicists, engineers, biochemists, and cell biologists.

As with almost every other scientific discovery, the discovery of gene regulation provided the means and impetus to assert human control over nature, in this case the genetic networks in cells. Starting with early proof-of-principle experiments that demonstrated how simple toggle switches and oscillators could be fashioned from repurposed bacterial genes and their control elements, today's synthetic genetic networks are beginning to approach the complexity of electronic ones.

These approaches could be applied to develop designer cells that seek and destroy pathogens in the human body, or to produce useful chemicals for medicinal or biofuel applications.

Uploading DNA-based circuits into cells so as to repair a diseased state, as shown in Figure 7.4, is one approach to personalized medicine, where a treatment is rationally designed based on the individual characteristics of a patient. In attempts to combat the rise of drug-resistant bacteria, viruses are being designed that can kill the bacterial pathogens. Bacteria, on the other hand, have been engineered to do our bidding and target and invade cancer cells. While these initial successes have only occurred in carefully staged laboratory conditions, the hope is that clinical applications are soon to follow.

Genetic manipulation of algae is being used to develop a new generation of biofuels. Algae are single celled organisms that live in an aquatic environment and can convert carbon from sugars or carbon dioxide into biomass. Synthetic biology approaches are being used to make algae that produce more biofuel, or that make the extraction process easier, for example by introducing mutations that decrease cell wall thickness and thereby make lipids and other biomolecules more accessible.

Antecedents to the synthetic biology revolution are biofuels produced from starch, sugar, animal fats, and vegetable oils, which are playing an increasingly large role in our energy landscape. Ethanol, a first-generation biofuel made primarily from corn, is one of the main biobased products produced worldwide and is present in more than 98 percent of the gasoline sold in the United States. Starch content,

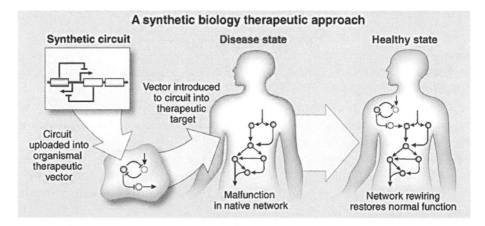

FIGURE 7.4 The uploading of synthetic DNA-based circuits into cells allows for more personalized medical treatment. Synthetic gene networks are uploaded into cells to therapeutically target the body's endogenous networks, causing a transition from disease to healthy state. Here, the uploaded network is a bistable toggle switch, which enables cellular memory with a network of two mutually repressible modules. SOURCE: W.C. Ruder, T. Lu, and J.J. Collins, 2011, Synthetic biology moving into the clinic, *Science* 333:1248, reprinted with permission from AAAS.

cellular morphology, and the particle size of ground corn kernels all affect ethanol yields. Ideas from biological physics have been instrumental in the development of today's ethanol processing capabilities, and technologies including scanning electron microscopy, laser confocal microscopy, small-angle X-ray scattering, and X-ray diffraction are revealing key morphological properties of maize starch that can be exploited for further process improvements. These technologies and genome editing have also led to advanced biofuels, which are produced from non-food biomass including agricultural waste, paving a future for more sustainable approaches to energy independence.

Genetic networks are one of many types of networks that are present inside cells. Metabolic networks connect many different enzymes and turn food molecules, which the cell takes up from its environment, into fuel needed to power all the machinery of the cell. Signaling networks relay information about the cellular environment and interface with both genetic and metabolic networks. On an even larger scale, populations of single celled organisms can be controlled by manipulating the interactions between individual cells.

Efforts to engineer new functionality in genetic networks have highlighted, in several cases, our lack of understanding of the underlying basic science. While there are qualitative pictures of the relevant regulatory interactions, and quantitative models for bits and pieces of the larger networks, there is no compelling theoretical framework within which to constrain the design problem. Although it is tempting to make analogies to electronic circuit design, which reaches extraordinary heights in modern computer chips, we still are missing important ingredients: We do not have a completely reliable quantitative description of the basic molecular interactions, which play the role of device physics in semiconductor circuits, and we just scratched the surface of the functional logic that genetic circuits can implement. To fashion synthetic biology into a true engineering discipline will require understanding the quantitative relationships between the microscopic interactions of the molecular parts and the emergent, cell-scale properties of the network. Again, basic scientific questions in biological physics are linked closely to opportunities for new technologies.

PREDICTING AND CONTROLLING EVOLUTION

The evolution of microbial and viral pathogens has played a central role in human history, and each of us has experienced the impact of microbial and viral evolution on our daily lives. Human health is increasingly threatened by the rise of antibiotic resistant bacterial pathogens, and by epidemics caused by viruses such as seasonal influenza, HIV, Ebola, Zika, and most recently SARS-CoV-2. This has led to intense efforts to infer the history of these pathogens, to predict their future, and even to manipulate the environment in ways that can control and alter these evolutionary processes. Methods originating in the biological physics community have played an important role in all of these efforts.

In recent years, widespread sequencing has improved our ability to understand and track emerging epidemics. By reconstructing phylogenies from this sequence information and overlaying them with the time and location of each sample, one can trace the transmission networks through which an epidemic spreads. One excellent example of the contribution of biological physicists to this endeavor is in the development of Nextstrain, a widely used analysis and visualization tool (see Figure 7.5). This tool has been used by researchers, journalists, policymakers, and the public to understand the spread of seasonal influenza, West Nile Virus, Zika, Ebola, and SARS-CoV-2. This has helped guide public health decisions by making it possible to evaluate the impact of policy decisions and epidemiological conditions. It also has provided important early insight into emerging threats.

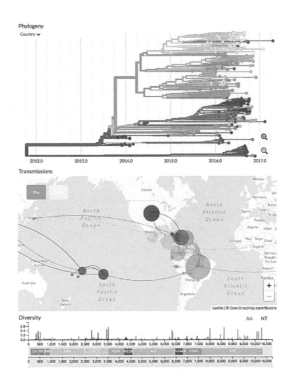

FIGURE 7.5 Widespread sequencing has improved our ability to understand and track emerging epidemics. The analysis and visualization tool Nextstrain shows us the genomic epidemiology of Zika virus as of October 2017 (live display at nextstrain.org/zika). The main interface consists of three linked panels—a phylogenetic tree, geographic transmissions, and the genetic diversity across the genome. SOURCE: From J. Hadfield, C. Megill, S.M. Bell, J. Huddleston, B. Potter, C. Callender, P. Sagulenko, T. Bedford, and R.A. Neher, 2018, Nextstrain: Real-time tracking of pathogen evolution, *Bioinformatics* 34:4124.

For example, one of the co-inventors of Nextstrain used these tools to show that community transmission of SARS-CoV-2 was occurring in the Seattle area in January 2020. Because these approaches are based on sequence information, they also can help us identify and track novel variants of concern based on their effects on phylogenetic structures.

Understanding microbial and viral evolution also is critical to efforts to predict these evolutionary processes. For example, rapid evolution in seasonal influenza allows the virus to evade the immune response to earlier strains. In order for vaccines to keep up with this, one must predict each spring how the virus will evolve over the following year, in order to have an appropriate vaccine available the following fall. Recent work from the biological physics community has introduced new approaches to use the statistical structure of influenza genealogies as the basis for inferring the relative fitness of different currently circulating strains, and hence to make predictions about their likely importance in the coming year.

The ability to predict evolution is also central to any attempt to control it. In the past few decades, there has been an increased appreciation for the fact that in many cases, disease treatments must consider not only their impact on a pathogen in its current form, but also how the treatment regime will influence the pathogen's evolution. A key example comes from the treatment of HIV, where early treatment regimens typically used antiviral therapies one at a time, which often led to the sequential evolution of resistance to each therapy, followed by treatment failure and eventually death. Analysis of kinetic experiments in the clinic established that the virus was reproducing rapidly despite the long time required for the disease to manifest. Furthermore, because viral replication occurs without proofreading (Chapter 2), mutation rates are high. Theoretical work from the then nascent biological physics community showed that, combining these observations, essentially all single- and double-point mutations were accessible to the virus on time scales that matter for treatment, but triple mutations are not. This means that there are almost no paths for the virus to evolve resistance to three different drugs simultaneously. This is the origin of the three-drug cocktails that revolutionized HIV treatment, making it possible to control the virus indefinitely and saving millions of lives.

Related considerations are at play in designing treatment regimes involving other antiviral or antimicrobial therapeutics, where analyzing the pharmacokinetics of the drug within an individual patient as well as the distribution of the drug across the larger population is essential in controlling the evolution of resistance. More recent work has also begun to consider the potential to exploit interactions between drugs to design dosing regimes that can reverse or severely limit resistance.

Inferring and controlling evolutionary dynamics also plays an important role in the study of cancer biology. Since any cancer is the result of a somatic evolutionary process, by using the tools of population genetics to analyze sequence variation within cancers we can make inferences about the effects of natural selection on putative driver and passenger mutations. This can shed important light onto the molecular mechanisms underlying cancer progression, and also has the potential to help in determining prognoses and in designing personalized treatment regimes based on the genotype of a particular cancer. Major investments in cancer sequencing have created enormous data sets for this purpose, and researchers from the biological physics community play an important role in designing methods, especially theoretical and computational methods, to exploit these new resources.

BIOMECHANICS AND ROBOTICS

Many people have envisioned a future where our lives are made easier by robots. In some ways, that future is now—many once dangerous and mind-numbing jobs in factories now are performed by special purpose robots. But what about robots with more general capabilities? Perhaps surprisingly for those outside the field, some of the biggest challenges are with everyday tasks, such as walking or running over complex, real world terrain. For the biological physics community, trying to understand why these problems are so hard is part of making precise what is meant by biological function, or specifying the physics problems that organisms must solve in order to survive (Chapter 1). In particular, the emerging focus on the physics of behavior (Chapter 1) creates the opportunity for deeper connections between robotics and the physics of living systems, moving beyond analogies and inspiration to testing common physical principles.

The mechanics of swimming, flying, and walking captures the imagination of scientists and the lay public alike, drawing appreciation for the physics of charismatic organisms such as water-walking insects, loping dogs, sidewinding rattlesnakes, flapping hawkmoths, and flocks of birds. The field makes connections across the vast expanse of field biology to theoretical physics, and provides numerous examples that are suited for teaching at all levels. Organisms' biomechanical capabilities also are inspiring the exploration and development of new materials. One example is the effort to create artificial dry adhesives mimicking the properties of gecko feet. Such adhesives would be valuable for their self-cleaning properties, ability to stick underwater, and ability to be placed and removed multiple times without losing their stickiness. Research in this area has yielded technologies that now are approaching commercialization.

At some level, all robots are inspired by the phenomena of life. To the extent that they are being built of very different materials, it is clear from the outset that one must transfer system-level principles rather than microscopic mechanisms. How far can this approach be taken? Among the possibilities are microrobots swimming through the bloodstream to fight disease; humanoid robots serving as caregivers; dog or snake-like robots searching for survivors in rubble; and self-navigating robots that drive us around or deliver our packages. There is a long and complex history, from ancient legends to successful modern factory automation. While the field is oriented toward practical applications, there is a theoretical side to robotics that is mathematically deep, drawing on differential geometry, dynamical systems, and more.

Some of the most successful recent developments are robots that incorporate mechanics into their control and morphology. These designs are based on the idea that effective locomotion in complex environments requires not only active controls but also well-chosen passive mechanical elements. This interplay emerged from studies of a number of different living systems, from cockroaches to kangaroos. Organisms tend to bounce in similar ways such that they can be described by a so-called spring loaded inverted pendulum (SLIP) template. Templates are essentially low order models that incorporate minimal dynamics displayed across groups of organisms. Dynamical systems integrating mechanics (nonlinear dynamics) and control have played an important role here in analyzing the important role of stability of such templates. An excellent example of how this interplay translates to robotics is the 6-legged robot RHex (see Figure 7.6E).

A critical idea here is that appropriate template dynamics can simplify control of movement, in essence offloading neural (and electronic) computation to the mechanics and body. This is in stark contrast to the philosophy of tightly controlled robots that have necessarily dominated robotics in factories. Indeed, the emergent aspects of living systems allow dynamics that are good in simple environments to on occasion be good in more complex terrain. Studies of cockroach response to perturbations on time scales faster than neuromuscular feedback loops coupled with SLIP directly led to RHex, which despite its simplicity displays impressive rapid running and complex environment performance. While RHex was commercialized in the 2000s, it has remained niche. However, the fruits of legged robotics work inspired by the SLIP template are now starting to appear in the real world, moving from laboratories to commercialization and deployment in real-world applications such as inspecting buildings. The doglike Spot robot (see Figure 7.6D), which takes advantage of both physical modeling and gauge theory principles, is one such example.

If walking is complicated, perhaps it would be easier to move without legs at all. Several groups in the biological physics community have focused their attention on the mechanisms of locomotion in snakes. In many cases, these organisms are moving on or through granular materials such as sand or soil, and the mechanics of such materials is itself an interesting and active physics problem. One common pattern of movement in such organisms is side-winding (see Figure 7.6C), which is quite effective over diverse terrain. Physicists study-

FIGURE 7.6 Organisms and physical robot models moving to the real world. (A) Four-legged locomotion in a red smooth saluki. (B) Six-legged locomotion in the cockroach *Periplaneta americana*. (C) Locomotion without legs, in the sidewinder rattlesnake. (D) Four-legged robot Spot. (E) Six-legged RHex. (F) Legless robot ModSnake. SOURCES: (A–C) https://en.wikipedia.org. (D) Image 1180640188, Harry Murphy/Sportsfile via Getty Images. (E) A.M. Johnson, M.T. Hale, G.C. Haynes, and D.E. Koditschek, 2012, "Autonomous Legged Hill and Stairwell Ascent," in pp. 134–142 in 2011 *IEEE International Symposium on Safety, Security, and Rescue Robotics (SSRR)*, http://ieeexplore. ieee.org/xpls/abs_all.jsp?arnumber=6106785; courtesy of Daniel Koditschek. (F) Courtesy of Howie Choset, Carnegie Mellon University.

ing actual sidewinders revealed a new "two wave" template that captures the sidewinding dynamics and reveals how phasing and modulation of the waves can lead to high-performance propulsion, maneuverability and stability. This description lends itself naturally to gauge theory formulation, importing ideas from the analysis of life at low Reynolds number (Chapter 1). Implementation of these principles in limbless robots has led to significantly improved mobility, illustrating the power of interaction among traditional biomechanics, robotics, and biological physics (see Box 7.2).

Efforts in the exploding field of soft robotics will certainly aid in robophysical modeling of living systems as well as lead to advances in robots that perform in real-world environments. In fact, concrete evidence for the promise of such systems is accumulating, as illustrated by the example of a soft, fish-like robot that recently explored the ocean at a depth of 3,000 m and will soon swim in the Mariana trench at depths of 10,000 m. This device takes advantage of biological-like propulsion (flapping), and is constructed from a soft elastomeric body and fins. This example demonstrates that advances in materials, sensors, actuators, and control principles can enhance the ability to instantiate principles discovered in

---

BOX 7.2
From Gauge Theories to Robophysics

Exploration of life at low Reynolds number (Chapter 1) led to many beautiful ideas, including the surprising appearance of gauge theories. While initially considered in relation to bacteria, many other organisms move in this regime where inertia is negligible, including sand-swimmers and many-legged macroscopic organisms.

At the same time that hidden gauge structures were being revealed, control theorists were developing geometric approaches to motion control. The control theorists' work is based on the idea that an enormous class of systems translate via oscillatory inputs—self-propulsion via self-deformation. Geometric phases arise naturally as accumulated net changes in positions or other global quantities despite local parameters returning to their initial values, and play a central role in understanding such movement.

Physicists and control theorists have now worked together to understand that their different formulations are equivalent, leading to the prospect of a gauge theory for robotics. This is leading to the articulation of principles by which diverse robots (swimming, stepping, crawling) can be controlled to gain life-like mobility in complex environments. Based on these studies, a few physicists have embraced the idea of using robots as physical models of living systems, essentially providing a "third way" (relative to analytic theory and digital computation) to recapitulate behaviors and predict new ones. This approach, dubbed "robophysics," forces one to focus on the mechanics, and thus the dynamical systems, that play a critical role in living systems.

The advent of low-cost additive manufacturing technologies, powerful actuators, sensors, and microcontrollers enables laboratory-scale synthesis of such models. Importantly, robophysical models can interact with real environments that are often impossible to model. Exploiting this, researchers have made significant progress toward robots capable of moving through challenging environments like sand. Reciprocally, the study of robot dynamics has provided insight into how living systems handle such challenges.

---

living systems, enabling engineered devices to move in hydro-, aero-, and terra-dynamic environments that are presently challenging for human-made systems.

NEURAL NETWORKS AND ARTIFICIAL INTELLIGENCE

One of the most consistent themes in the history of technology is the idea of building machines that emulate the extraordinary functionality of living systems. Perhaps flying like birds is the most ancient example of this. Building a machine that processes information in ways that mimic the function of the brain—artificial intelligence (AI)—is a more recent, but arguably even more grand idea. There have been important successes with pieces of this problem, such as acoustic signal processing that is modeled on the mechanics of the inner ear, or decomposing complex signals into positive parts, emulating the fact that neurons cannot generate negative action potentials. But since the turn of the 21st century, there have been startling developments which bring the grand dreams into sharper focus.

Emulating the brain depends on having computing hardware, but also on having a theoretical framework that defines more precisely what the brain does and how it does it. Systematic efforts in this direction began in the 1800s, with the formalization of "laws of thought" as mathematical logic. Twentieth-century discoveries about the dynamics of individual neurons, and the nature of their connections, seeded the idea that models of neural networks could embody computational functions. As described in Chapter 3, even simple models of individual neurons, if connected in arbitrary ways, can generate complex dynamics and have considerable computational power. Progress would require simplification.

The first great simplification in neural network architecture was to focus on layered structures, so that information flows from one layer to the next without feedback. This is loosely inspired by layers of processing in the brain itself. The "perceptron" architecture appears around 1960 (see Figure 7.7), and enthusiasm would wax and wane over decades. Today, perceptrons with many layers—deep

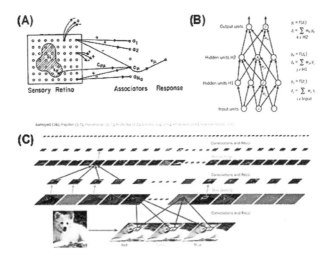

FIGURE 7.7 The "perceptron" model for neural network architecture, focused on layered structures inspired by the brain itself, first appeared around 1960 and now is the foundation for the first machines to achieve human level performance. Feed-forward networks, from the early perceptron to deep networks. (A) Inputs from the many pixels of a retina are summed with positive and negative weights by "associators," and responses are nonlinear functions of these summed inputs. (B) Generalization of (A) to multiple layers, making explicit the weights w_{ij} and nonlinearity $f(z)$. (C) A modern deep network classifying an image. Weights are organized in a convolutional structure, so that the same operation is applied to each small patch of the image. Nonlinear functions include both rectified linear (ReLU) and maximum pooling. Output units correspond to image classes; in this case, the unit with maximum output corresponds (correctly) to Samoyed dogs. SOURCES: (A) H.D. Block, 1962, The perceptron: A model for brain functioning, I. *Reviews of Modern Physics* 34:123. (B–C) Reprinted by permission from Springer: Y. LeCun, Y. Bengio, and G. Hinton, 2016, Deep learning, *Nature* 521:436, copyright 2015.

networks—are the foundation for the first machines that achieve human-level performance at tasks such as image recognition and game playing. The biological physics community and the physics community more generally have made important contributions to these remarkable developments.

Neural networks work because they are trained to perform particular tasks. The actual computations done by a particular network depend on the strengths of the connections between neurons, and a crucial part of the neural network idea is that these connections are plastic, as with real synapses in the brain. Learning rules for how synaptic strength changes can be modeled on what is known about synapses, but the engineering environment also allows going beyond the constraints of known mechanisms. This has been a productive dialogue, with more subtle forms of learning and adaption in the brain being applied to artificial systems and conversely the success of different learning rules driving re-examination of what is known about brains. More abstractly, the perceptron provided an important testing ground for theories of learning, especially for approaches grounded in statistical mechanics.

The push to develop truly deep networks was driven by classical theoretical results on the limitations of single layers, by experimental discoveries about the hierarchy of processing in the visual system, and by intuition. For image processing, there is an additional physics-based intuition that computations should be translation invariant, so that a small patch of the image will undergo the same transformations no matter where it appears. Taken together these ideas lead to convolutional networks, as in Figure 7.7C, which provided some of the first examples of human-level performance at classification and recognition of complex, real world images.

The influx of ideas from statistical physics to neural networks emphasizes that signal processing often involves an implicit model for the probability distribution of the incoming signals. This idea has been central to thinking about coding in real brains (Chapter 2) and the adaptation strategies that maintain the efficiency of these codes in varying environments (Chapter 4). In deep networks, it can be made concrete, (roughly) running the network backward to generate rather than process images. This has led to strikingly beautiful images, sparking the imagination. But it also has led to "deep fakes," which can be difficult to distinguish from real images, even of current events.

More recently, attention has turned to networks that have feedback and hence nontrivial dynamics. These recurrent networks are especially interesting for the processing of speech and language. In just a few short years, these networks have driven a revolution in our ability to interact with machines in natural language, and to use machines as language oracles, as in machine translation. A central problem is that real language has correlations on all scales. The relation between network architecture, correlation structure, and effectiveness in language understanding is a very active area of research. This work has strong overlap with ideas from the biological physics community about the physics of behavior and the emergence of long time scales in neural dynamics (Chapters 1 and 3).

As often is the case, the neural network revolution has drawn from basic science and is feeding back into basic science. At a practical level, deep networks are being deployed for the first steps of data reduction in many biological physics experiments. Deep networks also provide engineering solutions, of startling effectiveness, to problems ranging from protein structure prediction (Chapter 6) to the identification of genes relevant to disease. There are serious discussions about whether deep networks trained on image classification tasks, for example, provide models for the detailed responses on neurons in the visual system. More deeply, there is the problem of why these networks work: As models for the mapping of input to output, or for the dynamics of a process such language, they are enormously complex, sometimes with billions of parameters. This total number of parameters vastly exceeds the number of examples used in training the network, strongly violating our intuitions about "good" models. How is success possible in this under-determined regime? Can we make progress using the sorts of statistical physics methods that gave the first insights into neural network learning a generation ago (Chapter 4)? Does this influence thinking about the brain itself, which after all has roughly 100 trillion synapses? In specifying the problems that the brain is capable of solving, is it essential to include constraints on how much data is available from which to learn the correct solutions?

Neural networks today generally are implemented as simulations on digital computers. An alternative, which also goes back to the origins of these models in the biological physics community, is to build special purpose hardware, taking seriously the analog nature of computation in the nervous system. This is a rich field, which requires thinking about how basic mathematical operations carried out by neurons could be realized by semiconductor device physics. Among many other issues, such analog computations can be done at much lower power. This observation leads back to thinking about physical principles for brain organization, asking, for example, when it is more efficient to generate action potentials and when it is possible to compute with sub-threshold voltages. As a practical matter, special purpose "neuromorphic" chips have found application in niches where low power dissipation is especially important, starting with early touchpads and fingerprint sensors.

An important lesson from the long and complex history of neural networks and AI is that revolutionary technology can be based on ideas and principles drawn from an understanding of life, rather than on direct harnessing of life's mechanisms or hardware. Although it may take decades, it thus is reasonable to expect that principles being discovered today will inform the technologies of tomorrow.

PART III

REALIZING THE PROMISE

8

Education

Building a new scientific field is a multigenerational project. It is clear that realizing the promise of biological physics depends on what students are taught, and on how this material is presented. Developing effective educational strategies is vital for communicating the enticing intellectual opportunities of the field and for attracting talented aspiring scientists from the broadest possible cross-section of our society. The importance of this challenge is reflected in the fact that the majority of input the committee received from the community—voiced at the two town halls and in writing through the online platform—was about education. This input came from colleagues at all career stages—from senior faculty to beginning students—and from a wide range of institutions, including community colleges, primarily undergraduate institutions, and major research universities.

Science education is about much more than educating scientists. On the largest scale, a crucial responsibility of the scientific community is to contribute to the scientific literacy of the citizenry at large. A successful education transmits not just the importance of science for human health and the economy, but also a sense of wonder at the beauty and intricacy of our world. Science helps us to understand the world, but also holds that understanding to exacting standards, reminding us how difficult it can be to find convincing answers to important questions. Science is not just a foundation for technology and medicine, but part of human culture, and biological physics has a unique role to play in this larger cultural enterprise. The field combines the grandeur of the physicists' search for unifying principles with our human interest in ourselves; it brings extraordinary instruments that allow us literally to see what has never been seen, while engaging with the remarkable diversity of life on Earth; and it provides foundations for developments in technology and medicine that have revolutionized our lives. The unique appeal of combining the physicist's style of inquiry with the striking qualitative phenomena of life confers a special opportunity to attract a broader and more diverse community of students. This chapter explores how the emergence of biological physics fits into the culture of physics education, how biological physics can be integrated into the physics curriculum, and how this field can be leveraged to enhance the education of scientists more generally.

As we explore the educational challenges and opportunities created by the emergence of biological physics, it will be clear that some of these are internal to physics departments, while others involve collaboration between physics faculty and colleagues in other departments. Some are grounded in the frontier questions in our field, while others leverage the lessons of our field to explore foundational topics across disciplines. Some opportunities are in traditional classroom teaching, and others involve integrating teaching with research.

While the opportunities are inspiring, there are also barriers, both structural and perceived. Physics students who become interested in biological physics often wonder to what extent they need to "learn biology," something viewed as being outside of physics, taught in the equivalent of a foreign language. Conversely, biology students at most colleges and universities are required to have only minimal preparation in mathematics, so that physics is taught in a foreign language for them as well. Importantly, the scale of this cultural divide does not reflect the current state of the scientific enterprise: Biology today is a vastly more quantitative enterprise, more integrated with the mathematical and physical sciences, than it was a generation ago, and the curriculum for biology students has not kept pace. These problems will not be solved by making longer lists of courses from multiple departments and congratulating ourselves for our multi-disciplinarity. They demand a thoughtful approach to integrating biological physics into the fabric of physics education, and science education more generally, in ways that truly add value for all students.

Given the enormous variety of institutions and environments in which education takes place, there is no one-size-fits-all model for addressing the challenges and opportunities identified here, but there are some general principles on which to build these efforts. In defining biological physics, this report has emphasized that physics more generally is distinguished not by the objects or systems that are being studied, but rather by the kinds of questions that are being asked and by the kinds of answers that the physics community finds satisfying. Similarly, the teaching of physics is distinguished by a certain style and ambition: the focus on general principles, and the demonstration of how these principles are used to predict the behavior of particular systems; the sense that numbers matter, and that numerical facts about the world make sense in relation to one another and to the general principles; that one can construct instruments to measure these numbers, reliably; that much is understood by simplifying, and that sometimes even over-simplification is productive. Time and again, seemingly distant subfields of physics have been found to be connected deeply to one another, emphasizing that well-educated physicists need to develop the intellectual breadth that will prepare them to make and appreciate the next new connections that are discovered. These grand goals are not always achieved, but creating a new subfield of physics does not exempt the community from these aspirations. On the contrary, as physics expands it becomes even more important to transmit this core, unifying culture.

It is essential to acknowledge that any proposal to add something to the curriculum requires that something else be taken out, or at least compressed. This is painful, but crucial, and perhaps quite urgent. Typical core physics curricula today hardly require undergraduates to learn anything that happened after 1950, while modern biology and computer science focus on ideas and results from after 1950. Should we be surprised, then, to hear people speak of physics as the science of the past, while biology and computing are the sciences of the future? The findings, conclusions, and recommendations that emerge from this report address only part of this larger issue.

CURRENT STATE OF EDUCATION IN BIOLOGICAL PHYSICS

The current state of education in biological physics is largely a state of untapped opportunity. While a healthy community of biological physics researchers can be found in graduate schools and at the postdoctoral level, it is quite possible for today's undergraduate student to earn a degree in physics without ever encountering the physics of living systems. Exposure to the field, if it occurs at all, typically happens in an undergraduate's junior or senior year, making it difficult for students to engage deeply with the field before they graduate. At the same time, students in other fields, who sample physics only through a single introductory course, may get no hint about the relevance of physics to the phenomena of life.

These missed opportunities are more an artifact of history than a thoughtful analysis of the best path of study for today's students. Since the 1960s, many college physics programs have taken a narrower and more focused view of the subject, even as physics itself has become a much broader enterprise. A good illustration of this is provided by the table of contents of *Fundamentals of Physics*,[1] a textbook widely used for introductory physics courses. The book starts with Newtonian mechanics and builds the subject through electricity and magnetism toward the concluding chapters on nuclear and particle physics and the big bang. There is a short excursion into thermal physics, but the modern view of statistical physics does not make an appearance, nor does Brownian motion, despite its central historical role

---

[1] D. Halliday, R. Resnick, and J. Walker, 2013, *Fundamentals of Physics, Extended, 10th Edition*, John Wiley and Sons, New York.

as the proof that matter is composed of discrete atoms and molecules. There is no ef-fort to situate physics in relation to intellectual challenges outside this limited canon.[2] Such a structure reinforces the idea that there are two paths in the development of physics as a view of the natural world, one toward the very small (particles) and the other to the very large (the universe).

This tight focus on physics as an exploration of the very small and the very large leaves little room for students to experience the power of physics principles to illumi-nate the often-dramatic behaviors of systems at each of the many intermediate scales. It completely misses the idea that phenomena at intermediate scales are not merely a consequence of principles from the scale below, but may be a source of fundamental concepts and challenges in and of themselves. Indeed, the largest subfield of physics today, condensed matter physics, is defined by its focus on these emergent phenom-ena. While one might once have been able to think of this as quantum mechanics "applied" to macroscopic materials, it is now understood that condensed matter gives us concepts of great depth and broad applicability—order parameters, spontaneous symmetry breaking, scaling, the role of topology, and more. While many institutions do better, the typical introductory physics course gives little hint that one could learn anything fundamental by studying a block of metal, let alone a living cell.

The neglect of the living world, and its exploration by the physics community, continues into more advanced physics courses. Discussions of electric circuits and current flow seldom touch on the electrical dynamics of neurons; advanced mechanics courses seldom hint at the challenges of walking; optics courses rarely explain the principles of optical trapping or super-resolution microscopy; and quantum mechanics courses leave as mysteries the broad optical absorption bands of biological molecules, so different from atoms in gas phase but so central to the ability of life on Earth to capture the energy of the sun and to the rich colors that we experience every day. Statistical physics courses miss numerous opportunities to use the phenomena of life as illustrations of basic principles, while thermodynam-ics typically is presented without mentioning that experiments on animals played a key role in establishing the principle of conservation of energy. Fluid mechanics has slipped away from most core physics curricula, missing the opportunity to explore the surprising restoration of time-reversal invariance in the limit of large viscosity, and the profound implications of this for the movement of single cells. The end result is that physics students can easily get their undergraduate degrees without knowing that biological physics exists.

Of course, some physics programs offer undergraduates the opportunity to take special courses in biological physics or other subfields. However, this opportunity typi-

---

[2] This approach contrasts strongly with that taken in *The Feynman Lectures on Physics,* based on a course taught precisely 60 years ago (R.P. Feynman, R.B. Leighton, and M. Sands, 1963, *The Feynman Lectures on Physics,* Addison-Wesley, Reading, MA).

cally arises only after several years of studying Newtonian mechanics, electricity and magnetism, and quantum mechanics. One major obstacle in teaching biological physics earlier is that statistical physics typically is not offered before the junior year. Since the principles of statistical physics are so central to how physicists think about living systems, encountering these principles late in the physics curriculum means most students will have only a limited time window in which to engage with biological physics.

More subtly, conventional undergraduate courses on statistical and thermal physics emphasize non-interacting systems, such as the ideal quantum gases, which can be given an exact microscopic description. In contrast, much of modern statistical physics is concerned with how interactions among many degrees of freedom drive the emergence of qualitatively new macroscopic phenomena, and how these emergent phenomena can be described using models that ignore many microscopic details. The renormalization group explains how this simplification happens, and connects very concrete behaviors of real materials to more general and abstract theoretical principles. Monte Carlo methods make it possible to explore more complex, interacting systems, far beyond the ideal gases that are the focus conventional courses. These ideas are central to the physicist's exploration of life, both as theoretical methods and conceptual background.

In the same way that statistical physics provides much of the theoretical foundation for biological physics, modern optics is central to experimental biological physics. Optics itself has undergone revolutionary developments—from understanding the forces applied by light and the resulting invention of optical tweezers to the breaking of the diffraction limit to imaging, both recognized by Nobel Prizes—and this has been intertwined, beautifully, with developments in biological physics, as described in detail in Part I of this report. These connections, and even core ideas such laser-scanning imaging and super-resolution microscopy, are absent from the laboratory experience of most undergraduate physics students. There are missed opportunities both for the integration of biological physics into the curriculum and for the presentation of deep and fundamental physics.

In many cases ideas of great relevance for biological physics lie just beyond the bounds of traditional physics courses. Advanced mechanics courses typically do not point to the broader mathematical analysis of dynamical systems. Statistical physics courses, certainly at the undergraduate level, end before students can see Brownian motion as the primordial example of a stochastic process or realize that Monte Carlo simulation provides a path for exploring probabilistic models of systems well beyond thermal equilibrium. While physics teaching is properly focused on core subjects, all students would be well served by seeing that these subjects touch a wider variety of problems.

The problems identified in the undergraduate physics curriculum have a profound impact on efforts to grow the biological physics community. But these problems are more general, and have a much broader impact. The physics community today works on a range of problems that is much broader than what could be imagined when most of the current curriculum was solidified. The community has allowed a great chasm to develop between how active researchers think about physics and what is conveyed to the typical undergraduate. The emergence of biological physics is just one reason to think more deeply about the core physics curriculum.

Beyond the core curriculum, physics departments typically offer courses in the subfields of physics, both at the undergraduate and the graduate level. These courses play an important role in educating the students who will do research in these fields. Some departments also insist that graduate students take a number of these courses outside their research field in order to broaden their physics culture. It is important that while many areas of physics have strong connections to other disciplines, these courses are taught at a level suitable for advanced physics students. As an example, it is helpful for students interested in condensed matter physics to understand how their field is connected to areas of materials science, electrical engineering, and chemistry, but a multidisciplinary course that is built around these connections would not be an effective substitute for a course on condensed matter physics itself. It is even more important for students interested in biological physics to understand how their field is connected to many different parts of biology, but again exploring these connections cannot substitute for an advanced course on the physics of living systems itself. Such courses still are quite rare.

The structure of the undergraduate physics curriculum also influences how physics is viewed and understood by students and scientists in other disciplines. For the biological physics community the most important of these connections is with students and colleagues in the life sciences. For the vast majority of these students, their only interaction with physics is through an introductory "physics for life scientists" or "physics for premedical students" course. A traditional distinction between these courses and the introductory courses aimed at physicists and engineers is that the course for biologists does not make use of calculus, even when teaching mechanics (for which calculus was developed). In many cases, this course is required but not functionally prerequisite to other courses, and students therefore wait to take the course in their final year of undergraduate study. This structure completely misses the opportunity to convey how physics principles bear on the variety of biological problems that life science students confront, what the methods and concepts of physics have taught us about life's mechanisms, or more generally, how the physicist's perspective on the mathematical description of nature could guide further explorations.

In the same way that the phenomena of life are absent from the physics curriculum, the concepts and methods of physics are absent from the biology curriculum. In many ways, this gap is more surprising, since these concepts and methods have played such a crucial role in so many parts of biology. Addressing these issues requires appreciating the great breadth of the biological sciences, and situating the role of physics in the larger project of building a more quantitative biology. These topics are addressed in the next section.

In contrast to the situation for undergraduates, engagement with biological physics in graduate-level physics education is relatively strong. The National Center for Science and Engineering Statistics (NCSES) tracks the awarding of PhDs in the United States by field and subfield; since 2004 NCSES has tracked biological physics as a subfield of physics, with the results shown in Figure 8.1. Although many people have the sense that biological physics is a nascent or minor activity in the physics community, in fact, the number of students receiving PhDs and doing their thesis research in biological physics now is comparable to the numbers in well-established subfields, and this has happened in just 15 years. Biological physics today is producing the same number of new PhDs as did elementary particle physics in the years 2000–2005, and is growing.

In many physics departments, applicants to the PhD program are expected to articulate an area of interest. While this declaration is not binding, it does influence the admission process. Since many physics students receive their undergraduate degrees without learning that biological physics is a branch of physics, there is an obvious problem.

At the graduate level, biological physics education in physics programs—what is counted in Figure 8.1—coexists with a wide range of programs in the biological

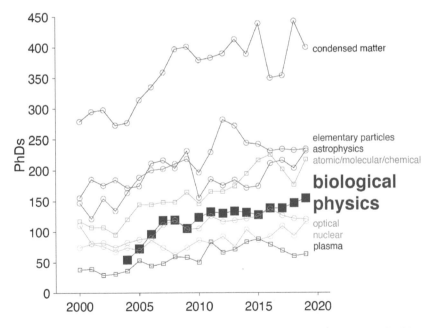

FIGURE 8.1 Monitoring the growth of biological physics as a subfield of physics in the 21st century. Doctoral degrees awarded in biological physics, compared with other subfields of physics. Data for 2010–2019. SOURCE: National Science Foundation, 2020, *Doctorate Recipients from US Universities: 2019*, NSF 21-308, Alexandria, VA, https://www.nsf.gov/statistics/doctorates/#tabs-1.

sciences that have some overlap with the field. It is important to keep in mind that biology is a much larger enterprise than physics, producing, for example, nearly five times as many PhDs per year, spread across many more distinct subfields. As emphasized throughout this report, many of these subfields—molecular biology, structural biology, cell biology, systems biology, neurobiology, and more—have had, and continue to have, important input from the ideas and methods of physics; many of these activities are identified as biophysics. Although it might be more accurate to view all of this activity as a continuum, the NCSES tracks the number of PhDs given in "biophysics (biological sciences)" as well as in "biophysics (physics)," as shown in Figure 8.2. Over the past decade, the number of PhDs granted in biophysics (biological sciences) has declined slowly, while the number in biophysics (physics) has increased, with the total remaining relatively constant. Thus, where physics students who became fascinated by the phenomena of life once saw themselves as becoming biologists, today they can retain their identity as physicists.

Finding: There has been considerable growth in the number of PhD students working in biological physics, so that the field now is comparable in size to well-established subfields of physics. This growth has occurred in less than a generation, and is continuing.

Finding: Biological physics remains poorly represented in the core undergraduate physics curriculum, and few students have opportunities for specialized courses that convey the full breadth and depth of the field.

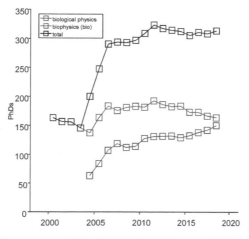

FIGURE 8.2 Doctoral degrees awarded in biological physics as subfield of physics, compared with biophysics as a subfield of the biological sciences. For clarity, data are shown as 2-year running averages. SOURCE: Data from National Science Foundation, 2020, *Doctorate Recipients from US Universities: 2019*, NSF 21-308, Alexandria, VA, https://www.nsf.gov/statistics/doctorates/#tabs-1.

Conclusion: The current physics curriculum misses opportunities to convey both the coherence of biological physics as a part of physics and its impact on biology.

STRENGTHENING BIOLOGICAL PHYSICS EDUCATION

The missed opportunities in biological physics education occur at every stage of the educational pipeline, from K–12 education to the launch of a scientific career. There are numerous opportunities to strengthen the effort, both within physics departments and at areas of intersection with other fields. This discussion of education is in the context of the first conclusion, from Part I of this report:

Conclusion: Biological physics, or the physics of living systems, now has emerged fully as a field of physics, alongside more traditional fields of astrophysics and cosmology; atomic, molecular, and optical physics; condensed matter physics; nuclear physics; particle physics; and plasma physics.

At research universities, the ideal is that teaching and research missions are aligned and synergistic. In that context, it is useful to recall our first recommendation, again from Part I:

General Recommendation: Physics departments at research universities should have identifiable efforts in the physics of living systems, alongside groups in more traditional subfields of physics.

Certainly a group of faculty who are active members of the biological physics research community will play a critical role in responding to the educational challenges identified here. On the other hand, these challenges arise in many different parts of the physics and biology curricula, and responses cannot be segregated. Universities, and their physics departments, cannot assume that a small group of biological physics faculty will solve these problems on their own, without engaging a broader range of colleagues and without perturbing physics and biology education more generally.

Nearly half of physics education in the United States happens outside the research intensive, PhD granting institutions. Quantitatively,[3] of the nearly 9,000 physics bachelor's degrees conferred in the United States in 2018, 45 percent were from institutions that are focused on undergraduate education and do not grant doctorates, and 15 percent of recipients started their educations at community col-

---

[3] P.J. Mulvey and S. Nicholson, 2020, "Physics Bachelor's Degrees: 2018," American Institute of Physics, https://www.aip.org/statistics/reports/physics-bachelors-degrees-2018.

leges, presumably taking their introductory physics courses at those institutions. Far beyond the number of students who receive physics degrees, nearly 250,000 students are enrolled in calculus-based introductory physics courses each year. These observations on the scale and breadth of physics education emphasize that integration of the physics of living systems into undergraduate physics education cannot be done solely by the relatively small number of faculty who identify as part of the biological physics research community.

Educational challenges do not have one-size-fits-all solutions, not least because the environment for teaching varies enormously across institutions. Faculty need to be empowered—and given the necessary resources—to develop curricula that are most appropriate for their institutions, and the recommendations that follow are meant to provide guidance and support rather than prescriptions. As a practical matter, the discussion begins with issues that can be addressed within physics departments, then moves outward to issues that engage educational institutions and their supporting agencies more broadly, and then to considering the integration of education with research. Throughout these efforts, it is crucial to ground education in biological physics firmly in the intellectual framework and principles of physics, even as we draw examples and inspiration from the fields with which it intersects.

Biological Physics in the Physics Curriculum

The different fields of physics often are represented by specialized courses aimed at advanced undergraduates or graduate students. These courses extend and reinforce the core physics curriculum and give students a view of where the subject is today. But it is difficult to imagine a student receiving a physics degree and not realizing that there is something called elementary particle physics, even if they never take a specialized course in the subject, and similarly for other well-established fields. In different ways, the results and goals of these different parts of physics are not just appended to the curriculum, but integrated into the core.

General Recommendation: All universities and colleges should integrate biological physics into the mainstream physics curriculum, at all levels.

This integration will have different implications for different groups of students. Some students will find themselves electrified by the field, sparking an interest in pursuing biological physics as part of their undergraduate degree (if the option is available) or in graduate school. For others, the field could act as a gateway, drawing their interest into physics from other disciplines. For still others, the field will form one part of their general physics education and help to deepen and inform their understanding of physics principles and other physics subfields. Serving the needs of all of these groups of students requires a multifaceted approach. Timing matters. Exposing students to biological physics too late in the undergraduate course of study can close off their options for specialization. Conversely, an early but narrow focus on biological physics risks compromising students' foundational knowledge of physics, and they will end up less well prepared to embrace and address the complexity of the living world. The goal is to give students paths for exploring the field in ways that reinforce, rather than sacrifice, the depth and breadth of a general physics education. We emphasize once again that there is no unique solution to these problems, and that workable solutions must be tuned to the context at each individual institution. It also is crucial that curricular innovation receive institutional support.

Students will not become better prepared to "do physics" in the more complex context of living systems by learning less physics. Indeed, biological physics is not the only part of physics now addressing phenomena of greater complexity. While physics continues to be characterized by the search for simplicity, the community now searches in more complex contexts, whether using machine learning in the search for elusive particles or in the analysis of images of living cells, perhaps using similar analysis methods; this trend toward the exploration of more complex phenomena is accelerating. Ideally, a physics degree will prepare our students for physics as it is practiced today, and as it is likely to be practiced in the next generation, starting from the beginning and continuing through the entire curriculum.

The Core Curriculum

The most straightforward way to expose physics students to living systems early in their education is to weave topics from biological physics into introductory courses. Standing waves literally come to life when explaining the physics of how *Escherichia coli* finds its middle via Min protein oscillations (Chapter 1); the physics of diffusion leads to fundamental limits on how well bacteria can sense their environment (Chapter 1); the statistical mechanics of two-level systems can be used to address single molecule experiments on ion channels (Chapter 2); and the mechanics of a bouncing pogo stick can rationalize aspects of animal movement, from cockroaches to kangaroos (Chapter 7). Using examples from living systems as a teaching tool in physics can provide early exposure to the field while simultaneously introducing key principles that form the foundation of a physics education.

Specific Recommendation: Physics courses and textbooks should illustrate major principles with examples from biological physics, in all courses from introductory to advanced levels.

Examples from biological physics can be used to teach topics in the core physics curriculum well beyond the introductory level. To revisit the list of missed opportunities from above, but now in positive terms: flying, swimming, and walking provide an engaging universe of examples of classical mechanics; the dynamics of neurons provide examples of electric circuits and current flow; optical trapping and super-resolution microscopy illustrate deep principles of electromagnetism and optics; and the broad optical absorption bands of biological molecules, which literally give color to much of our world, provide opportunities to build quantum mechanical intuition beyond the energy levels of isolated atoms. The concepts and methods of statistical physics, in particular, are illustrated by numerous phenomena from the living world, on all scales from protein folding to flocking and swarming, as seen in detail in Chapter 3. In an important counter to the impression that experiments on biological systems are messy, some of the most quantitative tests of simple polymer physics models, and the notion of entropic elasticity, have been done with DNA (see Figure 5.1).

For both the introductory and more advanced courses, the community of biological physicists has a special role to play in identifying good examples. But integrating these examples into the canon of physics teaching is a project that needs to be adopted by the broader community of physics faculty.

> **Conclusion:** There is a need to develop, collect, and disseminate resources showing how examples from biological physics can be used to teach core physics principles.

There have been several good starts in this direction, but the committee concludes that much more is needed.

As noted above, two topics in the core physics curriculum stand out for their great relevance to biological physics—statistical physics and optics. Unfortunately, these topics also stand out for the size of the gap between the typical presentation to undergraduates and our modern understanding of the subjects. The committee believes that these problems are important not just for the progress of biological physics, but for the progress of physics more generally.

> **Finding:** Current undergraduate courses in statistical mechanics often do not reflect our modern understanding of the subject, or even its full historical role in the development of physics. Among other neglected topics, Brownian motion, Monte Carlo simulation, and the renormalization group all belong in the undergraduate curriculum.

> **Finding:** Statistical mechanics courses typically come late in the undergraduate curriculum, limiting the window in which students can explore biological physics with an adequate foundation.

Specific Recommendation: Physics faculty should modernize the presentation of statistical physics to undergraduates, find ways of moving at least parts of the subject earlier in the curriculum, and highlight connections to biological physics.

Finding: Current treatment of optics in the undergraduate physics curriculum does not reflect modern developments, many of which have strong connections to biological physics. Among other neglected topics, optical traps and tweezers, laser scanning, nonlinear optical imaging modalities, and imaging beyond the diffraction limit all belong in the undergraduate curriculum.

Specific Recommendation: Physics faculty should modernize undergraduate laboratory courses to include modules on light microscopy that emphasize recent developments, and highlight connections to biological physics.

Specialized Coursework in Biological Physics

Beyond exposing students to biological physics in the core of their physics education, creating opportunities for interested students to delve more deeply raises a variety of additional considerations: What mathematics background and physics experience are needed before taking a specialized course on biological physics? What is the appropriate balance between biological physics coursework and general physics coursework for students who choose to specialize in this field? What are the appropriate roles for laboratory research experience, inquiry-based learning, and computational approaches? Should biological physics be offered as a separate major or track, or folded into the traditional physics program? Many of these questions arise for other fields of physics. While there are no universal answers to these questions, it is possible to identify a few guiding principles.

A well-educated physicist, regardless of specialization, is able to "think like a physicist" and make connections between different subfields of physics. One risk of creating an overly specialized program is that it can become obsolete, for example, as technology changes. Even when graduate-level courses in biological physics are offered, they are often rather narrow in scope. Emphasizing the approach to the living world through a physics mindset provides students with a flexible foundation that can later be adapted as fields and technologies evolve. Ensuring the breadth and depth of the general physics education is equally important for students who wish to pursue special study in biological physics as it is for those interested in any other subfield of physics. This foundation of general physics principles is crucial to a deep and productive exploration into the complexity of the living world.

Traditional biophysics or biological physics courses often are fragmented along lines defined by the subfields and history of biology. In practice, such a course might include protein structure but not the dynamics of neurons and networks, it may cover the mechanics of the cytoskeleton but not the collective behavior of flocks and swarms, and so on. To be consistent with the rest of the physics curriculum, teaching biological physics needs to be organized around conceptual questions and general principles—with diverse case studies as manifestations of those principles—rather than along a succession of disconnected biological topics. In addition to reinforcing the general physics culture, this approach will help physics students, who may not have extensive previous knowledge of biology, not to get lost in a sea of biological details.

> **Conclusion:** The great breadth of the field poses a challenge in teaching an introduction to biological physics for advanced undergraduates or beginning graduate students.

> **General Recommendation: Physics faculty should organize biological physics coursework around general principles, and ensure that students specializing in biological physics receive a broad and deep general physics education.**

An important part of the physicist's approach to nature, which also is central to the teaching of biological physics, is that our understanding is expressed in mathematical terms and tested in quantitative experiments. This interaction between theory and experiment can reach extraordinary precision, but physics crucially also is about simple approximate arguments. When something is finally understood, it is possible both to give order of magnitude estimates on the back of an envelope and to predict the results of detailed experiments, and the path to understanding often involves an interplay between these different approaches. Understanding also generates the ability to engineer new and often simplified systems, capturing the essence of what we see in nature; engineering in turn probes the limits of our understanding. In discussing the differences between physics and biology education, emphasis often is placed on the role of sophisticated mathematical analysis. This indeed is essential for physics, but focusing on this alone misses the roles of both quantitative experiment and simple models. Physics has a culture of quantitative measurement so pervasive that it can be taken for granted, and a taste for simple arguments that can feel more like art than science. But there are often explicit claims that biology is different, and that the complexity of life is both irreducible and irreducibly messy. Education needs to confront this problem, explicitly.

One of the basic conclusions from the vast array of work reviewed in Part I of this report is that the complex phenomena of life can be tamed, resulting in the sorts of reproducible, quantitative experiments that are the norm in the rest of physics. There are examples of this on all scales, from single molecules to populations of organisms, and in many cases, taming the complexity has involved building new instruments and introducing methods of data analysis that are grounded in more abstract theoretical principles. These results exemplify what is possible when the community holds to high standards for quantitative measurements and, more deeply, in the comparison between theory and experiment. Students need to be taught that the complexity of living systems is not an excuse to be satisfied with lower quality data, or with merely qualitative comparisons between theory and experiment, and that complexity itself is not an argument against the exploration of simple models.

Mathematical Methods

Physics has a special relationship to mathematics, and this is true for all parts of the discipline including biological physics. As with the core physics curriculum itself, there is a canon of mathematical methods for physics. Some of this is conveyed in a collection of courses, taught in the mathematics department and often designed for the first 2 years of undergraduate education, moving from single variable to multivariable calculus, linear algebra, and differential equations. But the more advanced parts of even the undergraduate physics curriculum draw on eigenfunction expansions, complex analysis, asymptotic approximation methods, Fourier methods, and more. Individual physicists differ in their relationship to this material, and this diversity of views is transmitted to the students. Physics departments might require their undergraduates to take particular advanced courses in applicable mathematics, they might offer their own courses on the mathematical methods of physics, or they might assume that more advanced methods are taught as part of physics courses; many institutions offer a mix of these approaches.

For students interested in deeper exploration of biological physics, what is missing from the conventional collection of mathematical methods is not so much particular topics as an understanding that these methods fit into larger and more generally applicable structures. As an example, physics students take courses on differential equations and advanced classical mechanics, but these typically stop short of introducing more general ideas about nonlinear dynamical systems. This is important because many of the dynamical systems relevant to the living world—from networks of biochemical reactions in a single cell to networks of neurons in the brain to interactions among species in an ecosystem—do not have the symmetries and conservation laws that are central to classical mechanics. This broader notion of nonlinear dynamical systems also is relevant for many other areas of physics, is connected to the foundations of statistical mechanics, and provides accessible examples of universality and renormalization. Physicists in general would benefit from knowing that these topics exist, just beyond the bounds of their traditional courses.

In a similar spirit, elementary statistical mechanics courses often do not emphasize that it is a fundamentally probabilistic description of the world. In fact, everything we observe—even the pressure that an ideal gas exerts on its container—is predicted to fluctuate, and the analysis of these fluctuations has been crucial at many stages in the development of the subject. Leaving fluctuations as an advanced topic, often beyond the core undergraduate course, also misses the opportunity to situate statistical mechanics in the larger context of probabilistic models and stochastic processes, many of which are relevant to our description of the living world. Indeed, the exploration of probabilistic models is a huge field, with applications to an ever growing array of problems, from economics to health care, machine learning, and more. Many of the concepts and methods in this field have their roots in statistical mechanics, with the Boltzmann distribution as the primordial example of a probabilistic model. All physics students would benefit from knowing that these connections exist, and the physics community as a whole would benefit from reclaiming some of this larger field, now belonging primarily to computer science and applied mathematics, where physicists have made many contributions.

Conventional boundaries for the mathematical methods of physics were established before computers became widely available. Today, the role of computation in the practice of science cannot be overstated, and biological physics is no exception. As is well known, physicists use computation in two very different ways. First is the simulation of models, using computing as an extension of theory, to explore phenomena that are not yet captured with analytic methods. Second is the analysis of data, using computing as an extension of experiment to extract meaning from ever larger data sets.

Simulation provides a path for students to explore theoretical questions even when they are not fully prepared for the analytic theory; as an example, some statistical physics courses now use Monte Carlo simulation for this purpose. Importantly, simulation can close the gap between more traditional physics problems, such as random walks, and biological physics problems, such as the "run and tumble" behavior of bacteria. The cost of making simulations more realistic can be very low, allowing students to go beyond the proverbial spherical cow while still seeing connections to simple models. Higher-level languages with relatively transparent syntax, such as MATLAB and Python, reduce the barriers to getting started on computational projects, and Jupyter notebooks provide a structure for collaboration and the sharing of resources online. While using these tools to help students explore more widely, confidently, and even playfully, it also is important to convey the lessons from generations of computational physicists: that simulations have something in common with experiments, and need to be analyzed carefully; that the ease of simulating complex models does not replace our search for principled, simplified descriptions, ultimately expressible with pen and paper; that simulation is one path to understanding rather than an end in itself.

Almost all fields of science are being revolutionized by the opportunity to gather "big data." While this often is presented as recent development, experimental high energy physics and cosmological surveys entered the big data era before it had a name. Today biological physics is following a similar path, with even modest experiments generating terabytes of data in an afternoon, and many experiments reaching the petabyte scale.

Conclusion: Biological physics, and physics more generally, face a challenge in embracing the excitement that surrounds big data, while maintaining the unique physics culture of interaction between experiment and theory.

Coda

Taken together, the recommendations above point toward a more general aspect of physics culture. Physics and physicists have played central roles in developments spanning the full range of science, technology, and policy, from plate tectonics and global climate to the semiconductor industry and the World Wide Web, from energy production to arms control. This engagement is an important part of physics culture, and is transmitted to students both informally and as side commentary in core courses and their textbooks. Despite the enormous impact that physics has had on our exploration of the living world, from basic science to the practice of medicine and care for public health, the phenomena of life remain largely absent from the broader notion of what physics is and what physicists do, that we transmit to our students. The recommendations here provide a path to closing this gap.

Biological Physics and Cross-Disciplinary Education

This report has emphasized biological physics as a branch of physics. This stands in contrast to the view of the field as the application of physics to biology, or as some interdisciplinary amalgam. The current state of the field emerged from rich interactions, over the course of a century, among distinct disciplines of physics and biology, as well as chemistry, psychology, and more. It is especially important to respect the richness of these interactions when teaching, while still conveying what is new and exciting about the physicist's perspective on the phenomena of life. A central theme in this history is that many scientists have made progress by crossing traditional boundaries between disciplines. These boundary-crossing events have many different outcomes: In some cases, an individual scientist changes fields; in other cases, attention is drawn to the opportunities for interdisciplinary collaboration; and finally, the boundaries can move, as has happened in biological physics. Students need to understand not so much the sociology of these boundary-crossing events, but rather that the ability to cross boundaries will expand their opportunities to formulate and solve problems that matter to them as individuals, to the scientific community, and to society as a whole.

One reaction to the complex history of biological physics, and to other examples of intellectual boundary crossing, is to emphasize that boundaries are artificial. Indeed, the phenomena in nature are not labeled intrinsically as being biology, chemistry, or physics. But faced with the same phenomena, biologists, chemists, and physicists will ask different questions, and expect different kinds of answers, as will applied mathematicians, computer scientists, and engineers. It is not reasonable to ask that these cultural differences be obliterated, any more than it would be reasonable to insist that all novels around the world be written in a single language. How should colleges and universities prepare students for a world in which their scientific interests will lead them to the edges of their chosen disciplines? What are the best ways to "translate" the culture of one discipline, making it accessible for others?

The analogy to language suggests that crossing boundaries between scientific disciplines is easiest if one starts early. Ideally, students would be exposed to biological physics at a pre-college level, potentially as early as middle school. In this early context, the interplay between physical laws and constraints and properties of living systems can motivate questions in both biology and physics. At the high school level, biological physics can be an integral part of both physics and biology education. This does not require a separate course; rather the relevance of physical considerations in understanding living systems can be taught as a part of high school biology, and high school physics courses can include examples drawn from biological systems, as described above for university courses. While the challenges of K–12 education are beyond the scope of the task for this report (Appendix A), the committee views them as crucial for progress in our field, and in science more broadly.

For most science students, their only contact with physics is through a single introductory course. Although sometimes referred to as "service courses," these courses offer a great opportunity. A meaningful goal is to convey to each student how the concepts and methods of physics provide productive tools for exploring the parts of the world that they find most interesting—this is the true service that can be performed in these courses. In addition, these courses can aspire to convey something of the beauty and grandeur of physics itself. The previous section emphasized the opportunities for integrating biological physics into introductory courses for physics students, and the same arguments apply even more strongly to physics courses for students in the life sciences. Examples from biological physics illustrate many core principles of physics more generally, and the notion that these principles are relevant to the phenomena of life is itself an important fact, one that can change a young student's view of the intellectual landscape. To make this work, the community needs to develop a catalogue of illustrative biological examples that serve as equivalents to the inclined plane, simple pendulum, planetary motion, and so on, emphasizing our earlier conclusion on updates to the curriculum:

> **Conclusion:** There is a need to develop, collect, and disseminate resources showing how examples from biological physics can be used to teach core physics principles.

Many widely used biology textbooks have zero equations. To the extent that mathematical analysis appears in the teaching of biology, it often focuses on the reliability of inference from limited data rather than on the phenomena themselves. This approach sends a clear message to students that numbers are irrelevant to the exploration of life, despite many counterexamples from the history of biology. By the end of the 20th century, it was clear to many people that this approach would not prepare students for the future of biology. This perspective was summarized in the *BIO 2010* report.[4]

BIO 2010 was motivated by the observation that the practice of biology was changing rapidly: Instead of studying individual genes, the field was moving to studying whole genomes; instead of probing single neurons in the brain, it was becoming possible to monitor large populations of cells; and more. At a purely practical level, the scale of biological data was reaching the point that individual scientists could no longer reason "by hand" about their results, but instead needed to formalize their data analysis in algorithmic and ultimately mathematical terms. More broadly, much of biomedical research was characterized as being at the interface between biology and the physical, mathematical, and information sciences. *BIO 2010* emphasized that revolutionary changes in the research environment should drive comparably profound changes in teaching. In particular, it was necessary to push biology students to develop more quantitative skills, and to be sure that biology courses draw on these skills throughout the curriculum. Making these changes would require new resources, both from the federal government and from individual academic institutions, and collaboration among faculty from multiple departments.

How far has the community come in responding to *BIO 2010*? Almost all research universities now have visible programs in areas that can be described as "quantitative biology," although exactly what this means is different at different institutions, and the extent to which these programs are accessible to undergraduates also varies. Interestingly, many institutions have programs in biophysics

---

[4] National Research Council, 2003, *BIO 2010: Transforming Undergraduate Education for Future Research Biologists*, The National Academies Press, Washington, DC.

that anticipated the goals of *BIO 2010* by decades. These programs have played an important role in helping students who have strong physics preparation engage with the frontier of biological research. In particular, biophysics programs have been a major source of students working on X-ray crystallography, nuclear magnetic resonance, and cryogenic electron microscopy approaches to the structure of biological molecules, well before these approaches merged into structural biology.

Even with the growth of quantitative biology programs, the basic requirements for traditional biology undergraduates remain light in mathematics and the physical sciences, and this has consequences for how more advanced biology students engage with central topics in the field. Thus, modern molecular biology and biochemistry courses are built around molecular structures, but little can be said about the experimental methods that make it possible to visualize these structures. Similarly, core neurobiology courses describe the central role of ion channels in brain function, but typically do not make reference to the equations that describe the dynamics of these molecules or to the deeply quantitative analyses that led to their discovery. Despite dramatic changes in the practice of biology as a science, this curriculum continues to send the message that mathematical approaches are optional rather than integral to our exploration of life.

> **Conclusion:** There still is room to improve the integration of quantitative methods and theoretical ideas into the core biology curriculum, continuing the spirit of *BIO 2010*. This remains crucial in preparing students for the biomedical sciences as they are practiced today, and as they are likely to evolve over the coming generation.

While the *BIO 2010* report referred broadly to the physical, mathematical, and information sciences, physicists in particular have a crucial role to play in the education of more quantitative biologists. The business of physics is the development of instruments for the quantitative observation of the natural world and the development of mathematical structures that rationalize these data, allowing understanding, prediction, and design. Physicists have been in the forefront of collecting and analyzing large data sets, and in building collaborations that create capabilities far beyond what can be accomplished by single investigators. Because of these traditions, physics departments have developed a substantial infrastructure for quantitative education—in the laboratory, at the blackboard, and at the computer. The emergence of biological physics as a branch of physics has made clear how all of this can be brought to bear on the phenomena of life.

Integrating the principles of biological physics into the education of more quantitative biologists needs to happen at many levels, and at all stages of education. As noted above, the traditional model has been that biology students see physics only in an introductory course, and attention has therefore been focused on improving these courses. But students will not develop a more quantitative approach to the life sciences if equations never appear in their subsequent biology courses. Diverse educational experiments at institutions around the country provide examples for how to proceed:

- Physics, chemistry, and biology faculty can collaborate to offer an integrated introduction to the natural sciences, providing an alternative to separate courses in the individual disciplines.
- In the tradition of courses on the mathematical methods of physics, departments can offer courses on mathematical methods in biology, or in subfields of biology.
- Laboratory courses can introduce experimental methods from physics, the building of experimental apparatus, and the physics culture of connecting theory to experiment, while remaining focused on biological systems.
- Intense summer courses can bring together students from across the divide between physics and biology. Sustained over many years, these programs can help to close the gap between the disciplines.

What all these approaches to the education of quantitative biologists have in common is that they require collaboration among faculty from multiple departments. As was noted in the *BIO 2010* report, such collaborative teaching often faces administrative obstacles.

Conclusion: The biological physics community has a central role to play in initiatives for multidisciplinary education in quantitative biology, bioengineering, and related areas.

General Recommendation: University and college administrators should allocate resources to physics departments as part of their growing educational and research initiatives in quantitative biology and biological engineering, acknowledging the central role of biological physics in these fields.

It is widely appreciated, following the arguments of *BIO 2010*, that the education of quantitative biologists must engage faculty in multiple departments. We would add that real progress requires departing from the model in which courses taught by one department are in "service" to the educational agenda of another department. University and college administrators must create funding structures that support genuine collaboration and equal partnership among all relevant faculty.

Coupling Education and Research

In the academic world, research and education are linked. Yet, only a small fraction of institutions require undergraduates to engage with research as part of their degree program, although many more offer research opportunities. Larger, research-intensive institutions, both universities and national laboratories, have summer programs to welcome visiting students who might not have comparable opportunities at their home institutions. It is generally agreed that engaging with research is enormously beneficial for undergraduates, and in some cases it is not an exaggeration to say that such experiences are life changing.

Biological physics research groups have a special role to play in the ecosystem of undergraduate research experiences. Many experimental groups in the field are small and focus on "table top" experiments, providing a more intimate community for young students. Theorists work on problems ranging from data analysis through simulation to abstract theory, providing opportunities for students to enter with varying levels of background knowledge. The same research groups can appeal to students planning a range of undergraduate majors, not just physics and biology but also chemistry and many fields of engineering. Seeing directly, in the laboratory and at the blackboard, how the physicist's approach illuminates the phenomena of life can reignite students' interests in physics itself.

Finding: Meaningful engagement with research plays a crucial role in awakening and maintaining undergraduate student interest in the sciences.

Conclusion: Biological physics presents unique opportunities for the involvement of undergraduates in research at the frontier of our understanding, offering more intimate communities through smaller research groups and providing opportunities for students to enter with varying levels of background knowledge and from a range of undergraduate majors.

The widespread enthusiasm for engaging undergraduates in research has been supported by federal funding agencies for a very long time. The National Science Foundation supported an Undergraduate Research Participation Program starting in 1958; cut from the federal budget in 1982, it was revitalized in 1987 as Research Experiences for Undergraduates (REU). The REU program has touched tens of thousands of students, and by now, almost every federal agency has mechanisms aimed specifically at supporting undergraduate involvement in research projects.

An unintended consequence of broader support for undergraduate research experience is that this experience has become a de facto requirement for admission into highly ranked doctoral programs. As a result, the community needs to attend not just to the total amount of support for such programs, but to the equality of opportunity in access to these programs.

It is important to emphasize that meaningful student engagement in research builds on the foundation provided by the core curriculum. If the goal is to have students get involved in science as soon as possible, ideally in the summer after their first year of undergraduate study, then it is necessary to ask if the introductory courses are preparing our students properly for this experience. The recommendations above reveal additional considerations for the goals of introductory courses. Do laboratory modules instill a taste for measurement that prepares students to move toward the frontier of the subject? Do homework exercises hone the theoretical and computational skills that provide a starting point for engaging with data and models beyond our current understanding? These are challenging questions, and the path to more satisfying answers will necessitate resources beyond those that are currently allocated.

Institutions vary widely in the resources that they bring in support of introductory undergraduate physics courses. It is not unusual for these courses to involve a single faculty member lecturing to hundreds of students, perhaps with problem-solving sessions led by graduate student teaching assistants; institutions without robust doctoral programs may not be able to staff these smaller group discussions. In physics, perhaps more than in biology, one course builds on another, so that inequalities of investment in introductory courses are amplified with time.

> **Conclusion:** Equality of opportunity for students to engage with physics, including biological physics, depends on high-quality introductory courses, emphasizing the interconnectedness of education and research.

> **Finding:** Current models for support of undergraduate research perpetuate a sharp distinction between the core curriculum (education) and the development of the scientific workforce (research). This extends to the fact that science and education are overseen by different standing committees in Congress.

> **Conclusion:** Support for the development of the scientific workforce will require direct federal investment in the core of undergraduate education, especially at an introductory level.

These observations lead to our recommendations about the integration of teaching and research:

> **Specific Recommendation: Universities should provide and fund opportunities for undergraduate students to engage in biological physics research, as an integral part of their education, starting as soon as their first year.**

Specific Recommendation: Funding agencies, such as the National Institutes of Health, the National Science Foundation, the Department of Energy, and the Department of Defense, as well as private foundations, should develop and expand programs to support integrated efforts in education and research at all levels, from beginning undergraduates to more senior scientists migrating across disciplinary boundaries.

As explained above, many agencies have programs that support the engagement of undergraduates in research, and so these recommendations could be read simply as a plea for expansion of these programs. While this would help, the committee feels strongly that new programs are needed for more effective *integration* of teaching and research. Students need to see the connection between the core of their curriculum and the advancing frontiers of science not just in summer laboratory sojourns but in the classroom as well. While not all science students will become scientists, all will benefit from making these connections. Such efforts will take different forms in different institutions and for different groups of students.

POSTDOCTORAL TRAJECTORIES

The education of professional scientists does not end with the award of the PhD. Postdoctoral positions, once a brief and luxurious pause between graduate school and the responsibilities of a university faculty position, have become the destination for half of all new physics PhDs[5] and a near absolute requirement for advancement to faculty and independent research positions. Postdoctoral periods also have become longer, so that what was once a transitional period is becoming a substantial phase of career and life; a corollary is that postdoctoral fellows are becoming a larger part of the scientific workforce. These trends are especially strong in the biomedical sciences, where they have been identified by prominent commentators as among the "systemic flaws" in the research enterprise.[6] The situation in the physics community is different, but might not be better. In many areas of theoretical physics, for example, postdoctoral appointments are limited to 3 years, but it is common for people to have multiple postdoctoral sojourns before arriving at a faculty position. Young biological physicists are influenced by both cultures.

Because biology is a much larger enterprise than physics, many new PhDs in biological physics will move to postdoctoral positions in biology departments or to basic science departments at medical schools. On the one hand, this is a sign

---

[5] P. Mulvey and J. Pold, 2019, *Physics Doctorates: Initial Employment*, American Institute of Physics, https://www.aip.org/statistics/reports/physics-doctorates-initial-employment-2016.

[6] B. Alberts, M.W. Kirschner, S. Tilghman, and H. Varmus, 2014, "Rescuing US Biomedical Research from Its Systemic Flaws," *Proceedings of the National Academy of Sciences U.S.A.* 111(16):5773–5777.

of success for the field, and creates opportunities for productive exchange of ideas and expertise. On the other hand, the dispersal of young biological physicists adds to the problems of maintaining coherence in the field, discussed at many points in this report. For the individuals involved, the differences in culture surrounding postdoctoral positions in physics and biology can be a source of anxiety and can also make it difficult to return to a physics environment, if desired, at the next step in their careers.

Physics has a tradition of treating postdocs as budding independent investigators rather than merely as skilled labor. To maintain this tradition, fellowships that provide competitive salaries and some degree of independence are essential. These issues bridge the challenges of education and those of supporting the field more generally, addressed in Chapter 9.

> **Conclusion:** Accelerating young researchers to independence is critical to empowering the next generation of biological physicists. As in other fields of physics, independent, individual fellowships are an effective mechanism.

To maintain coherence as postdoctoral fellows move to a wide variety of research environments requires support for their attendance at events that bring them into contact with the broad biological physics community. In this respect, an important role is played by institutions such as the Aspen Center for Physics and the Kavli Institute for Theoretical Physics, which have hosted many programs on topics in biological physics alongside those in better established subfields of physics. Such gatherings promote the exchange of ideas and formation of new independent collaborations, as well bringing promising postdoctoral fellows into contact with senior colleagues from outside their immediate circle of mentors.

The growth in biological physics also has attracted numerous scientists transitioning from other subfields of physics at the postdoctoral level. This again is a testament to the excitement and promise of the field. However, unlike students who enter biological physics as undergraduate or graduate students, postdoctoral scholars have limited time to learn, and often have more competing demands on their time. Intensive summer schools play an important role as "crash courses" that help these young scientists learn essential background quickly, and sample the frontiers of the subject. There are successful examples of this at the locations of well-known theoretical physics schools, such as École de Physique des Houches, Intitut d'Études Scientifiques de Cargèse, and the University of Colorado Boulder. There is an independent tradition of summer schools in different parts of experimental biology, and several of these have evolved in response to the influx of ideas, methods, and young people from physics; examples include the Marine Biological Laboratory, Cold Spring Harbor Laboratory, and Friday Harbor. These different courses span the range from "pure" physics courses to fully interdisciplinary courses to biology courses where physicists are welcome. In some cases, these courses have run for decades, and their alumni have grown into leaders in both physics and biology. Continued financial support and community engagement with these courses is important.

Biological physics also is remarkable for the diversity of potential career opportunities, in both academic and nonacademic settings. While postdoctoral fellows in some areas of physics largely confine their searches for academic jobs to physics departments, biological physicists often face a bewildering array of academic job options. There are opportunities in physics departments, but these are not proportional to the size of the field. Parts of what this report describes as the broad field of biological physics have their natural home in chemistry departments, and at several institutions topics outside the more traditional areas of physics are found in applied physics departments; theorists may find homes in applied mathematics departments. Many departments in engineering schools (not just bioengineering) are more and more deeply connected to problems in the life sciences, in ways that often resonate with the biological physics community. Finally, there are opportunities in the many different kinds of biology departments that one finds at universities, and in the even wider range of departments and research centers found at medical schools.

There are practical challenges for postdoctoral fellows in preparing for this broad range of opportunities. Different departments give different weight to publications in different journals, for example, and still have varying attitudes about e-print archives (although these views are converging). Problems that physicists find closely connected might be the focus of different departments or programs in the biological sciences, or different groups in the same department, and dividing research effort among these problems could weaken the case in the eyes of each specialized group. Senior scientists have a responsibility to attend to these issues, often case by case.

Career opportunities for postdoctoral scholars outside of academia are varied and broad, and include positions in biotech and pharmaceutical industries as well as quantitative analysis and data science positions in an increasingly wide range of industries. As in academia more generally, there is considerable room for improvement in how young biological physicists are introduced to these opportunities.

SUMMARY

Bringing the physicists' style of inquiry to bear on the phenomena of life has a unique appeal. Beyond the intellectual opportunities of the field itself, biological physics provides a path for communicating the excitement of physics more broadly, for attracting talented scientists from the widest cross-section of our society, and for contributing to the scientific literacy of the society at large. But realizing these opportunities requires rethinking how we teach physics, biology, and science more generally. Today, physics students interested in biological problems can retain their identity as physicists, and, in less than a generation, the number of such students has become comparable to those in well-established subfields of physics. On the other hand, even though biological physics has emerged as a coherent subfield of physics, the phenomena of life are largely absent from the core undergraduate physics curriculum. Similarly, while methods and concepts from physics have played a central role our modern view of life, the principles of biological physics are largely absent from the core biology curriculum. These are symptoms of the large gap that has developed between the practice of science and the education of undergraduates. This chapter has examined these issues, leading to a series of interlocking findings, conclusions, and recommendations about:

- the integration of biological physics into the core physics curriculum;
- the need for modernization of the physics curriculum;
- courses on biological physics for advanced physics students;
- the special role of biological physics in building a more quantitative biology; and
- integration of education and research, and support for this integration.

Uniting these issues is a concern for the communication of general physics culture—an emphasis on general principles, on the power of both mathematical sophistication and simple arguments, and on the benefits of holding high standards both for quantitative measurements and for the interaction between experiment and theory. Beyond their formal education, young biological physicists face a remarkable diversity of potential career opportunities, but also must navigate the cultural differences they will encounter along these paths. Attention of senior scientists to these differences, as well as the availability of postdoctoral opportunities that provide some degree of independence, are essential for helping young biological physicists realize the promise of the field.

9

Funding, Collaboration, and Coordination

This chapter reviews the current state of research funding, infrastructure, and coordination in biological physics and outlines opportunities to structure future investments for greatest impact. Funding touches nearly every aspect of the scientific endeavor. From a student's earliest forays into a field through their emergence as an independent investigator; from the seed of an idea through the fits and starts of theory-building and experimentation; from an area of focus through the birth of a field, funding is essential. Consistent funding makes possible the time and tools to enable discovery, supports the environments for knowledge to flourish, and maintains the continuity that allows the best ideas to grow and bear fruit. Competitively awarded scientific funding both rewards past successes and recognizes the potential within people, groups, and ideas, with an eye toward cultivating fertile intellectual grounds that will enrich science and society for generations.

Biological physics is central to the missions of an astonishing array of research funders and stakeholders. Government agencies seeking to advance fundamental understanding of our world turn to biological physics for insights into the physical underpinnings of life. Others look to biological physics to elucidate processes that could form the basis for exciting new applications in medicine, energy, engineering, and more. Governments, foundations, and the private sector all have much to gain from the basic mechanisms uncovered and the ideas sparked by advances in the field. These opportunities exist across the full range of the field, along all axes: from the dynamics of single molecules to the collective behavior in large communities of organisms, from theory and experiment, from research by individual investigators and by larger groups. Many funding agencies have successfully targeted particular parts of this multidimensional space, often through connections with topics nominally outside of biological physics.

The existence of multiple funding sources for biological physics lends a degree of robustness to the system, and gives the community an opportunity to observe the best practices of different agencies. On the other hand, the diversity of funding sources fragments the field, obscuring its coherence. Individual investigators have adapted to this funding landscape, but the analysis which follows argues that the time is right for the funding landscape to adapt to the field.

CURRENT FUNDING FOR RESEARCH AND EDUCATION

Overview and Methodology

The principal funding for research in biological physics comes from three federal agencies: the National Science Foundation (NSF), the Department of Energy (DOE), and the National Institutes of Health (NIH). Beyond this there is support from the Department of Defense (DoD) agencies through the Defense Advanced Research Projects Agency (DARPA), the Army Research Office (ARO), the Office of Naval Research (ONR), and the Air Force Office of Scientific Research (AFOSR). Furthermore, several philanthropic foundations are invested in the field, prominently the Alfred P. Sloan Foundation, the Burroughs Wellcome Fund (BWF), the Gordon and Betty Moore Foundation, the Howard Hughes Medical Institute (HHMI), the Simons Foundation, and the Kavli Foundation. The analysis here makes use both of publicly available data, and responses to queries from the committee as described in Appendix C.

The diversity of funding sources creates challenges in assembling a global view of support for the field, and it seems prudent to begin by enumerating some of the resulting caveats. First, there can be a challenge even in finding biological physics projects in an agency's categorization of its own funding programs, and the committee adopted different approaches with different agencies, as described in detail below. Second, different agencies have different approaches to multiyear grants. At NIH, for example, these are listed in publicly available data as multiple annual grants, while analogous databases at NSF list a single grant. For these reasons, broad trends over multiple years are more instructive than year-by-year comparisons.

A third distinction is that support is provided in different ways. All the agencies support individual investigators and small groups, but DOE in particular plays an enormous role in supporting large infrastructure, often in the form of user facilities that are accessible by peer-reviewed proposal at no cost to the researcher. These facilities address the needs of the scientific community as a whole. Biological physics is a fraction of this user base, sometimes highly visible but not easy to quantify. NSF, NIH, and others provide user facilities to a lesser extent, sometimes in partnership with DOE. User facilities are important enough to the community that they are discussed in a separate section below.

Figure 9.1 provides a somewhat crude summary of total spending over the past decade. As should be clear from the previous remarks, uncertainties are large, especially beyond NSF and NIH, where public databases make it possible to drill down to the level of individual grants to check on what is being included as support for biological physics. Exploring these issues involves looking more closely at what these numbers mean for each of the many agencies that support the field, and then stepping back for some perspective. This sets the stage for addressing the challenges which emerge from the analysis. Finally, the chapter concludes with a survey of the state of user facilities.

Agency by Agency Analysis

In fiscal year 2020, federal research and development funding in the United States reached an estimated \$164 billion.[1] Congress appropriates this funding through 12 annual appropriations bills, supporting science and engineering re-

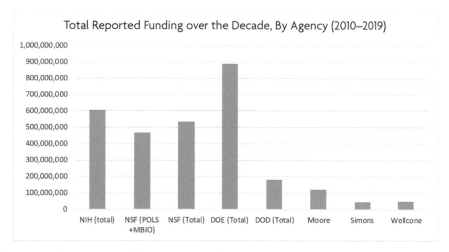

FIGURE 9.1 Aggregate spending on programs that include biological physics, across funding agencies, over the decade 2010–2019. As explained in the text, totals for NIH and NSF reflect detailed searches through full databases of grants. Total for DOE is from congressional budget documents, including several programs that overlap biological physics but are much broader; real spending on the field is much less. Totals for other agencies drawn from reports provided by agency representatives in response to committee's queries (Appendix C).

---

[1] Excluded from this total is research directed specifically at the SARS-CoV-2/COVID-19 crisis.

search that advances national objectives aligned with each agency's mission. NSF has a uniquely broad mission—"to promote the progress of science; to advance the national progress of health, prosperity, and welfare; and to secure the national defense; and for other purposes"—while other agencies can be (much) more focused (Appendix D). Further subdivided, agency directorates and programs have missions and objectives that contribute specifically to the agency's overarching mission, and funding is awarded accordingly.

National Science Foundation

NSF supports biological physics in part through the Physics of Living Systems (PoLS) program within the Physics Division (PHY) of the Mathematical and Physical Sciences (MPS) Directorate. This program has its origins in the early 2000s, and although it remains small—with a budget five times smaller than the NIH support for the field—it has played a key role in supporting the emergence of the field as a branch of physics. Biological physics also has been supported through the Physics Frontier Centers (PFC) program, again led by PHY but with significant contributions from other divisions and directorates. Uniquely among the funding programs surveyed here, NSF/PHY has supported the physics community's exploration of life across all scales, from single molecules to populations of organisms, including both theoretical and experimental work, and has embraced the field as a part of physics more broadly.

There is a much longer history of supporting biophysics through NSF's Biological Sciences Directorate (BIO); today much of this support is through the Molecular Biophysics (MBIO) cluster within the Molecular and Cellular Biosciences (MCB) division. As the name suggests, MCB is a large division that funds work across a broad range of molecular and cellular biology, with clusters focused on Cellular Dynamics and Function, Genetic Mechanisms, and Systems and Synthetic Biology, alongside Molecular Biophysics. Physicists working on the brain also are supported by the Neural Systems program within Integrated and Organismal Systems, and by the program on Collaborative Research in Computational Neuroscience (CRCNS), which is part of the Information and Intelligent Systems Division of the Computer and Information Science and Engineering Directorate and funded jointly by NIH; this is not included here due to the complexities of joint funding. Note that all NSF awards can be found at https://www.nsf.gov/awardsearch, which is the source for all data shown here.

In Figure 9.1, what is counted as total NSF funding for biological physics combines the PoLS program, the PFCs with a clear focus on the field, and MCB awards identified as "molecular biophysics." This total for the decade is broken out year by year in Figure 9.2, and further divided into the regular awards from PoLS and MCB and the PFCs.[2] The partnership between PHY and BIO in supporting biological physics has been a powerful catalyst for growth and played a critical role in establishing the United States as a world leader in the field.[3] More generally, NSF has managed to provide much better support for the field by coordinating across even its largest administrative divisions.

> **Finding:** The Physics of Living Systems program in the Physics Division of NSF is the only federal program that aims to match the breadth of biological physics as a subfield of physics.

Department of Energy

Of the U.S. federal science and technology funding agencies, the DOE Office of Science provides the largest amount of funding for physical sciences research broadly, and is second to NSF in providing physical sciences research funding to universities and colleges. At DOE, the Basic Energy Sciences (BES) program

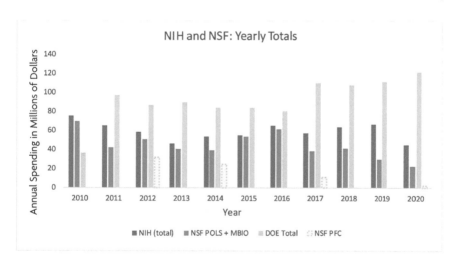

FIGURE 9.2 Annual spending on programs that include biological physics, by agency. As with the caveats to Figure 9.1, these budgets considerably exceed actual spending on biological physics. in particular, for the Physics Frontier Centers, 2012 includes $28 million to the Kavli Institute for Theoretical Physics; only ~30 percent of Kavli Institute programs are in biological physics, and some of these are supported by funds outside the main NSF Physics Frontier Centers award.

---

[2] The PFC competition happens only once every 3 years, but (as noted above) multiyear NSF awards are listed as belonging to a single year. If not separated, this would produce an artificial oscillation in apparent annual funding.

[3] Many NSF awards are co-funded by different programs. In this accounting the full award amount is attached to the lead program, and there is no double counting.

provides the majority of physical sciences research funding, including funding for projects in condensed matter and materials physics, chemistry, geosciences, and the "physical biosciences," which have substantial overlap with what we identify in this report as biological physics. A major component of the BES program are shared research facilities based at DOE national laboratories and open to researchers from all disciplines. These user facilities serve more than 16,000 scientists and engineers each year, and include X-ray, neutron, and electron beam scattering sources. The enormous impact of these facilities on biological physics is discussed more fully below, along with DOE support for advanced scientific computing.

The interaction between basic science and the agency mission has a long history at DOE and its precursor agencies; for example, the study of biological responses to radiation led to the discovery of very general DNA repair mechanisms; the pursuit of more efficient solar energy conversion led to many discoveries about the molecular events in photosynthesis; and more. DOE also supported "theoretical biology" at the Los Alamos National Laboratory at a time when those words were very unpopular, helping to advance research from protein dynamics to immunology and more; much of that work is connected directly to modern research in biological physics. The Human Genome Project began with DOE efforts in the mid-1980s, before being fully launched in 1990 through a memorandum of understanding between DOE and NIH. For some perspective on these efforts, a 2011 report estimated the $3.8 billion federal investment in the Human Genome Project generated $796 billion in economic impact, providing more than 300,000 jobs in the genome-driven industries that emerged.[4]

Beyond shared research facilities, the BES program provides funding for biological physics research through its Chemical Sciences, Geosciences, and Biosciences (CSGB) Division. Under CSGB, there are three foci for investment: Fundamental Interaction, Photochemistry and Biochemistry, and Chemical Transformations, each of which is further subdivided. Photochemistry and Biochemistry is focused on "research on molecular mechanisms involved in the capture of light energy and its conversion into chemical and electrical energy through biological and chemical pathways" and is furthered subdivided into Solar Photochemistry, Photosynthetic Systems, and Physical Biosciences. Significant parts of the work on photosynthesis described in Chapter 1 are supported by these programs, as are efforts to build artificial systems that mimic this and other biological functions. In addition to funding through BES, the DOE Biological Environmental Research (BER) program provides funding for biological physics research through its Biological Systems Science Division (BSSD). The BER mission is "to support transformative science and

[4] S. Tripp and M. Grueber, *Economic Impact of the Human Genome Project*, 2011, Battelle Memorial Institute, Columbus, OH, https://www.battelle.org/docs/default-source/misc/battelle-2011-misc-economic-impact-human-genome-project.pdf.

scientific user facilities to achieve a predictive understanding of complex biological, Earth, and environmental systems for energy and infrastructure security, independence, and prosperity," and BSSD is organized to support research that "integrate[s] discovery- and hypothesis-driven science with technology development on plant and microbial systems relevant to national priorities in energy security and resilience." Through BSSD's Biomolecular Characterization and Imaging Science effort, DOE funds efforts to study structural, spatial, and temporal relationships of metabolic processes to better understand environmental and biosystems design impacts on ecosystems from the atomic to the microbial and plant scales.

The funding reported for DOE in Figures 9.1 and 9.2 represent totals for CSBG and BSSD, and were obtained from congressional budget documents.[5] These programs have strong overlap with biological physics, but also clearly support much work outside the field. Indeed, DOE has made substantial efforts to integrate different disciplines, and to create a continuum of support from basic science to applications, all in pursuit of its mission. The simple sums reported here thus substantially over-estimate the support for the field, but on the other hand leave out the support for shared facilities described later in the chapter (see "User Facilities").

Rather than trying to dissect the budget, the committee asked DOE staff for their views (Appendix C). They responded that,

> Biophysics research funded by DOE [Office of Science] has traditionally focused on molecular and cellular biophysical topics, such as structural biology and enzyme kinetics. However, biophysics as a discipline is now expanding to encompass a broader range of techniques that include both experimental and theoretical tools to measure phenomena related to quantum biology, nucleic acid interactions, protein biosynthesis, cell membrane fluid dynamics, cell-cell interactions within and between microbiomes, as well as many more.

In a second direction, BES mission-directed programs draw from understanding of microscopic mechanisms of energy transduction but are increasingly looking at "using biology as a blueprint for the design and synthesis of self-regulating, resilient materials that incorporate predetermined functionality and information content approaching that of biological materials."

The DOE Office of Science supports an Early Career Research Program (ECRP) for universities and national laboratories. Awardees are all within 10 years of receiving a PhD, and can receive $150,000 per year at universities and $500,000 per year at national laboratories for 5 years to launch their careers. A number of ECRP funding opportunities in BES and BER have focused on biological physics topics in recent years.

---

[5] U.S. Government Publishing Office, "Budget of the United States Government," https://www.govinfo.gov/app/collection/budget/2023.

Finding: The United States has had a long-standing role as a leader in the area of biological physics at the molecular scale. Crucial support for this effort comes from DOE investment in programs and user facilities.

DOE's capabilities in "large project team science" are well known as essential to support of the U.S. effort in elementary particle physics. As particle physics began to connect with astrophysics and cosmology, these efforts expanded, including through partnerships with the National Aeronautics and Space Administration (NASA). As noted above, the Human Genome Project was joint effort of DOE and NIH, while the follow-up National Plant Genome Initiative has involved DOE, NSF, and the U.S. Department of Agriculture (USDA). More recently, DOE has explored bringing its team science expertise to bear on explorations of the brain, under the umbrella of the national BRAIN initiative, and this has involved connections with both NSF and NIH.[6]

National Institutes of Health

NIH has no single program geared specifically for supporting work in biological physics, but it has made substantial investments in the field. Funding comes from multiple Institutes and represents a major source of individual research support for the community. Using NIH's RePORTer tool[7] to search the term "physics" produced 572 funded projects totaling $234,958,323 for the 2019 fiscal year alone, though a closer inspection of the results for this and other fiscal years revealed that many of these projects did not appear related to biological physics. In contrast, searching for projects led by scientists whose primary departmental affiliation was physics or biophysics yielded results that were far more relevant and consistent. Thus, the committee tracked NIH support for biological physics based on departmental affiliation, while support from other agencies was tracked in other ways. This approach certainly undercounts biological physics researchers who do not have primary appointments in physics or biophysics departments, and may include some scientists who would not identify with the definition of the field adopted in this report, but seems a reasonable proxy.

Over the decade 2010–2019, NIH made approximately 170 awards per year to principal investigators (PIs) whose primary affiliations are in physics or biophysics departments, with a total budget of roughly $60 million annually. About 84 percent of these awards were single PI grants. The National Institute of General Medical Sciences (NIGMS) accounted for the largest share, about 40 percent of overall funding, but the National Cancer Institute (NCI); the National Heart, Lung,

---

[6] National Institutes of Health, "Brain Research Through Advancing Innovative Neurotechnologies®," https://braininitiative.nih.gov.

[7] National Institutes of Health, "RePORTER Version 2020.9," https://reporter.nih.gov.

and Blood Institute (NHLBI); the National Institute of Biomedical Imaging and Bioengineering (NIBIB); and the National Institute of Neurological Disorders and Stroke (NINDS) each awarded more than 100 grants over the decade, and awards from 17 other institutes form a long tail, as shown in Figure 9.3. While this funding has been a key contributor to biological physics, it represents only about 0.2 percent of all NIH awards.

While NIH grants are funded by institutes, they are reviewed by study sections, which are (largely) standing committees with responsibility for particular areas of science. There are roughly 200 of these study sections, on topics ranging from macromolecular structure to the organization and delivery of health services. It is remarkable that PIs whose primary affiliations are in physics or biophysics departments received their grants through 75 different study sections (Appendix E). On one hand, this means that the impact of the physicists' approach to life is felt throughout a large part of the NIH portfolio. On the other hand, it means that biological physics proposals need to be tuned to the interests of these very different groups.

Finding: NIH provides strong support for many individual investigators in biological physics, through multiple institutes and funding mechanisms.

The committee notes, more explicitly, that there are several study sections traditionally thought of as "biophysics" study sections, but these only serve a rather narrow slice of the biological physics community. This includes traditional structural biology and single molecule research, some works on theories and models connected

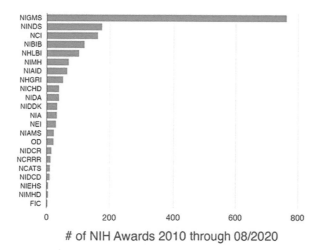

FIGURE 9.3 Number of NIH awards to individual investigators, by institute.

very closely to experiments, and some neuroscience. This leaves large segments of the biological physics community having to navigate the study sections across many different subject areas to find which ones are friendly to their flavor of scientific approach. This has the effect of scattering the field and obscuring its coherence.

NIH awards grants in several broad categories, including research awards (R**), training grants (T**) that support stipends and tuition for students and postdoctoral fellows, and career development awards (K**) that include grants to aid the transition from postdoctoral fellow to independent investigator; for a complete list of relevant funding mechanisms see Appendix E. Of the 4,841 K awards (K01, K08, K22, K25, K99) NIH made across all institutes and disciplines in the past 10 years, about 2 percent (104 awards) went to PIs in physics or biophysics departments. Note that this is 10 times larger than the fraction of research awards flowing from NIH into the biological physics community. This is a strong endorsement of the idea that the community is producing young scientists who are exceptionally well prepared to meet the intellectual challenges posed by the phenomena of life, whether they end up identifying as biological physicists, physical biologists, biologists, or medical scientists.

Department of Defense

Biological physics researchers and projects also are supported by and integral to a variety of programs at DoD. This funding comes through DARPA, ARO, ONR, and AFOSR; where available, annual funding levels over the decade are included in Figure 9.4, while a broader set of DoD agencies contribute to the decadal total in Figure 9.1. These agencies' engagement with the field—evident in both their funding allocations and in the thoughtful comments they provided to the committee—reflects the strong potential biological physics holds for advancing mission-driven basic science as well as practical applications in a variety of domains (for descriptions of these agencies' missions, see Appendix D). Overall, the amount of support these agencies provide to biological physics research is substantial, though it is qualitatively different from the support provided by NIH and NSF. Most grants at the DoD agencies support biological physics as a part of larger, multidisciplinary collaborations, so that the budgets quoted in Figures 9.1 and 9.4 usually are shared across many disciplines. Nonetheless, DoD's enthusiasm to bring biological physics researchers into these larger projects is a marker of the field's impact.

> **Finding:** DoD agencies have highlighted multiple areas where the interests of the biological physics community intersect their missions.

Concretely, in response to the committee's request for data on the support for biological physics, representatives of several DoD agencies called out multiple

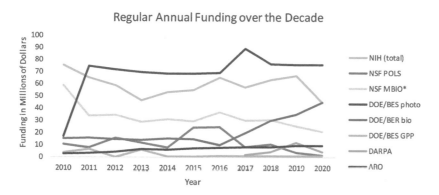

FIGURE 9.4 Funding in regular programs, by agency, broken down by program if that information was readily available. NOTES: DOE BER = DOE/BER/BSSD Biomolecular Characterization and Imaging Science; DOE BES photo = DOE/BES/CSGB Photochemistry and Biochemistry; NSF MBIO = NSF/MCB/ Molecular Biophysics; NSF POLS = NSF/PHY/Physics of Living Systems. DARPA reported awards just for the years shown. ONR did not report an annual breakdown, but the total decadal awards of $96 million are included in the DoD heading in Figure 9.1.

areas where the field intersects their mission: soft robotics; bio-inspired materials and autonomous systems; computational neuroscience and sensorimotor control; radio-bio; insect brains; soft-matter circuit design; locomotion; non-equilibrium active matter; physics-based models of stochasticity in populations; inter-cellular communication; agent-based models; and more. While distributed across many different funding programs in multiple DoD agencies, this broad spectrum of topics connects with a large swath of the biological physics community.

In the mid-2000s, the federal government created a new agency, the Intelligence Advanced Research Projects Agency (IARPA), parallel to DARPA but under the auspices of the Director of National Intelligence. Although not a part of our full survey, we note that IARPA supports a major project to uncover principles of data representation and computation in real brains, with the goal of exporting these principles to artificial systems. As part of this effort, IARPA contributed to one of the largest efforts to date in connectomics (Chapter 3).

Private Foundations

Beyond the federal funding agencies, several private foundations in the United States have made, and continue to make significant contributions to the support of biological physics and allied efforts. Examples include the Alfred P. Sloan Foundation, BWF, the Gordon and Betty Moore Foundation, HHMI, the Simons Foundation, and the Kavli Foundation. These different organizations have used a wide range of funding mechanisms. Starting in the mid-1990s, for example, the Sloan

Foundation supported the establishment of six centers for theory in neuroscience. As described in several sections of Part I, this was a moment when a new generation of ideas and methods from physics were having an impact on thinking about the brain, and the physics community was beginning to appreciate the challenges of real neural networks. The young people who passed through the Sloan centers formed an important part of the nucleus for a more quantitative, theoretically oriented exploration of the brain, with substantial connections to the biological physics community.

Not long after the Sloan initiative, BWF started a program to support centers for research and graduate education at the interface of the biological and physical sciences, and this evolved into the Career Awards at the Scientific Interface (CASI), which provides $500,000 grants to postdoctoral fellows, nominally for 5 years, with the intention that support is carried into the initial years of their junior faculty positions. While not aimed solely at biological physics, a number of young people in the field have been supported by the CASI program, which also provides community and mentorship for the grant recipients, who now number nearly 200. Importantly, the CASI program has taken a broad view of the field, and grant recipients have worked on problems ranging from the dynamics of single molecules to the behaviors of large populations of organisms.

The Gordon and Betty Moore Foundation has two major initiatives that overlap the biological physics community, in Marine Microbiology and Symbiosis, and in Aquatic Systems. Separately it makes grants not tied to specific initiatives. In response to the committee's query, Moore Foundation staff identified 64 grants related to the physics of living systems during 2010–2019, including investigator awards, multidisciplinary awards, and support for postdoctoral scholars in larger research communities.

HHMI is a major supporter of biomedical research in the United States, both at the Janelia Research Campus and through the appointment of HHMI investigators based at universities and medical schools around the country; total expenditures in fiscal year 2020 were $822 million. In this broad portfolio, one can find many people and projects that connect strongly to the physics of living systems community. On the other hand, only a handful of the more than 250 HHMI investigators have appointments in physics or applied physics departments, which is roughly the same as the representation of biological physics among NIGMS grantees. There is stronger representation of physicists among the HHMI professors, who are selected for their integration of research with undergraduate teaching. HHMI also has supported fellowship programs for PhD students coming from abroad, and from underrepresented groups in the United States.

The Simons Foundation has multiple programs that overlap with the interests of the biological physics community. Since 2017, the foundation has partnered with the Division of Mathematical Sciences at NSF to support four NSF-Simons Research Centers for Mathematics of Complex Biological Systems. The Simons Foundation also supports several large-scale collaborations, including on the Origin of Life, the

Principles of Microbial Ecology, and the Global Brain, all of which have substantial participation from the biological physics community. These collaborations fund multiple investigators around the world, often with associated programs for postdoctoral fellows and to support the transition from fellow to independent investigator, similar to the BWF/CASI program. The Simons Investigator program provides stable, longer term support for faculty working in theoretical physics and astrophysics, theoretical computer science, and mathematics. For several years, there was a young investigator program for mathematical modeling of living systems, but now the theoretical physics program explicitly includes biological physics. At the recently established Flatiron Institute, two of the five centers—the Center for Computational Biology and the Center for Computational Neuroscience—have strong overlap with the physics of living systems.

> **Finding:** Private foundations have supported programs that engage the biological physics community, often before such programs become mainstream in federal agencies, and have explored different funding models.

Synthesis

Looking at funding agency by agency matches the way in which grants are awarded, and the way in which Congress supports science. But there are other axes along which to decompose the funding of biological physics that highlight the challenges of the current funding environment more clearly.

Scale of Support for Individual Investigators

The discussion above emphasized individual investigator grants because the majority of support for biological physics at both NSF and NIH is in this form (see Figure 9.5). But looking more closely at these individual awards uncovers dramatic differences. At NIH, single investigator and small group grants (R** research awards) typically have 3- to 5-year terms, with a median budget of roughly $300,000 per year, stably over the decade; roughly 10 percent of grants are early career awards. For NSF/PoLS, the median budget for individual investigators over this period was under $125,000 per year. Figure 9.6 shows the full range of award sizes to individual investigators in biological physics at both NSF and NIH.[8]

---

[8] Echoes of these differences between NIH and NSF can be seen in the differential treatment of the physical and biological sciences within NSF itself. NSF's major young investigator award, the CAREER, is a 5-year grant of $400,000 for proposals funded through the Physics Division (PHY), but $500,000 for those funded through the Biological Sciences Directorate (BIO). See National Science Foundation, *Faculty Early Career Development Program (CAREER)*, NSF 20-525, https://www.nsf.gov/pubs/2020/nsf20525/nsf20525.htm.

NSF and NIH Funding Dollars Compared by Grant Type

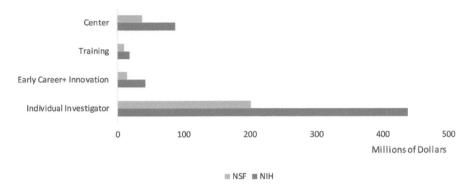

FIGURE 9.5 Support for biological physics at the National Science Foundation (NSF) and the National Institutes of Health (NIH), by category.

The contrast between NSF and NIH award sizes is striking. The typical (median) NIH award has an annual budget larger than all but a few percent of NSF awards. It is important that the NSF grants in our field are not just smaller than NIH grants, but small in absolute terms. Appendix F estimates the bare minimum grant size needed to support a faculty member working with one PhD student;

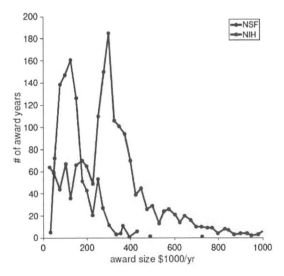

FIGURE 9.6 The full distribution of award sizes to individual investigators in biological physics at the National Science Foundation (NSF; blue) and the National Institutes of Health (NIH; red). NIH data extracted via the RePORTer tool as described in the text; annualized NSF awards estimated as described in Appendix E.

this minimal budget is essentially equal to the median NSF annual award. While many members of the biological physics community manage to do outstanding science with less than this minimal budget, this low level of funding is not healthy for the field.

The small grant sizes at NSF hold back the development of biological physics in many ways, and generate hidden costs. NSF has been unique in embracing biological physics as a field of physics, in all its breadth. If individual investigators cannot support their research programs through NSF, they are driven to other funding agencies that fragment the field in various ways and de-emphasize connections to the rest of physics. Smaller grant sizes mean that many investigators need multiple grants to support their research programs, which require multiple reviewers and multiple reports, creating a cascade of burdens on the community that are largely unaccounted for. If particular grant programs cannot meet scientists' needs, eventually they stop applying, giving the impression that community demand is shrinking, and making it more difficult for program officers to argue for greater resources.

> **Finding:** NSF award sizes for individual investigators in biological physics have reached dangerously low levels, both in contrast to NIH and in absolute terms.

The committee emphasizes that this is not a criticism of NSF, which in fact has done a commendable job of stretching limited resources through partnerships across programs and divisions. Rather, NSF is caught between its budgetary constraints and its goal of supporting the field in its full breadth.

Centers and Research Communities

In many areas of science, from elementary particle physics to global climate, there are crucial projects that require large teams of investigators. These big efforts capture the public imagination, as did their historical precursors. There is a sense that progress more generally depends on supporting a "moonshot for X," with argument more about the correct X than about the moonshot model. It is worth remembering that the most successful examples of "big science" did not start big, and for most of their history were as small as they could possibly be. The search for gravitational waves, for example, began with individual investigators, and grew only gradually to the point where the eventual discovery was reported by more than 1,000 authors.

Even when a single project does not require a large team, however, there can be benefits from more collective support for groups of investigators. In condensed matter physics and atomic physics, for example, "table top" experiments benefit enormously from shared resources in local communities, and the same is true for biological physics. Theorists seldom write papers with large numbers of authors, but nonetheless congregate to share ideas, and communal mentorship of junior theorists is common. These research communities enhance the productivity of their individual members. Support for such research communities thus is important, not as an alternative to individual investigator grants but as a complement to them. Mechanisms for funding these groups need to support intellectual infrastructure as well as laboratory infrastructure.

There are several current models for support of biological physics research communities. At NSF, of the 17 PFCs that have been launched since the program began in 2001, three are focused on problems in biological physics, and several others have overlap with the field. Also at NSF, of the 19 currently active Materials Research Science and Engineering Centers (MRSEC), more than one-third host interdisciplinary research groups working on living or biomimetic systems. These examples are important in part because they demonstrate the ability of biological physics programs to succeed in open competition across all areas supported by the NSF Divisions of Physics and Materials Research. NSF and the Simons Foundation also have partnered to support four Centers for the Mathematics of Complex Biological Systems, as noted above, all of which have substantial participation from the biological physics community. Many of the institutions that host these different Centers are linked through NSF's Physics of Living Systems Student Research Network, which also connects to a number of institutions internationally. Although this network has not yet reached all relevant U.S. institutions, it provides a model for how to share the benefits of strong local community support much more widely. More generally, Center grants often have a strong mandate for the integration of research and education, including reaching students outside the host institution.

NIH also supports centers and larger collaborations. There are Program Project/Center grants (P01, P30, and P50), but these now tend to be more specialized and are not offered across the full range of topics supported by NIH. There are cooperative agreements (U01) to support larger group efforts, and a number of biological physics projects can be found inside these agreements. As discussed below, the NIH training grant programs provide for institutional support of doctoral students, and in some cases postdoctoral fellows, working in a broad area, and as such serve some of the same functions as Centers in the NSF model. Perhaps the most radical experiment undertaken by NIH was the establishment of 12 Physical Sciences in Oncology Centers. Run through NCI, these centers incorporate the physical sciences into cancer research and have been quite successful in combining biological and physical sciences. Their focus has largely shifted, however, toward tool building and engineering and away from the search for deeper physical and mathematical principles.

Conclusion: As in many areas of science, there is a challenge in maintaining a portfolio of mechanisms to fund the spontaneity of individual investigators, the supportive mentoring environments of research centers, and the ambitious projects requiring larger collaborations.

Support for Education and Workforce Development

Both NSF and NIH provide direct support for graduate education, through different mechanisms. The Graduate Research Fellowship Program (GRFP) was the first program established by NSF and now appoints just over 2,000 new fellows each year, across all fields of science. In 2020, 136 new fellows were appointed in physics and astronomy, with a distribution across subfields, including the Physics of Living Systems, that roughly mirrors the distribution of PhDs (see Figure 8.1). The GRFP also provides a model in which individual students are given considerable freedom and agency, with fellowship support attached to them as individuals rather than to their mentors or their doctoral program.

The NSF PoLS program has established a student research network connecting multiple institutions both within the United States and internationally. These grants (included in the analysis above) aim not at core funding but rather provide supplementary funding to enable student exchanges, summer schools, conference attendance, and other activities that enrich the experience of doctoral students. There are smaller and more focused fellowship programs sponsored by DoD (National Defense Science and Engineering Graduate Fellowship Program) and DOE (Office of Science Graduate Student Research Program), and both have some overlap with biological physics. As found in Chapter 8, however, there are some sharp distinctions between support for education and research that limit these agencies.

NIH has a very different approach to the support of graduate education, with a large program of "training grants" that fund multiple students at single institutions. For many doctoral programs in the biomedical sciences, broadly defined, these training grants are a major pillar of support. Perhaps surprisingly, fully 14 percent of NIH grants awarded to physics and biophysics departments over the past decade have been training grants, but this represents only 0.5 percent of the more than 35,000 NIH training grants in total. The tiny fraction of NIH training grants that support doctoral education in biological physics is not commensurate with the impact that physics has had on the biomedical sciences. It is even inconsistent with the larger fraction of NIH career development (K) awards to physicists, as described above. This would not be a problem if NSF had analogous programs for the support of physics students more generally, but in the absence of such programs this represents a significant gap.

Finding: Physics programs do not have the stable, programmatic support for PhD students that is the norm in the biomedical sciences.

Theory as an Independent Activity

In physics, theory and experiment are partners in exploring the world. The relationship between theoretical and experimental physics is complex, but there is little doubt that theoretical physics has an independent identity. Correspondingly, there is a long tradition of federal agencies supporting theory as an independent activity. At NSF, for example, in the Division of Physics there are separate, parallel programs for theory and experiment in atomic, molecular, and optical physics; elementary particle physics; gravitational physics; nuclear physics; and particle astrophysics and cosmology. Also at NSF, in the Division of Materials Research the condensed matter physics program supports experimental condensed matter physics while the materials theory program supports theoretical condensed matter. One can find similar structures at DOE, and to a lesser extent in the DoD agencies. Importantly, independent funding for theory has never inhibited theory/experiment interaction.

> **Finding:** Physics has a unique view of the relationship between theory and experiment, and in many fields of physics this is supported by separate programs funding theorists and experimentalists. This structure does not exist in biological physics.

Adding Things Up

The physicists' view is that the numbers describing the world fit together into some coherent framework. It seems natural, in this spirit, to ask if the total funding for biological physics, across all the sources discussed here, fits together with other measures of the activity and vitality of the field. But what sets this scale? One well-defined anchor is the number of students each year who receive their PhD for research in the field, as described in Chapter 8 (see Figure 8.1). At a minimum, supporting the field means supporting these students during their thesis research, continuing to support the fraction who move on to postdoctoral positions, and supporting the community of university faculty and professional researchers who provide mentorship for these young scientists. Appendix F provides estimates for this minimal level of support.

Taking a very strict view of the field's boundaries, corresponding to research that would be carried out by students receiving their PhDs in Physics, minimal support is in the range of $83 to $92 million per year. This is vastly larger than what is available, for example, through programs in the Physics Division of NSF. With a slightly broader view of the field, the minimum level of support comes close to the total expenditures of all relevant agencies (see Figure 9.4). These minimal levels do not include the costs of research facilities, equipment and supplies, technical or administrative support staff, travel for collaboration, and so forth.

Finding: Total support for biological physics is barely consistent with the minimum needed to maintain the current flow of young people into the field. This approximate balance of needs and support leaves significant gaps, and provides little room for new initiatives.

Conclusion: Biological physics is supported by multiple agencies and foundations, but this support is fragmented, obscuring the breadth and coherence of the field. It is precariously close to the minimum needed for the health of the field.

RESPONDING TO CHALLENGES AND OPPORTUNITIES

One of the major challenges facing the biological physics community is the substantial mismatch between the community's intellectual activity and the structure of current funding programs. This should be understood in the context provided by the initial conclusion from Part I of this report:

Conclusion: Biological physics, or the physics of living systems, now has emerged fully as a field of physics, alongside more traditional fields of astrophysics and cosmology; atomic, molecular, and optical physics; condensed matter physics; nuclear physics; particle physics; and plasma physics.

In contrast, much of the support for the field continues to follow a model of biophysics as only the application of physics to biology. This perspective constrains what the funding agencies perceive as relevant, and reinforces an organization of the field around the classical subdivisions of the biological sciences. The biological physics community has been successful in adapting to this funding environment, but it now is time for the funding environment to respond more fully to the community.

General Recommendation: Funding agencies, including the National Institutes of Health, the National Science Foundation, the Department of Energy, and the Department of Defense, as well as private foundations, should develop and expand programs that match the breadth of biological physics as a coherent field.

This section explores how this broad recommendation can be implemented across the many relevant agencies and across other dimensions of research support. Discussion on major user facilities is deferred to the "User Facilities" section below.

It is important that the current funding structures are not only mismatched to the state of the field, but also fail to maximize the opportunities for agencies to advance their missions. As an example, there is a set of interlocking questions about collective behavior in groups of organisms, the emergence and persistence of ecological diversity in multispecies groups, and, on longer time scales, the evolution of these populations. Taken together these questions are crucial for the missions of multiple agencies. The biological physics community has sharpened these questions and made progress toward answers by adopting the unifying language of statistical mechanics and dynamical systems, both in theoretical work and in the design and analysis of new experiments, as described in Part I. But mapping these questions into the current funding programs fragments the community: If the multispecies groups are bacteria living inside the human gut, then the problem is relevant for NIH, while bacteria living in soil are relevant for DOE; if the question asked is not about bacteria but about plants in the rainforest, then connections are drawn to climate science and the mission of the National Oceanic and Atmospheric Administration or the ecology programs of NSF, while interest in cooperative behaviors is a basic science problem connected to the practical problem of coordinating multiple autonomous vehicles and the mission of DoD. All of these agency missions would be advanced by more coherent and coordinated support for the biological physics community's attack on these problems. More deeply, the community's discovery of conceptual connections among these seemingly different problems offers opportunities for each of these agencies to benefit by seeing its mission in the broadest possible terms.

National Science Foundation

The committee's concerns about support for biological physics at NSF can be stated simply: NSF does not have enough resources to accomplish a mission that includes biological physics. As emphasized above, NSF is the only agency to recognize biological physics in all its breadth, as a field of physics. The launch of a Biological Physics program within the Physics Division, a program that evolved into the current Physics of Living Systems program, not only provided an important source of support but also a marker that there is new physics to be found in the phenomena of life. But, as noted above, the level of funding for these programs has not kept pace with the growth of the field.

> **Specific Recommendation: The federal government should provide the National Science Foundation with substantially more resources to fulfill its mission, allowing a much needed increase in the size of individual grant awards without compromising the breadth of its activities.**

National Institutes of Health

A substantial component of support for biological physics research in the United States comes from NIH. As emphasized above, these grants derive from a wide range of different institutes and 75 different study sections, testimony to the breadth of the field. None of these study sections, however, is devoted to biological physics itself. This fragments the field, and misses intellectual opportunities that cut across the historical sub-divisions of biology, as described in Part I of this report. More subtly, young investigators especially are pushed toward defining their work in relation to the communities represented by the study sections, thus working against the emergence of biological physics as a field of physics.

> **Finding:** Support for the physics of living systems is scattered widely across NIH, making it difficult for investigators to find their way and obscuring the coherence of the field.

> **Specific Recommendation: The National Institutes of Health should form study sections devoted to biological physics, in its full breadth.**

NIH study sections are established via a process called ENQUIRE, through the Center for Scientific Review (CSR). ENQUIRE integrates data and input from stakeholders to determine whether changes in study section focus or scope are needed to facilitate the identification of high impact science, with special consideration of emerging science. Clusters of study sections are formed based on scientific topics (instead of CSR managerial units); review via the ENQUIRE process is systematic, data-driven, and continuous—roughly 20 percent of CSR study sections are evaluated each year, and every study section is evaluated every 5 years.

Department of Energy and Other Agencies

DOE, along with most of the federal agencies involved in the support of science, has a concrete mission (see Appendix D), and there is an important challenge in connecting the frontiers of biological physics research with these different missions. Importantly, DOE often views its mission not only in terms of particular domains of science and technology, but also in bringing its expertise in "big science" to bear on problems well outside of energy. An example of this is the recent series of workshops sponsored jointly by DOE and NIH addressing the feasibility of larger scale projects in connectomics, mapping the synaptic connections among neurons, perhaps even in entire brains. Several imaging modalities, including various forms of electron microscopy, synchrotron X-ray computed nano-CT or pytchography, and various forms of optical microscopy, are under evaluation as foundational technologies, and this effort could require investments on the order of several hundred million dollars. Such maps would provide a structural scaffolding on which to place the growing body of experiments that monitor the electrical activity of many neurons simultaneously, also a problem where measurement technologies are advancing rapidly. These efforts have had substantial input from the biological physics community, both in the development of experimental methods and in the articulation of theories within which one might make sense out of such large data sets. More broadly, the physics community has tremendous experience with projects on this scale, from the construction of large instruments to making the data accessible and integrating experiment with theory. If DOE is to be the path for large-scale brain projects in the physics tradition, it will be important to engage the broader biological physics community from the very start.

As DOE looks to partnerships with other agencies to explore a broader view of its mission, it is important to have successful models for such partnerships in support of biological physics. An example is the program for Collaborative Research in Computational Neuroscience (CRCNS), funded jointly by NSF and NIH, which has supported a number of groups in the biological physics community. In this program, NSF takes the lead in soliciting and managing the review of proposals, while representatives of several participating NIH Institutes engage with the process to identify proposals recommended for funding that would fit in their larger programs. Since its inception as an NSF/NIH effort, CRCNS has expanded to include partnerships with DOE and with agencies in France, Germany, Israel, Japan, and Spain; the procedures leading to joint funding necessarily have become more complex, but are well managed. The joint support of the Center for Quantitative Biology by NSF and the Simons Foundation provides a related but very different model.

> **Finding:** Biological physics has benefited from funding programs that are shared across divisions within individual federal funding agencies, between agencies, and between federal agencies and private foundations.

The problem of connecting biological physics to agency missions also exists in the DoD agencies. In these agencies, the individual programs and program officers have considerable autonomy, however. Furthermore, there is a tradition of DoD agencies, at various times, seeing their mission in the broadest possible terms. As described above, DoD officials see a wide range of connections to topics being explored by the biological physics community. Importantly, many of these connections involve not a simple translation of biological mechanisms into engineering contexts, but rather the emulation of deeper physical principles that enable the extraordinary functionality of living systems.

Specific Recommendation: The Department of Defense should support research in biological physics research that aims to discover broad principles that can be emulated in engineered systems of relevance to its mission.

In addition to supporting a broad spectrum of activities that engage the biological physics community, DoD agencies also offer models of different funding structures. In particular, the Multidisciplinary University Research Initiatives (MURI) Program can support mid-sized collaborations of five to seven PIs with budgets of $3–$5 million over several years. Work on this scale is becoming increasingly important in biological physics, as frontier experiments often require combinations of new technologies and produce data on a scale where strong integration with theory and analysis is necessary starting from the design of the experiments. A similar scale of project can be supported through the U01 and U19 mechanisms of NIH, but these are less common, making them something of a rarity in the biological physics community.

Conclusion: There is an opportunity for DoD agencies to use the MURI Program to support biological physics, and for NSF and NIH to expand their support of these mid-sized collaborations.

Industrial Research Laboratories

As described in Part II of this report, biological physics has strong overlap and interaction with many fields that are relevant for the pharmaceutical and biotechnology industries—structural biology, systems biology, molecular design, synthetic biology, and more. At various times, research laboratories sponsored by these industries have thus been very supportive of the field.

In the 1980s and 1990s, a very different set of industrial research laboratories provided strong support for the emergence of biological physics as a part of physics. AT&T Bell Laboratories, IBM Research, the NEC Research Institute, and others did not have biology programs in their research laboratories, simply because biology was irrelevant to their business. But all of these companies had biological physics groups that grew out of their basic physics efforts, looking at problems ranging from cooperativity in hemoglobin to neural coding and computation and collective behavior in flocks and swarms. In parallel, Exxon Research supported a large effort in soft matter physics, and many of the scientists from this group would move into biological physics. All these laboratories offered investigators the opportunity to explore the physics of living systems at a time when very few academic physics departments in the United States had identifiable groups in the field. There were paths from this basic scientific work into more applicable results, including foundational work on neural networks, but—as with much of the work in these laboratories—the support for science was not tied to immediate application.

Pharmaceutical and biotechnology research has grown, but from the point of view of the physics community as a whole, the perceived golden age of industrial research labs is over. What has emerged instead are research laboratories supported by new industries such as Microsoft, Google, and Facebook. There also are new structures, intermediate between industrial research and private foundations, such as the Chan-Zuckerberg Initiative and Open AI, as well as new non-profit institutes such as the Chan-Zuckerberg Biohub and the Allen Institutes. There are many connections between the activities of these laboratories and the interests of the biological physics community, from understanding how the identities of cells emerge from patterns of gene expression to the physics of behavior, and this landscape continues to evolve rapidly. There are many opportunities both for technological progress and for fundamental discovery.

Specific Recommendation: Industrial research laboratories should reinvest in biological physics, embracing their historic role in nurturing the field.

Support for Education and Workforce Development

One advantage of support being so widely distributed is that the biological physics community samples the different approaches adopted by the different agencies. Nowhere are these differences more apparent than in the support for graduate education. As noted above, NSF, DOE, and DoD all have graduate fellowship programs, but these are individual fellowships and support only a tiny fraction of the PhD students who are funded through research grants by these agencies. NIH, in contrast, has a very large program of "training grants" that provide substantial support for the infrastructure of graduate education in the biomedical sciences. In the current fiscal year, these T32 grants are active at 150 institutions across the country, with a total budget of roughly $750 million per year. While there is widespread appreciation of the impact that physics has had on the mission of NIH, only 0.5 percent of training grant funds go to the biological physics community. To put this in perspective, a 3.5 percent increase in the T32 budget would provide enough funds to support *all* physics PhD students currently working on biological physics (Appendix F).

The training grant model has many advantages, not just for biological physics. Support is collective, and thus encourages the building of local communities, whose importance is emphasized above. With training grant support, students are empowered to pursue new and exciting directions. Programs can be judged not only on the collection of research directions available to students, but on their effectiveness in mentoring and placement of students after their degrees.

Specific Recommendation: Federal funding agencies should establish grant program(s) for the direct, institutional support of graduate education in biological physics.

It is important that programs for the support of graduate education in biological physics be matched to the culture of physics education more broadly. A number of T32 requirements, for example, are not well aligned with the typical structure of physics doctoral programs. More subtly, training grants typically impose significant restrictions on the support of international students. As discussed in Chapter 10, for the United States to maintain its position of scientific leadership will require renewed commitment to welcoming talented students from all over the world.

Specific Recommendation: Federal agencies and private foundations should establish programs for the support of international students in U.S. PhD programs, in biological physics and more generally.

NIH has multiple funding mechanisms that emphasize career development beyond graduate education. The same T32 training grant programs that fund PhD students also can fund postdoctoral fellows; there are Institutional Research and Academic Career Development Awards that support communities of postdoctoral fellows engaged with modest amounts of teaching at minority serving institutions alongside their research activities; and there are Pathway to Independence programs to support individuals in the transition from postdoctoral fellow to independent investigator (K99/R00). As noted above, the biological physics community has done very well in the competition for these individual fellowships, but there is room for improvement of the community's representation in the institutional grants. NIH also supports postdoctoral fellows through the Kirschstein National Research Service Awards, but these are not focused as explicitly on the independence of the fellows.

Many NSF divisions have programs for the direct support of individual postdoctoral fellows, for example in Mathematical Sciences, Astronomy and Astrophysics, Earth Sciences, and Biology. Similarly, NASA supports the Hubble Postdoctoral Fellows. As an example, the NSF Astronomy and Astrophysics Fellows, together with the NASA Hubble Fellows, support a bit less than 10 percent of the new PhDs awarded each year in astronomy and astrophysics. While these programs are small in comparison to the total budget for the field, the fact that fellows are appointed through a national competition attaches considerable prestige to the awards, and the impact on the community extends beyond the direct financial support. Perhaps more importantly, individual fellowships give the fellows a significant degree of independence, while the selection process allows judgements to be made about the quality of the mentoring environment in which they will be immersed. As explained in Chapter 8, this balance is especially important for young scientists developing their own perspectives on the next generation of challenging problems in biological physics.

> **Conclusion:** Accelerating young researchers to independence is critical to empowering the next generation of biological physicists. As in other fields of physics, independent, individual fellowships are an effective mechanism.

In addition to examples of how such fellowships are implemented at NSF and NIH, there are examples such as the BWF Career Awards at the Scientific Interface, which has had substantial engagement with the biological physics community, as discussed above.

Support for Theory

As more and more of the living world becomes accessible to large-scale, quantitative experiments, the biology community and the agencies that support research in the biological sciences have understood the need to support new approaches to data analysis. For the physics community, however, theory is more than data analysis (see also Chapter 8). The history of physics shows that the most compelling data analysis methods often are grounded in more general theoretical principles, sometimes articulated long before the relevant experiments were possible. Theory thus engages productively with questions that may not be relevant for today's experiments, but help to set the agenda for the next generation of experiments. Conversely, a compelling analysis of today's experiments can raise new theoretical questions, not about how best to fit the data but whether these data provide hints of new principles. Close examination of the examples presented in Part I of this report shows that biological physics has made substantial progress through this more subtle interplay between theory and experiment. Nonetheless, many members of the community report that finding funding for theoretical work that reaches beyond today's experiments remains challenging.

The agencies responsible for support of physics more generally have established mechanisms for the support of theory as an independent activity, as noted above. Much could be gained by extending these mechanisms to encompass biological physics, or the physics of living systems. Not only would it be easier to support forward-looking theoretical research, but independent funding mechanisms would play a role in unifying the community of theorists and highlighting the role of theory in the search for a physicist's understanding of life.

> **Specific Recommendation: Federal agencies and private foundations should develop funding programs that recognize and support theory as an independent activity in biological physics, as in other fields of physics.**

Coda: The Usefulness of Basic Science

In October 1939, just after the start of World War II, *Harper's* magazine published an essay on the usefulness of useless knowledge.[9] It was an astonishing moment at which to offer a vigorous defense of human curiosity and intellectual exploration as the engines of progress. But the argument was prescient. Although the pre-war world had been transformed by automobiles, airplanes, and radio, the pace of change would accelerate dramatically during and after the war. In the industrialized world today, our lives are noticeably more dependent on advanced technologies than even a decade ago, and there is no end in sight.

Every piece of technology in the modern world has at its foundation remarkable developments in basic science. The path from science to useful technology can be long, and requires its own unique innovations, but without the scientific foundation, none of this is possible. An example from 1939 was that radios required the 19th-century unification of electricity and magnetism, and the resulting prediction of electromagnetic waves. Today, computer chips would not exist without quantum mechanics to describe the behavior of electrons in semiconductors. It would not have been possible to make effective vaccines against COVID-19 so rapidly without a generation of work on protein structure and on the basic mechanisms of gene expression and replication in cells.

While it is not so difficult to trace back in time from useful technology to foundational scientific discoveries, it is much harder to predict which discoveries or even which areas of research will lead to useful technology. Investments in microwave spectroscopy around 1950 led to the maser, which led to lasers, which led to dramatic improvements in eye surgery (Chapter 7), among other medical applications. But who would have argued, in 1950, that support for a small group of microwave spectroscopists was important for the future of ophthalmology?

All of this matters because much of the justification for substantial government spending on science is that these investments will have impact on our lives. More concretely, NIH support for research on biological physics ultimately is justified by the fact that physicists' explorations of the phenomena of life have a profound impact on human health. As emphasized throughout this report, this is not merely "physics applied to medicine." Similarly, DoD support for the field is justified by the impact that biological physics has on relevant technology. Again this impact comes in large part not from the application of physics to well-posed problems outside the field, but from the discovery of basic physical principles that govern living systems and can be emulated in engineered systems. These paths for impact are reviewed in detail in Chapter 7.

---

[9] A. Flexner, 1939, "The Usefulness of Useless Knowledge," *Harper's* (179):544–552, reprinted with a companion essay by R. Dijkgraaf, 2017, Princeton University Press.

General Recommendation: To maintain the flow of concepts and methods from biological physics into medicine and technology, the federal government should recommit to the vigorous support of basic science, including theory and the development of new technologies for experiments.

To implement this recommendation it is essential that not only NSF but also the mission-driven agencies contribute to the support of basic science. Existing funding mechanisms often follow a pattern of siloed utilitarianism, focusing only on areas of funding where single agencies and even individual programs within these agencies are confident of their expertise or sovereignty. This style of funding can lead to the premature drawing of boundaries between what is relevant and what is not, limits the ability of agencies to advance their missions, and in the end can even inhibit the progress of basic science by fragmenting the field along these boundaries. History teaches us that the best investments in society's future often involve trying to answer today's most basic scientific questions.

USER FACILITIES

User facilities are substantial pieces of research infrastructure that are too large, too complex, or too costly for a single university or company to maintain. Examples relevant to biological physics research include synchrotron light sources; cryogenic electron microscopes; high-performance computing facilities; neutron-scattering sources; high magnetic fields; and specialized facilities for nanoscience, imaging, and genomics. These facilities and resources are generally available for research access on the basis of competitive peer review.

Many are physically located at the DOE National Laboratories and are supported directly by the DOE Office of Science. Other facilities and funders relevant to the biological physics community include, among others:

- The Advanced Imaging Center at Janelia Research Campus of HHMI, supported in part by the Gordon and Betty Moore Foundation;
- The New York Structural Biology Center, supported by a consortium of New York–based universities, the Simons Foundation, and NIH;
- The National High Magnetic Field Laboratory at Florida State University, supported by NSF along with DOE through facilities at Los Alamos National Laboratory; and
- The Pacific Northwest Center for Cryo-EM (PNCC), supported by NIH along with DOE through the Pacific Northwest National Laboratory.

The federal facilities are a huge national investment in science that support tens of thousands of research projects each year. As technologies evolve, evolution of the facilities can provide a dramatic democratization of frontier science.[10,11]

To gauge trends and needs with regard to the use of such facilities for biological physics research, the committee solicited information via a questionnaire distributed both to facilities directly and also to the DOE Office of Science (Appendix C). The responses reflect the key role of user facilities in enabling cutting-edge biological physics work and also suggest some important opportunities for improvement.

The facilities reviewed here categorize their users as belonging to Biology and Life Science, Chemistry, Earth and Environmental Science, Engineering, Materials, or Physics. Figure 9.7 shows the proportion of users identified as Biology and Life Science at each of the facilities. Three centers primarily support the life sciences, but several other user facilities that might have been seen as giving primary support to the physical sciences in fact support much broader communities. The four X-ray sources report biology and life science use ranging from 25 to 40 percent; the high magnetic field laboratory 24 percent, the three advanced computing centers 5 to 13 percent. The Environmental Molecular Sciences Laboratory (EMSL) has a quarter of its users in biology, and more than a third in environmental sciences. A common thread is that the genesis of these powerful tools has come from decades of investment in physics-based technologies, often without a particular vision that they would impact our exploration of the living world. When these tools are found to be useful in a new field, a community of users learns to exploit them.

Finding: Large-scale physical tools, particularly those for imaging and advanced computing and data, are an important part of the infrastructure supporting thousands of researchers exploring the living world.

Engagement of the biological physics community with user facilities goes beyond instruments to include high-performance computing. As an example, of the approximately 60 peer-reviewed Innovative and Novel Computational Impact on Theory and Experiment (INCITE) awards supported annually by the Argonne and Oak Ridge Leadership Computing Facilities, there have annually been 5 to 10 in the category "Biological Sciences: Biophysics."

---

[10] E. Hand, 2020, "Cheap Shots," *Science* 367(6476):354–358, https://www.science.org/doi/10.1126/science.367.6476.354.

[11] C. Tachibana, 2020, "Democratizing Cryo-EM: Broadening Access to an Expanding Field," Technology Feature, *Science.com*, https://www.science.org/content/article/democratizing-cryo-em-broadening-access-expanding-field.

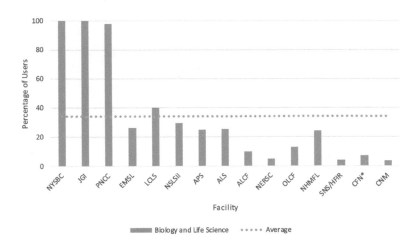

FIGURE 9.7 Distribution of users categorized by the facilities as "Biology and Life Science," one of six categories. Facilities are grouped into broad categories. Genomics: New York Structural Biology Center (NYSBC) and Joint Genome Institute (JGI). Microscopy: Pacific Northwest Center for Cryo-EM (PNCC) and Environmental Molecular Sciences Laboratory (EMSL). Light sources: LINAC Coherent Light Source (LCLS), National Synchrotron Light Source (NSLSII), Advanced Photon Source (APS), and Advanced Light Source (ALS). High-performance computing: Argonne Leadership Computing Facility (ALCF), National Energy Research Scientific Computing Center (NERSC), and Oak Ridge Leadership Computing Facility (OLCF). National High Magnetic Field Laboratory (NHMFL). Spallation Neutron Source and High Flux Isotope Reactor (SNS/HFIR). Nanoscience: Center for Functional Nanomaterials (CFN) and Center for Nanoscale Materials (CNM).

The facilities report generally standardized methods of access (application and peer review), and each has multiple ways of getting community input for facility development. Most of them provide a "concierge service" for new users and they are incentivized to get this right. They invariably offer specialized training, courses and summer schools, run conferences bringing together their user communities, and host student interns. PNCC notes the centrality of training, "One of our center goals is to train cryo-EM researchers toward independence, with a focus of providing hands-on training opportunities, that are in short supply amongst many existing cryo-EM workshops."

Although the raw data do not offer a detailed accounting of facility use by biological physics researchers, respondents offered insightful comments on the role of biological physics in the evolving interests and needs they are seeing among their user communities. Of growing interest, for example, is the ability to image with high resolution over large fields of view and to perform multi-modal studies. EMSL noted its investments "in the development of multi-modal analytical capabilities to enable the spatio-temporal imaging of cellular, communal, and systems level organisms (microbial, fungal, plant, for example) to facilitate the prediction and control of biological systems for beneficial purposes." This highlights an opportunity for the intersection of the biological physics community's approach to collective behavior and ecological diversity (Chapter 3) with approaches to related problems in the environmental sciences community. As noted elsewhere, partnerships between funding bodies will be instrumental in advancing this exchange.

Often the large facilities themselves act as agents to bring research communities together and help those partnerships. For example, past partnerships between DOE and NIH to make dedicated beam-lines and support protein characterization played an important role in advancing structural biology. Advanced Photon Source (APS) users deposit more protein structures into the Protein Data Bank than any other facility in the world. APS noted:

> Recently, there has been increased interest in using various imaging techniques (microtomography, X-ray microscopy, etc.) to study biological samples, in part due to improved instrumentation at the light sources. With the advent of new high-brightness sources being designed for facility upgrades (both at the APS and at the LBNL Advanced Light Source in the United States) that allow for smaller X-ray focal-spot sizes, the trend toward the use of [these] imaging techniques in the biological sciences is likely to increase substantially.

User facilities are not always based around large instruments. The five DOE nanoscience centers, for example, provide advanced instrumentation to a very wide research community, though the individual instruments might not be beyond the financial reach of some university laboratories. Their value is often in providing a professional concierge service to a science user who wants the tool without the overhead of managing it, and may also need a few weeks of access rather than full-time ownership. Sometimes the proximity of multiple tools matters, and perhaps also a straightforward connection to large-scale data analysis and computing. This model offers an opportunity to rapidly democratize advances in, for example, optical imaging, NMR imaging, force microscopies, and electron diffractive imaging. These tools are often of interest to many disciplines, of course, but most facilities are managed by organizations that have a predominantly physical sciences agenda, especially DOE. It also is critically important to invest in the development of next generation technologies (see Box 9.1).

While DOE facilities are generous in welcoming users from many different communities, the process of commissioning new facilities and upgrades is strongly confined by agency mission need. As an example, when modern synchrotrons were developed in the 1990s the case for them barely mentioned what has been one of their most productive contribution to science, medicine, and the economy: protein structure.

Finding: DOE has become a major sponsor of research in biological physics, especially through facilities, without acknowledging the field's supporting contribution to the DOE mission.

BOX 9.1
Investments in Technology Development

   While support for PIs, collaborations, and new centers is substantial, investments in technology development are lagging. For example, although NIH and DOE made significant investments in cryo-electron microscopes in 2018 and 2019, few investments have been focused strategically for technology advancements to improve research outputs for biological research. Smaller, cheaper, lower energy microscopes designed specifically for biological research; sample preparation advancements including high throughput microfluidics approaches that reflect the reduced protein inputs needed for cryo-electron microscopy (cryo-EM) as compared to X-ray crystallography; and overall major increases in overall throughput could not only improve the return on investment and productivity of the research but also expand access to cryo-EM. An important model dedicated to this challenge now exists in the United Kingdom. Originally funded by the UK government at £103 million, the Rosalind Franklin Institute in Didcot, United Kingdom, will focus on cross-cutting themes in automation, imaging, data handling, and artificial intelligence applied to biological science. In the United States, some foundations have recently begun to address some of the technology development needs, as demonstrated by the $2.4 million grant awarded by the Gordon and Betty Moore Foundation in May 2020 for miniature laser phase plates for "next generation cryo-EM" and the investment of the Simons Foundation for the Simons Machine Learning Center located at the Simons Electron Microscopy Center. Scaled, focused federal cryo-EM technology development investments to develop next generation cryo-EMs with improvements including less shielding and no need for field emission guns that deliver electrons suffused in sulfur hexafluoride gas would be beneficial. An emerging NIH/DOE collaboration around connectomics for the BRAIN initiative promises to develop next generation imaging technologies and broaden their access while tackling a moonshot project of the mouse connectome and leaving a legacy of a new user facility for the community.

Specific Recommendation: Congress should expand the Department of Energy mission to partner with the National Institutes of Health and the National Science Foundation to construct and manage user facilities and infrastructure in order to advance the field of biological physics more broadly.

A challenge in implementing this recommendation is that, in the language that DOE uses to describe users of its facilities, biological physics simply does not exist. If you are studying the structure of a protein, for example, you are a "biology or life sciences" user, no matter what question you are asking. This enforces the view that physics provides tools, but the phenomena being studied are outside of physics, belonging to another discipline. Obviously, this view has not stopped DOE from making enormous contributions to our exploration of life, and to many problems that fit squarely under the umbrella of biological physics as defined in this report. Nonetheless, it is worrisome that the nation's largest source of funding for the physical sciences does not recognize more explicitly that the living world is a source of profound physics problems.

Finally, one major user facility does not fit the usual model, but has had a substantial impact nonetheless: the Kavli Institute for Theoretical Physics (KITP), which describes itself as a "user facility for theorists." KITP was founded in 1979 (as ITP), supported by NSF and housed at the University of California, Santa Barbara. The model was to have a small permanent faculty, a community of postdoctoral fellows, and a steady stream of visitors. This model was common to many centers for theoretical research, both in the United States and around the world, but what made ITP a user facility was that it solicited proposals from the community for programs that would organize larger groups of visitors for extended periods. By any measure, this has been very successful, and is widely emulated. Now in its fifth decade, KITP efforts have expanded, and to keep pace with this expansion funding for its core programs now comes not just from NSF but from the Kavli Foundation, NASA, NIH, the Gordon and Betty Moore Foundation, and other sources. ITP's first forays into biological physics were in the mid-1980s, and accelerated in the 21st century. An important component of these programs has been that they highlight, for the physics community as a whole, that biological physics has a place alongside other subfields as part of an integrated effort in theoretical physics. As the field grows, there is room for more experimentation with support for theory communities, where considerable impact is possible at relatively low cost.

10

Building an Inclusive Community

Race, gender, and immigration are topics in the background of almost all policy discussions in the United States, but recent events have made these topics more urgent. This report, written in 2020–2022, comes at the end of a 4-year period in which U.S. government policies toward immigrants, including PhD students in the sciences, shifted substantially; rhetoric and perceptions shifted even more. There is constant discussion of international students, and scientific exchange with international collaborators, as threats to national security and U.S. intellectual property. During this period hate crimes against ethnic minorities also increased,[1] and police violence against members of the Black community created a national focus on the problems of racism not seen since the peak of the Civil Rights Movement in the 1960s. Highly publicized episodes of sexual harassment, including in the scientific community, brought renewed attention to the challenges of achieving equal opportunity for women. All of these issues obviously reach far beyond the field of biological physics, and far beyond the scientific community. This chapter is not intended as a comprehensive review of these problems. Rather we provide some sampling of the issues, conditioned by the experiences of our particular community and especially by the moment at which we write. In many cases, we echo and reinforce the conclusions reached in previous reports, hopefully doing justice to this previous work.

A central challenge for the scientific community is to be welcoming and supportive of people coming from all parts of society. To do this requires overcoming prejudices inherited from the larger culture, and addressing our policies directly to

---

[1] See, for example, the regular reports on *Hate & Extremism* from the Southern Poverty Law Center, https://www.splcenter.org/issues/hate-and-extremism.

societal injustices. Professional scientists are the stewards of great resources—access to high-quality education and the opportunity for aspiring scientists to pursue careers devoted to their intellectual passions. The community's stewardship can be judged not only on whether it is productive, but on whether it is just. International engagement, race, and gender are different axes along which success can be judged, and these are taken in turn. We emphasize at the outset that the experience of each group in our society is unique, and that history creates particular responsibilities in achieving a just relationship with each group. At the same time, there are universal themes in the search for social justice. In making recommendations we try to balance the universal and the particular, again recognizing that our discussion is far from comprehensive.

INTERNATIONAL ENGAGEMENT

Science is an international activity. For many years, the scientific community in the United States held a position of great privilege on the world stage. There was a widespread sense that our nation was among the best, if not the best place to launch a scientific career. Admission to U.S. graduate programs was sought after by aspiring scientists around the globe, and the path from PhD to full scientific independence was enviably short. These conditions led to a steady flux of young scientists into to the country, and they were joined by many children of immigrants who had come seeking a broader sense of economic opportunity and freedom from persecution. In many ways, the growth of the United States as a global scientific power during the 20th century is intertwined with the history of immigration.

Many of the benefits of openness to the world are quantifiable. In 2018, the economic impact of international students in the United States was estimated to be roughly $41 billion, supporting 458,000 U.S. jobs.[2] Also as of 2018, immigrants who came to the United States as international students had founded 21 startup companies that are still held privately but with capitalization over $1 billion; more than half of such successful startups have been founded by immigrants.[3] More than one-third of doctoral degrees in the sciences and engineering from U.S. institutions are earned by temporary visa holders, and the fraction is even higher for physics, having been above 40 percent consistently since 1998.[4] Roughly one-third of all

---

[2] National Association of Foreign Student Advisors, 2020, *Losing Talent 2020: An Economic and Foreign Policy Risk America Can't Ignore*, Washington, DC, https://www.nafsa.org/sites/default/files/media/document/nafsa-losing-talent.pdf.

[3] National Center for Science and Engineering Statistics, 2019, "Doctorate Recipients from U.S. Universities: 2019," NSF 21-308, National Science Foundation, Alexandria, VA, https://ncses.nsf.gov/pubs/nsf21308.

[4] From data collected by the American Physical Society and the Integrated Postsecondary Education Data System (IPEDS) of the National Center for Education Statistics (NCES), *Physics Degrees Earned by Temporary Residents*, https://www.aps.org/programs/education/statistics/temp-residents.cfm.

Nobel Prize winners in Physics who were affiliated with U.S. institutions at the time of the award were immigrants.

Finding: Science in the United States has long benefited from the influx of talented students and scientists from elsewhere in the world.

Finding: International students have made substantial contributions to the economy of the United States.

The pipeline of talented students from around the world who flock to the United States to study science now is under threat. The United States began restricting visas in 2018, and in September 2020, the Department of Homeland Security issued a Notice of Proposed Rulemaking to change F, J, and I visa aliens from "duration of status" to a fixed period of 4 years, not long enough to complete most U.S. PhD programs.[5] Thanks in part to pressure from the scientific community, this last proposal now has been withdrawn. Nonetheless, international student enrollment has suffered. The Department of State funds the Open Doors report annually to provide information relating to foreign students and scholars in the United States and U.S. citizens studying abroad.[6] The 2020 Open Doors report shows that international student enrollment dropped for the fourth consecutive year, ending 50 years of nearly exponential growth. National Center for Science and Engineering Statistics (NCSES) data show that international graduate student enrollment dropped starting in 2017 after several years of increases. A survey of 49 of the largest physics PhD programs at U.S. institutions found a nearly 12 percent decline in international applications between 2017 and 2018.

The American Physical Society (APS) conducted a survey of international students who were accepted into U.S. physics graduate programs but declined these offers:[7] 32 percent expressed a belief that the United States is "unwelcoming to foreigners," 21 percent said they have better educational opportunities outside the United States, and 20 percent said they have better long-term employment opportunities outside the United States. A more extensive survey, also from the APS,

---

[5] Department of Homeland Security, 2020, "Establishing a Fixed Time Period of Admission and an Extension of Stay Procedure for Nonimmigrant Academic Students, Exchange Visitors, and Representatives of Foreign Information Media," *Federal Register* 85(187):60526–60598, https://www.federalregister.gov/documents/2020/09/25/2020-20845/establishing-a-fixed-time-period-of-admission-and-an-extension-of-stay-procedure-for-nonimmigrant.

[6] J. Baer and M. Martel, 2020, Fall 2020 *International Student Enrollment Snapshot*, Department of State and Institute of International Education, Washington, DC, https://opendoorsdata.org/research-briefs/fall2020snapshot.

[7] T.W. Johnson, 2019, "The United States Is Losing the Ability to Attract International Students Due to Visa Obstacles," *APS News* 28:11.

discovered that at least 40 percent of international early career scientists who chose to come to the United States to study and/or work believe that the U.S. government's current response to research security concerns makes their decision to stay in the United States long term less likely or much less likely.[8] These perceptions and the steep decline in the pool of international applicants put the U.S. physics enterprise at high risk of no longer attracting the top students in the world.

Finding: Applications to U.S. physics graduate programs from international students have decreased since 2016.

Finding: Many international students find the United States unwelcoming and feel that they have better opportunities outside the United States.

Beyond attracting students, scientific exchange has been an instrument of diplomacy, even in challenging times. Today, however, concerns over economic security relating to systematic capture of intellectual property, trade secrets, and advanced technologies by China have given rise to a number of policies and practices that may significantly restrict international collaboration. In particular, the Department of Justice launched a "China Initiative," and in February 2020, the Federal Bureau of Investigation confirmed that it was conducting 1,000 active investigations involving allegations of intellectual property theft by China.[9] A number of researchers in the United States have been accused of conflicts of interest as a consequence of their Chinese collaborations, and some have been removed from their academic posts. The vast majority of scientists targeted in this way are of Asian ethnicity.[10] A recent database established by the *MIT Technology Review*[11] shows that the China Initiative has "strayed from economic espionage and hacking cases to 'research integrity' issues, such as failures to fully disclose foreign affiliations on forms." Their reporting and analysis further showed that "the climate of fear created by the prosecutions has already pushed some talented scientists to leave the United States and made it more difficult for others to enter or

---

[8] American Physical Society, 2021, *Research Security Policies & Their Impacts: Key Results of APS Member Survey*, APS Government Affairs, https://www.aps.org/policy/analysis/upload/APS-Research-Security-Survey-Key-Findings-2021.pdf.

[9] A. Silver, 2020, "Scientists in China Say US Government Crackdown Is Harming Collaborations," *Nature* 583:341–342, https://www.nature.com/articles/d41586-020-02015-y.

[10] J. Mervis, 2020, "'Has It Peaked? I Don't Know.' NIH Official Details Foreign Influence Probe," *Science*, June 22, https://www.science.org/content/article/has-it-peaked-i-don-t-know-nih-official-details-foreign-influence-probe.

[11] E. Guo, J. Aloe, and K. Hao, 2021, "The US Crackdown on Chinese Economic Espionage Is a Mess. We Have the Data to Show It," *MIT Technology Review*, December 2, https://www.technologyreview.com/2021/12/02/1040656/china-initative-us-justice-department.

stay, endangering America's ability to attract new talent in science and technology from China and around the world." In the same spirit, a survey by the University of Michigan's Association of Chinese Professors found that nearly two-thirds of respondents "don't feel safe" in their academic positions, while a national survey found that academic scientists of Chinese heritage are four times more likely to fear government intrusions than their non-Chinese colleagues.[12] There is a danger that normal components of academic interaction and scientific collaboration are being criminalized.[13] These concerns are reinforced in a recent analysis by the leadership of the American Physical Society.[14]

> **Finding:** Discussions of U.S. policy toward international students and scientists are being driven by concerns about national and economic security.

In response to these developments, NSF commissioned a study by the JASON group to explore the value of openness and the concerns about security.[15] While finding some basis for increased concern, the report emphasizes that the scale of the problem remains poorly defined and cautions against over-reaction. Many of the issues involving the conduct of U.S.-based researchers in their interactions with foreign governments and institutions could be addressed through broader and clearer disclosure requirements. Openness in scientific exchange needs to be supported with openness about individual commitments and potential conflicts, and education about the underlying ethical considerations.

The U.S. government has a well-established structure for restricting the flow of information—classification. But the threshold for classification is high, to preserve the benefits of scientific exchange even on matters that have the potential to impact national security. There are mechanisms by which universities and research institutions handle "controlled unclassified information," such as preserving patient privacy in medical research, but the JASON report cautions against expanding these to create new boundaries around fundamental research activities. In response to the rhetoric that surrounds these issues, it is essential to insist on compelling evidence, rather than anecdotes, in guiding policy.

---

[12] J. Mervis, 2021, "U.S. Academics of Chinese Descent Organize and Speak Out—with Caution," *Science*, November 2, https://www.science.org/content/article/u-s-academics-chinese-descent-organize-and-speak-out-caution.

[13] G. Wilce, 2021, "Crackdown on Spying Damages US Science, Says Chinese–Born Physicist," *Physics* 14(63), https://physics.aps.org/articles/v14/63.

[14] P.H. Bucksbaum, S.J. Gates, Jr., R. Rosner, F. Hellman, J. Hollenhorst, B. Balantekin, and J. Bagger, 2021, "Current US Policy on China: The Risk to Open Science," *APS News*, August 9, https://www.aps.org/publications/apsnews/updates/china-risk.cfm.

[15] JASON, 2019, *Fundamental Research Security*, JSR-19-2I, The MITRE Corporation, McLean, VA, http://www.nsf.gov/JASON_Security_Report.

Conclusion: The open exchange of people and ideas is critical to the health of biological physics, physics, and the scientific enterprise generally. This exchange has enormous economic and security benefits.

General Recommendation: All branches of the U.S. government should support the open exchange of people and ideas. The scientific community should support this openness by maintaining the highest ethical standards.

As this report was in its final review, plans were announced to end the China Initiative. As is clear from the discussion and recommendation above, the Committee views this as a step in the right direction. It is important to emphasize, however, that this one step does not solve the general problem, and that open exchange of people and ideas needs to be supported more broadly.

An important step in implementing this recommendation would be to reaffirm Presidential Directive NSDD189,[16] which states that fundamental research is defined as research that is meant to be published in the open literature and that the products of fundamental research remain unrestricted to the maximum extent possible; if control of particular fundamental research is required for national security, the mechanism is classification. To deal with concerns about potential conflicts of interest arising from international collaboration, the federal agencies need to adopt clear and uniform rules regarding disclosure, rules which academic and research institutions can apply effectively to their faculty and staff without restricting their intellectual freedom. Regarding visa policy, making the F-1 visa dual intent would provide a pathway for international students to stay in the United States after receiving their degrees, thus capturing for the U.S. economy some of our investment in their education and accumulated expertise. These observations echo and extend those articulated in the recent decadal survey of atomic, molecular, and optical physics.[17] A 2020 report on safeguarding the bioeconomy, which is more focused on security issues and places its discussion in a larger economic context, nonetheless comes to similar conclusions and recommendations.[18]

Although concerns about engagement with China may seem tied to the current political situation, it is worth remembering that China was long an exception to general U.S. immigration policy. The Chinese Exclusion Act was passed in 1882,

---

[16] National Security Decision Directives (NSDD) Reagan Administration, 1985, *National Policy on the Transfer of Scientific, Technical and Engineering Information*, NSDD 189, Federation of American Scientists and Intelligence Resource Program, https://fas.org/irp/offdocs/nsdd/nsdd-189.htm.

[17] National Academies of Sciences, Engineering, and Medicine, 2020, *Manipulating Quantum Systems: An Assessment of Atomic, Molecular, and Optical Physics in the United States*, The National Academies Press, Washington, DC.

[18] See, especially Recommendation 4-1, p. 354 in National Academies of Sciences, Engineering, and Medicine, 2020, *Safeguarding the Bioeconomy*, The National Academies Press, Washington, DC.

and its provisions were not fully repealed until the Immigration and Nationality Act of 1965. The modern experience of Chinese students applying to U.S. PhD programs in physics began only in 1979, with the launch of CUSPEA (China-U.S. Physics Examination and Application). More generally, we know that our openness to scientific talent is not equally distributed across the globe. The American Physical Society has several programs to foster collaboration between U.S. physicists and those in the developing world.[19] For 30 years the Abdus Salam International Centre for Theoretical Physics (ICTP) has hosted a Diploma Program designed to help students who take their first degrees in the developing world make the transition to PhD programs in the United States or Europe, and related programs are growing at ICTP partner institutes in Brazil, China, Mexico, and Rwanda. Cooperative agreements with these institutes, in the spirit of the original CUSPEA agreement, provide a path for selection of exceptional students from a much broader range of backgrounds than might otherwise find their way to PhD programs in the developed world.[20]

While it is essential that the United States reaffirm its openness to international scientific exchange through changes in both rhetoric and policy, restoring the nation's privileged status in attracting PhD students from around the world will require special attention. Students need to feel welcome and, concretely, they need to be supported. Scientific exchange needs to be viewed once again as a positive instrument of diplomacy rather than as a danger. As explained in Chapter 9, there often is a mismatch between the existing funding structures and the needs of international students, leading to an effective bias against these students in the admissions process. This creates an opportunity.

Specific Recommendation: Federal agencies and private foundations should establish programs for the support of international students in U.S. PhD programs, in biological physics and more generally.

RACE AND ETHNICITY

Restrictions on international students damage U.S. science by restricting the pool of talent entering the field. The dramatic inequalities faced by racial and ethnic minorities that persist within our society also diminish the pool of talent. Throughout this period, the core funding for pre-college education has been the responsibility of local governments. This has allowed dramatic differences in the

---

[19] American Physical Society, 2022, "Programs, International Affairs, Developing Countries," https://www.aps.org/programs/international/programs/index.cfm.

[20] An example is the joint MSc program between the South American Institute for Fundamental Research (ICTP-SAIFR) in Brazil and the Perimeter Institute in Canada.

support for schools in districts serving different socio-economic groups, exacerbating other racial and ethnic disparities in our society. Although the U.S. Supreme Court issued the landmark *Brown v. Board of Education* decision in 1954, abolishing racial segregation in public schools, the principles of equal financial support for education of students from all backgrounds and communities continue to be litigated on a state-by-state basis, even today. Inequalities of educational opportunity have been reinforced by racially discriminatory economic practices, even in policy initiatives that, on average, raised the standard of living for poorer Americans.[21]

In the same way that the progress of U.S. science has been entwined with immigration policy, changing attitudes and policies in the treatment of different ethnic groups also have had impact. Over the course of the 20th century, Chinese, Japanese, and Jewish families in the United States all faced explicit discrimination—in university admissions, in housing, in employment, and more. As treatment of these groups became more just, the children of these families took their places among our nation's scientific leadership. As just one measure of this, nearly half of the individuals who received the Nobel Prize in Physics while affiliated with U.S. institutions come from these once persecuted groups. This is a reminder that opening access to a more diverse community can unleash talent on a grand scale. The search for justice is aligned with the search for scientific excellence.

Discrimination based on race and ethnicity has a long history. The scientific community's clearest window into this legacy today is the dramatic underrepresentation of Black and Latinx Americans in our community.[22] Underrepresentation is not by itself an indicator of injustice, but the case for systemic racism in the United States does not depend on the fraction of physics degrees awarded to different groups. The challenge for our community is to welcome, support, and nurture talented young people from around the world and from U.S. citizens of all ethnic groups. A recent report from the American Institute of Physics reviews the ways in which the community falls short of this goal with respect to African Americans.[23]

Inequalities of opportunity have an especially large impact on physics education. Serious engagement with physics requires some level of mathematical maturity, and students from underresourced high schools have less access to the

---

[21] See, for example, I. Katznelson, 2005, *When Affirmative Action Was White: An Untold History of Racial Inequality in Twentieth Century America*, W.W. Norton, New York.

[22] From data (2014–2018) collected by the American Physical Society, the U.S. Census Bureau, and the Integrated Postsecondary Education Data System (IPEDS) of the National Center for Education Statistics (NCES), https://www.aps.org/programs/minorities/resources/statistics.cfm.

[23] M. James, E. Bertschinger, B. Beckford, T. Dobbins, S. Fries-Britt, S.J. Gates, M. Ong, et al., 2019, *The Time Is Now. Systemic Changes to Increase African Americans with Bachelor's Degrees in Physics and Astronomy*, https://www.aip.org/diversity-initiatives/team-up-task-force.

teachers who can nurture this maturity:[24] students in schools with large minority populations have only a 50 percent chance of being taught by math and science teachers who are fully qualified for their assignments. This is not a failing of the teachers, but of the school administrators, who preferentially assign less qualified teachers to minority-serving schools, as well reflecting a lack of resources needed to attract highly qualified teachers into minority-serving districts. These differences in resources at the high school level continue when we look at minority-serving undergraduate institutions.

> **Finding:** Physics education is layered, with one layer building strongly on the one below. Inequality of access or resources is compounded.

> **Finding:** Recent data indicate that while the number of Black students earning physics bachelor's degrees is growing, the percentage has not increased.

> **Finding:** Historically Black Colleges and Universities have played a crucial role in the scientific and professional education of Black Americans.

> **Finding:** The total number of physics bachelor's degrees awarded by Historically Black Colleges and Universities has shrunk.

> **Conclusion:** Inequalities of educational opportunity continue to limit the accessibility of physics education for Black students.

Although the experience of each group is unique, one can find related problems for all of the underrepresented ethnic groups in our community. Parallel to the role of HBCUs for Black students are the broader collection of Minority Serving Institutions (MSIs) and Tribal Colleges and Universities (TCUs).[25]

There is a strong connection between our specific concerns regarding the education of underrepresented groups and our general concerns about the lack of proper support for core undergraduate education as part of scientific workforce development. As described in Chapter 8, it is not reasonable to expect that problems in this core can be solved by supplemental programs alone.

---

[24] L. Darling-Hammond, 2001, Inequality in teaching and schooling: How opportunity is rationed to students of color in America, pp. 208–233 in *The Right Thing to Do, the Smart Thing to Do Enhancing Diversity in the Health Professions—Summary of the Symposium on Diversity in Health Professions in Honor of Herbert W. Nickens, M.D.*, B.D. Smedley, A.Y. Stith, L. Colburn, and C.H. Evans, eds., National Academy Press, Washington, DC.

[25] For details see Office of Civil Rights, "Minority Serving Institutions Program," Department of the Interior, https://www.doi.gov/pmb/eeo/doi-minority-serving-institutions-program.

General Recommendation: Federal agencies should make new resources available to support core undergraduate physics education for underrepresented and historically excluded groups, and the integration of research into their education.

This recommendation goes beyond the boundaries of biological physics, but as we have emphasized in Chapter 8 the educational issues in our field are intertwined with those of physics more generally. While there are a variety of programs that support research experiences for undergraduates, especially those from historically excluded groups, this recommendation emphasizes the need to integrate teaching and research. We should not assume that a lack of resources for core educational programs, especially at minority serving institutions, can be compensated by pulling students away from their home institutions for relatively short visits to wealthier research intensive environments. Thus, while there are successful examples of partnerships between HBCUs or MSIs and major research universities, including programs organized and supported by the APS and the Biophysical Society, this recommendation emphasizes the need to support core educational programs.

Finally, there is ample evidence that the lived experience of students from minority groups is very different from that of the majority. It is essential that new programs be grounded in this experience.

Specific Recommendation: Recognizing the historical impact of Historically Black Colleges and Universities, Minority Serving Institutions, and Tribal Colleges and Universities, faculty from these institutions should play a central role in shaping and implementing new federal programs aimed at recruiting and retaining students from underrepresented and historically excluded groups.

GENDER

Race is not the only axis along which societal prejudices influence participation in the scientific community. It is well known that women continue to be underrepresented in the sciences, and this gap is particularly large in physics.[26]

Finding: The fraction of women who take a high school physics course is almost equal to the fraction of men, but women comprise only ~25 percent of students in the most advanced high school courses.

---

[26] A.M. Porter and R. Ivie, 2019, *Women in Physics and Astronomy*, American Institute of Physics, https://www.aip.org/statistics/reports/women-physics-and-astronomy-2019.

Finding: After steady growth for a generation, the fraction of bachelor's degrees in physics earned by women plateaued in 2007 at ~20 percent. The fraction of PhDs in physics earned by women has continued to grow, now matching the fraction of bachelor's degrees.

In contrast, women account for roughly 40 percent of new PhDs awarded in astronomy and chemistry and more than 50 percent of PhDs awarded in biology. But even within physics, there are substantial variations. Notably, female physics students are twice as likely as their male colleagues to do their thesis research in either astrophysics or biological physics. It would be useful to understand the origins of these differences; anecdotally the difference is more about the attitudes that women encounter upon entering these fields than about the subject matter,[27] but this deserves further study.

Although the problem of gender in the sciences has many dimensions, data on high school, bachelor's, and doctoral programs focuses our attention on the experience of women in courses near the end of high school and the beginning of college. Recent work surveying the performance of more than 10,000 students over a decade shows that the women who enroll in calculus-based introductory physics courses received final grades that are statistically indistinguishable from their male counterparts, but their self-assessments are significantly lower.[28] In one study, the women's low self-assessment was (perhaps surprisingly) not coupled with a perception that they were less included in the course or the student community. This example emphasizes the subtlety of the problems.

Specific Recommendation: In implementing this report's recommendations on introductory undergraduate education and its integration with research, special attention should be paid to the experience of women students.

In exploring different approaches to building a more inclusive educational environment, it is important that the physics community has access to significant, specialized resources. In 2005 the APS launched *Physical Review Physics Education Research*, alongside the other *Physical Review* journals that address different subfields of physics.[29] Among other topics, this journal publishes a steady stream of papers on gender in physics education.

---

[27] T. Feder, 2021, "Why Does Biophysics Attract a Disproportionate Number of Women?," *Physics Today*, https://physicstoday.scitation.org/do/10.1063/PT.6.5.20210607a.

[28] M. Dew, J. Perry, L. Ford, W. Bassichis, and T. Erukhimova, 2021, "Gendered Performance Differences in Introductory Physics: A Study from a Large Land-Grant University," *Physical Review Physics Education Research* 17(1):010106, https://link.aps.org/doi/10.1103/PhysRevPhysEducRes.17.010106.

[29] Biological physics is grouped with statistical, nonlinear, and soft matter physics in *Physical Review E*.

Even more disturbing than the pattern of enrollment in physics courses are reports of sexual harassment. In surveys of undergraduate women in physics, 75 percent report that they have experienced some form of harassment.[30] This is intolerable. Throughout the scientific community, and academia more generally, there is movement toward more concrete policies creating accountability, such as channels for anonymous reporting, and measures for enforcement, for these unacceptable behaviors. These are understood procedures that are well established elsewhere that need to continue and be supported vigorously by the community.

TOWARD BROADER ENGAGEMENT

Finally, the committee's findings, conclusions, and recommendations regarding the human dimensions of science apply in large part to all areas of physics, and in many cases to the scientific community more generally. There is a sense, however, that biological physics has a special role to play in welcoming a broader community.

Conclusion: The biological physics community has a special opportunity to reach broader audiences, leveraging human fascination with the living world to create entrance points to physics for a more diverse population of students and for the general public.

This impression—that the physics of life should provide a more accessible introduction to physics—is shared quite widely, and was repeated several times in the input from the community, but the committee is unaware of data that could make this claim precise. It seems best to end on the optimistic note that our community senses an opportunity to reach a broader audience, even if many details remain to be determined.

---

[30] L.M. Aycock, Z. Hazari, E. Brewe, K.B.H. Clancy, T. Hodapp, and R.M. Goertzen, 2019, "Sexual Harassment Reported by Undergraduate Female Physicists," *Physical Review Physics Education Research* 15(1):010121, https://link.aps.org/doi/10.1103/PhysRevPhysEducRes.15.010121.

Appendixes

A

Statement of Task

The committee will be charged with producing a comprehensive report on the status and future directions of physics of the living world. The committee's report shall:

1. Review the field of Biological Physics/Physics of Living Systems (BPPLS) to date, emphasize recent developments and accomplishments, and identify new opportunities and compelling unanswered scientific questions as well as any major scientific gaps. The focus will be on how the approaches and tools of physics can be used to advance understanding of crucial questions about living systems.
2. Use selected, non-prioritized examples from BPPLS as case studies of the impact this field has had on biology and biomedicine as well as on subfields of physical and engineering science (e.g., soft condensed matter physics, materials science, computer science). What opportunities and challenges arise from the inherently interdisciplinary nature of this interface?
3. Identify the impacts that BPPLS research is currently making and is anticipated to make in the near future to meet broader national needs and scientific initiatives.
4. Identify future educational, workforce, and societal needs for BPPLS. How should students at the undergraduate and graduate levels be educated to best prepare them for careers in this field and to enable both life and physical science students to take advantage of the advances produced by BPPLS?

The range of employment opportunities in this area, including academic and industry positions, will be surveyed generally.

5. Make recommendations on how the U.S. research enterprise might realize the full potential of BPPLS, specifically focusing on how funding agencies might overcome traditional boundaries to nurture this area. In carrying out its charge, the committee should consider issues such as the state of the BPPLS community and institutional and programmatic barriers.

B

Recommendations

The list below contains each recommendation made in the report, both general and specific.

EMERGENCE OF A NEW FIELD

General Recommendation: Realizing the promise of biological physics requires recognition that is distinct from, but synergistic with, related fields, both in physics and in biology. In colleges and universities it should have a home in physics departments, even as its intellectual agenda connects profoundly to efforts in many other departments across schools of science, engineering, and medicine. (Part I)

General Recommendation: Physics departments at research universities should have identifiable efforts in the physics of living systems, alongside groups in more traditional subfields of physics. (Part III, Chapter 8)

Specific Recommendation: The biological physics community should support exploration of the full range of questions being addressed in the field, and assert its identity as a distinct and coherent subfield embedded in the larger physics community. (Part I)

EDUCATING THE NEXT GENERATION

General Recommendation: All universities and colleges should integrate biological physics into the mainstream physics curriculum, at all levels. (Part III, Chapter 8)

Specific Recommendation: Physics courses and textbooks should illustrate major principles with examples from biological physics, in all courses from introductory to advanced levels. (Part III, Chapter 8)

Specific Recommendation: Physics faculty should modernize the presentation of statistical physics to undergraduates, find ways of moving at least parts of the subject earlier in the curriculum, and highlight connections to biological physics. (Part III, Chapter 8)

Specific Recommendation: Physics faculty should modernize undergraduate laboratory courses to include modules on light microscopy that emphasize recent developments, and highlight connections to biological physics. (Part III, Chapter 8)

General Recommendation: Physics faculty should organize biological physics coursework around general principles, and ensure that students specializing in biological physics receive a broad and deep general physics education. (Part III, Chapter 8)

General Recommendation: University and college administrators should allocate resources to physics departments as part of their growing educational and research initiatives in quantitative biology and biological engineering, acknowledging the central role of biological physics in these fields. (Part III, Chapter 8)

Specific Recommendation: Universities should provide and fund opportunities for undergraduate students to engage in biological physics research, as an integral part of their education, starting as soon as their first year. (Part III, Chapter 8)

Specific Recommendation: Funding agencies, such as the National Institutes of Health, the National Science Foundation, the Department of Energy, and the Department of Defense, as well as private foundations, should develop and expand programs to support integrated efforts in education and research at all levels, from beginning undergraduates to more senior scientists migrating across disciplinary boundaries. (Part III, Chapter 8)

SUPPORTING THE FIELD

General Recommendation: Funding agencies, including the National Institutes of Health, the National Science Foundation, the Department of Energy, and the Department of Defense, as well as private foundations, should develop and expand programs that match the breadth of biological physics as a coherent field. (Part III, Chapter 9)

Specific Recommendation: The federal government should provide the National Science Foundation with substantially more resources to fulfill its mission, allowing a much needed increase in the size of individual grant awards without compromising the breadth of its activities. (Part III, Chapter 9)

Specific Recommendation: The National Institutes of Health should form study sections devoted to biological physics, in its full breadth. (Part III, Chapter 9)

Specific Recommendation: Congress should expand the Department of Energy mission to partner with the National Institutes of Health and the National Science Foundation to construct and manage user facilities and infrastructure in order to advance the field of biological physics more broadly. (Part III, Chapter 9)

Specific Recommendation: The Department of Defense should support research in biological physics research that aims to discover broad principles that can be emulated in engineered systems of relevance to its mission. (Part III, Chapter 9)

Specific Recommendation: Industrial research laboratories should reinvest in biological physics, embracing their historic role in nurturing the field. (Part III, Chapter 9)

Specific Recommendation: Federal funding agencies should establish grant program(s) for the direct, institutional support of graduate education in biological physics. (Part III, Chapter 9)

Specific Recommendation: Federal agencies and private foundations should establish programs for the support of international students in U.S. PhD programs, in biological physics and more generally. (Part III, Chapter 9)

Specific Recommendation: Federal agencies and private foundations should develop funding programs that recognize and support theory as an independent activity in biological physics, as in other fields of physics. (Part III, Chapter 9)

General Recommendation: To maintain the flow of concepts and methods from biological physics into medicine and technology, the federal government should recommit to the vigorous support of basic science, including theory and the development of new technologies for experiments. (Part III, Chapter 9)

HUMAN DIMENSIONS OF SCIENCE

General Recommendation: All branches of the U.S. government should support the open exchange of people and ideas. The scientific community should support this openness by maintaining the highest ethical standards. (Part III, Chapter 10)

General Recommendation: Federal agencies should make new resources available to support core undergraduate physics education for underrepresented and historically excluded groups, and the integration of research into their education. (Part III, Chapter 10)

Specific Recommendation: Recognizing the historical impact of Historically Black Colleges and Universities, Minority Serving Institutions, and Tribal Colleges and Universities, faculty from these institutions should play a central role in shaping and implementing new federal programs aimed at recruiting and retaining students from underrepresented and historically excluded groups. (Part III, Chapter 10)

Specific Recommendation: In implementing this report's recommendations on introductory undergraduate education and its integration with research, special attention should be paid to the experience of women students. (Part III, Chapter 10)

C

Queries to Funding Agencies

In support of Chapter 9, the committee queried representative of the following funding agencies and foundations:

Agouron Institute
Burroughs Wellcome Fund
Department of Defense, Air Force Office of Scientific Research
Department of Defense, Army Research Office
Department of Defense, Defense Advanced Research Projects Agency
Department of Defense, Office of Naval Research
Department of Energy, Office of Science
Gordon and Betty Moore Foundation
Howard Hughes Medical Institute
National Institutes of Health
National Science Foundation, Biological Sciences Directorate
National Science Foundation, Physics Division
Simons Foundation

In addition, the committee made direct inquiries to staff at the following user facilities:

Advanced Light Source, Lawrence Berkeley National Laboratory (LBNL)
Advanced Photon Source, Argonne National Laboratory (ANL)
Center for Functional Nanomaterials, Brookhaven National Laboratory (BNL)

Center for Integrated Nanotechologies, Los Alamos National Laboratory (LANL)

Center for Nanophase Materials Sciences, Oak Ridge National Laboratory (ORNL)

Center for Nanoscale Materials, ANL

Cornell High Energy Synchrotron Source, Cornell University

Department of Energy High Performance Computing[1]

Environmental Molecular Sciences Laboratory, Pacific Northwest National Laboratory (PNNL)

Joint Genome Institute, LBNL

LINAC Coherent Light Source, SLAC National Accelerator Laboratory

Molecular Foundry, LBNL

National High Magnetic Field Laboratory, Florida State University

National Nanotechnology Coordinated Infrastructure

National Synchrotron Light Source II, BNL

New York Structural Biology Center

Pacific Northwest Center for Cryo-EM, PNNL

Spallation Neutron Source and High Flux Isotope Reactor, ORNL

Stanford-SLAC Cryo-EM Center, SLAC National Accelerator Laboratory

Stanford Synchrotron Radiation Laboratory, SLAC National Accelerator Laboratory

The main text of the committee's request is below. The first two sets of questions on funding and research training and large-scale activities and partnerships were sent to the agencies and foundations listed above, while the section on large-scale user facilities was sent to all of the listed user facilities.

As you may know, once every ten years the National Academy of Sciences is charged with assessing the state of physics research and education in the United States. This year, for the first time, the Decadal Survey includes an assessment of Biological Physics, or the Physics of Living Systems (PoLS). On behalf of the survey committee, I am writing to ask your help in collecting data from your agency/foundation.

As an agency/entity that supports PoLS research, the committee requests that [name of agency or foundation] provide important information that will allow us to address our charge. We kindly ask you to complete the following questionnaire to enable the committee to assess the state of funding in the field.

The committee understands that it is difficult to define the boundaries of our field. We have chosen to take a broad view, with the understanding that there are many subfields where interesting and exciting discoveries are being made. The committee only asks that

---

[1] Oversees the Argonne Leadership Computing Facility (ALCF), the National Energy Research Scientific Computing Center (NERSC), and the Oak Ridge Leadership Computing Facility (OLCF).

you outline the parameters you used when querying databases for funded researchers or project funding opportunities within PoLS.

We would very much appreciate it if you could provide information from your agency/entity to the committee, and we request that you send us this information by July 10, 2020. Please feel free to contact either me or the National Academies staff, by email or telephone if you have any questions. Please also let us know as soon as possible if your schedule constraints will not allow a response by the date above so that we can accommodate your responses in our program.

FUNDING OF RESEARCH AND TRAINING

In order to understand the impact of federal agency and foundation funding on PoLS research, the committee is seeking the answers to the following questions from your organization:

1. What are the various funding mechanisms or programs at your agency/foundation related to how the approaches and tools of physics can be used to advance understanding of crucial questions about living systems, or how biology inspires or generates new paradigms in physics?
2. What new directions or emerging areas has your agency/foundation identified for the establishment of new programs in PoLS research? What are the major gaps that these funding opportunities address?
3. What is the absolute number of dollars spent on PoLS research each year over the past decade at your agency/foundation?
4. What is the total number of research grants given to PoLS each year for the past 10 years? Within that total, what fraction is awarded to individual investigators (PIs), co-PIs, or multiple PIs, and what is the average funding per PI?
5. What is the distribution of grant size each year for the past 10 years?
6. What is the distribution of investigators who receive PoLS grants, with relation to their primary academic/institutional departments. If available, please provide this distribution as percentages.
7. Over the past decade, how many awards and what average dollar amount per award go to grantees each year as a function of the grantee's time past PhD. We can divide this into three time spans: 5 years after PhD, 10 years after PhD, and beyond 10 years.
8. Over the past decade, how many awards and what average dollar amount go to grantees each year as a function of gender, race, and ethnicity?
9. Over the past decade, how many awards and what average dollar amount go to training grants and fellowships in POLS each year and what fraction goes to PhD students versus postdoctoral fellows?

10. Please give examples of how the PoLS projects funded by your organization have impacted biology and biomedicine, as well as subfields of the physical and engineering sciences.

LARGE-SCALE ACTIVITIES AND PARTNERSHIPS

Because PoLS research is inherently cross-disciplinary, the committee is interested in understanding how funding is distributed between large-scale research and single-PI groups. To this end, we pose the following questions to the funding agencies.

1. If your agency supports centers or large-group efforts in PoLS science, what is the size of the awards per year for each center relative to the total budget spent on PoLS each year?
2. Which areas in PoLS research are part of these large efforts?
3. Briefly describe the extent (% of total funding, $ amount) to which your agency supports interdisciplinary large-scale activities and partnerships that include PoLS science.
4. If appropriate, please briefly describe the extent of industrial participation (% of total funding, $ amount) in large scale PoLS awards and how this has changed year to year over the past 10 years.

LARGE-SCALE USER FACILITIES

Many of the scientists engaged in PoLS research require use of the instrumentation and resources available at agency- and non-agency-funded large-scale user facilities. Examples include synchrotrons, nanoscience centers, and high-performance computing. In order to better understand the role user facilities play in PoLS research, the committee poses the following questions:

1. What types of resources or instrumentation are available to researchers at your user-facility (choose from the answers below, or specify):
   o Synchrotron
   o Nanoscience center
   o High-performance computing
   o Electron microscopy
   o Light microscopy
   o Mass spectrometry
   o Nuclear magnetic resonance (NMR)
   o DNA sequencing, synthesis, and/or modification
   o Other (please specify)

2. For scientists using the instruments and resources, what is the breakdown of their associated fields of study?
3. What mechanisms does your facility use to solicit community input for future funding decisions for large-scale user facilities?
4. What mechanisms do you have to solicit proposals for access to facilities from scientists representing new areas of research, and how are proposals that represent new areas of research supported through the review process?
5. What new technologies or user facility capabilities would be of interest to the PoLS research community?

Is there anything else the committee should consider in terms of the current and future funding landscape of PoLS research?

D

Agency Missions

NATIONAL SCIENCE FOUNDATION

"To promote the progress of science; to advance the national health, prosperity, and welfare; and to secure the national defense; and for other purposes."[1]

DEPARTMENT OF ENERGY

"The mission of the Energy Department is to ensure America's security and prosperity by addressing its energy, environmental and nuclear challenges through transformative science and technology solutions."[2]

NATIONAL INSTITUTES OF HEALTH

"NIH's mission is to seek fundamental knowledge about the nature and behavior of living systems and the application of that knowledge to enhance health, lengthen life, and reduce illness and disability."[3]

---

[1] National Science Foundation, *The NSF Statutory Mission*, https://www.nsf.gov/pubs/2014/nsf14002/pdf/02_mission_vision.pdf.

[2] Department of Energy, "Our Mission," https://www.energy.gov/about-us.

[3] National Institutes of Health, "Mission and Goals," https://www.nih.gov/about-nih/what-we-do/mission-goals.

DEPARTMENT OF DEFENSE

"Our mission is to provide the military forces needed to deter war and ensure our nation's security."[4]

---

[4] Department of Defense, "About," https://www.defense.gov/About.

E

Details Regarding National Science Foundation and National Institutes of Health Grants

To assess the current scope and structure of funding for biological physics in the United States, the committee gathered data from publicly accessible databases at the National Science Foundation (NSF) and the National Institutes of Health (NIH), and distributed a questionnaire to a wide range of centers, institutions, foundations, and other funding agencies and stakeholders (Appendix C). This effort generated valuable insights into the amount of funding flowing into our field, the types of projects and training efforts being supported, and the priorities of different funders. The headline funding over the decade 2010–2020 is shown, by agency, in Figure 9.1. This appendix focuses on details at NSF and NIH.

NIH STUDY SECTIONS

NIH awards to principal investigators with their primary affiliations in physics and biophysics departments were reviewed by 75 different standing scientific study sections, 237 different special emphasis panels, 3 National Cancer Institute subcommittees, and 5 training/career/workforce development subcommittees. The following is a list of the relevant standing scientific study sections:

- AIDS Molecular and Cellular Biology Study Section (AMCB)
- Anterior Eye Disease Study Section (AED)
- Arthritis and Musculoskeletal and Skin Diseases Special Grants Review Committee (AMS)

- Atherosclerosis and Inflammation of the Cardiovascular System Study Section (AICS)
- Auditory System Study Section (AUD)
- Basic Mechanisms of Cancer Therapeutics Study Section (BMCT)
- Basic Neuroscience of Aging Review Committee (NIA-N)
- Biochemistry and Biophysics of Membranes Study Section (BBM)
- Biodata Management and Analysis Study Section (BDMA)
- Bioengineering, Technology and Surgical Sciences Study Section (BTSS)
- Bioengineering of Neuroscience, Vision and Low Vision Technologies Study Section (BNVT)
- Biology of the Visual System Study Section (BVS)
- Biomaterials and Biointerfaces Study Section (BMBI)
- Biomedical Computing and Health Informatics Study Section (BCHI)
- Biomedical Imaging Technology A Study Section (BMIT-A)
- Biomedical Imaging Technology B Study Section (BMIT-B)
- Biomedical Imaging Technology Study Section (BMIT)
- Biomedical Library and Informatics Review Committee (BLR)
- Biophysics of Neural Systems Study Section (BPNS)
- Cancer Etiology Study Section (CE)
- Cancer Immunopathology and Immunotherapy Study Section (CII)
- Cell Death in Neurodegeneration Study Section (CDIN)
- Cellular, Molecular and Integrative Reproduction Study Section (CMIR)
- Cellular and Molecular Biology of Glia Study Section (CMBG)
- Cellular and Molecular Biology of Neurodegeneration Study Section (CMND)
- Cellular and Molecular Biology of the Kidney Study Section (CMBK)
- Cellular and Molecular Immunology—A Study Section (CMIA)
- Cellular and Molecular Technologies Study Section (CMT)
- Clinical Molecular Imaging and Probe Development (CMIP)
- Clinical Neuroscience and Neurodegeneration Study Section (CNN)
- Clinical Translational Imaging Science Study Section (CTIS)
- Cognitive Neuroscience Study Section (COG)
- Communication Disorders Review Committee (CDRC)
- Development—2 Study Section (DEV2)
- Developmental Biology Subcommittee (CHHD-C)
- Developmental Therapeutics Study Section (DT)
- Digestive Diseases and Nutrition C Subcommittee (DDK-C)
- Drug Discovery and Molecular Pharmacology Study Section (DMP)
- Enabling Bioanalytical and Imaging Technologies Study Section (EBIT)
- Gene and Drug Delivery Systems Study Section (GDD)
- Imaging Technology Development Study Section (ITD)

- Instrumentation and Systems Development Study Section (ISD)
- Intercellular Interactions Study Section (ICI)
- International and Cooperative Projects—1 Study Section (ICP1)
- Lung Cellular, Molecular, and Immunobiology Study Section (LCMI)
- Macromolecular Structure and Function A Study Section (MSFA)
- Macromolecular Structure and Function B Study Section (MSFB)
- Macromolecular Structure and Function C Study Section (MSFC)
- Macromolecular Structure and Function D Study Section (MSFD)
- Macromolecular Structure and Function E Study Section (MSFE)
- Medical Imaging Study Section (MEDI)
- Membrane Biology and Protein Processing Study Section (MBPP)
- Microbiology and Infectious Diseases B Subcommittee (MID)
- Microbiology and Infectious Diseases B Subcommittee (MID-B)
- Microscopic Imaging Study Section (MI)
- Minority Programs Review Subcommittee A (MPRC-A)
- Modeling and Analysis of Biological Systems Study Section (MABS)
- Molecular and Integrative Signal Transduction Study Section (MIST)
- Molecular Genetics A Study Section (MGA)
- Molecular Neuropharmacology and Signaling Study Section (MNPS)
- Nanotechnology Study Section (NANO)
- Neural Basis of Psychopathology, Addictions and Sleep Disorders Study Section (NPAS)
- NeuroAIDS and other End-Organ Diseases Study Section (NAED)
- Neurobiology of Learning and Memory Study Section (LAM)
- Neurodifferentiation, Plasticity, and Regeneration Study Section (NDPR)
- Neuroscience and Ophthalmic Imaging Technologies Study Section (NOIT)
- Neurotechnology Study Section (NT)
- Neurotransporters, Receptors, and Calcium Signaling Study Section (NTRC)
- NIDR Special Grants Review Committee (DSR)
- NST-2 Subcommittee (NST-2)
- Nuclear and Cytoplasmic Structure/Function and Dynamics Study Section (NCSD)
- Oral, Dental and Craniofacial Sciences Study Section (ODCS)
- Prokaryotic Cell and Molecular Biology Study Section (PCMB)
- Radiation Therapeutics and Biology Study Section (RTB)
- Sensorimotor Integration Study Section (SMI)
- Skeletal Biology Structure and Regeneration Study Section (SBSR)
- Synapses, Cytoskeleton and Trafficking Study Section (SYN)

NIH FUNDING MECHANISMS

NIH offers research support through a wide variety of different mechanisms, each with a defined "activity code." From 2010 through August 2020, 1,770 NIH awards received by recipients whose primary departmental affiliation was Physics or Biophysics came through 47 of these different mechanisms. These are listed here, in larger categories, with number of awards in parentheses for individual programs. Note that in keeping with NIH reporting practices each award represents a single funding year such that, for example, a 5-year grant is counted as five awards.

Innovator and Pioneer Awards: 22 awards, $26,777,038

- DP1: Pioneer Award (15)
- DP2: New Innovator Award (7)

Research Training and Fellowships: 147 awards, $17,879,386

- F30: Kirchstein Predoctoral award for dual degree (5)
- F31: Kirchstein Predoctoral award (43)
- F32: Kirchstein Postdoctoral award (30)
- F99: Predoctoral to Postdoctoral Transition award (2)
- R25: Education Project Grant (35)
- R36: Dissertation Award (3)
- T32: Kirchstein Institutional Pre/Postdoctoral Training award (26)
- T34: Kirchstein Institutional Undergraduate Training award (3)

Career Development Awards: 105 awards, $14,995,424

- DP5: Early Independence Award (1)
- K01: Mentored Research Scientist Career Development Award (14)
- K08: Mentored Clinical Scientist Research Career Development Award (1)
- K22: Career Transition Award (22)
- K25: Mentored Quantitative Research Career Development Award (21)
- K99: Pathway to Independence Award, phase 1 (28)
- R00: Pathway to Independence Award, phase 2 (18)

Program Project/Center Awards: 77 awards, $87,485,518

- P01: Program Project (10)
- P30: Center Core Grant (3)

- P41: Biotechnology Resource Grants (18)
- RF1: Multi-Year Funded Research Project Grant (1)
- RL5: Linked Education Project (linked to U54) (7)
- U19: Multi-Project Cooperative Agreements (with NIH Intramural) (8)
- U54: Specialized Center-Cooperative Agreements (11)
- UF1: Multi-Year Funded Research Cooperative Agreement (3)
- UG3: Exploratory/Developmental Cooperative Agreement Phase I, large budget (4)
- UH2: Exploratory/Developmental Cooperative Agreement Phase I (1)
- UH3: Exploratory/Developmental Cooperative Agreement Phase II (4)
- UL1: Linked Specialized Center Cooperative Agreement (linked to U54) (7)

Research Grants: 1,293 awards, $438,145,936

- R01: Research Project Grant (950)
- R03: Small Grant Program (14)
- R15: Academic Research Enhancement Awards (48)
- R21: Exploratory Developmental Research Grant (157)
- R33: Second Phase of Exploratory Developmental Research Grant (4)
- R35: Maximizing Investigator's Research Award (MIRA) (45)
- R37: Method to Extend Research in Time (MERIT) Award (29)
- R56: High Priority, Short-Term Project Award (4)
- R61: First Phase of Exploratory Developmental Research Grant (2)
- RC1: Challenge Grant (5)
- RC2: High Impact Interdisciplinary Science (1)
- U01: Cooperative Agreement (with NIH Intramural) Award (34)

Conference Grants: 5 awards, $70,100

- R13: NIH Support for Conferences and Scientific Meetings (5)

Resource and Shared Instrumentation Grants: 13 awards, $7,449,944

- S10: Shared Instrumentation Grant (12)
- U24: Resource-Related Research Cooperative Agreement (1)

Diversity/Capacity Building Grants: 82 awards, $21,129,177

- SC1: Support of Competitive Research Award (SCORE), undergrad (18)
- SC2: Support of Competitive Research Award (SCORE), grad (21)
- SC3: Support of Competitive Research Award (SCORE), post-doc (36)
- TL4: Kirchstein Undergraduate Training award for URM Institutions (7)

NSF FUNDING LEVELS

To survey the state of funding for biological physics at NSF, the committee used the publicly accessible (advanced) award search tool.[1] Selecting for the Physics of Living Systems program and start dates after January 1, 2010, produces a list of 416 awards, many of which are co-funded with other programs. Of these, 60 are in support of conferences, workshops, and summer schools, and 16 are single grants in support of centers or large research networks. This leaves 340 awards to individual investigators or small groups.

As an aside, NSF also has a "collaborative research" mechanism, which involves making multiple parallel grants to individual investigators working together, sometimes bound only by a loose theme. This is in contrast to center grants, which involve one large award to a group. The committee concluded that individual components of the collaborative research grants, which are reported as distinct awards, are meaningfully grouped with the individual investigator awards.

As noted in the main text, NSF attaches all award funding to the initial award date, even as it accumulates in increments. For the 224 awards with end dates before December 31, 2020, however, all of the funds have been awarded and one can annualize the award amount. Figure E.1 shows the cumulative distribution of these annualized awards. Although there is a small tail of large awards, the mean annual award is just over $150,000 per year and the median is $122,600 per year. One can also make a distribution over award years, which would be more comparable to how NIH reports its data, and this is shown in Figure 9.6 of the main text.

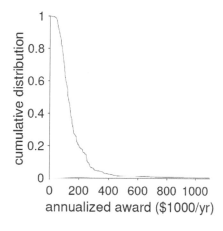

FIGURE E.1 Cumulative distribution of annualized awards to individual investigators and small groups through the Physics of Living Systems program at the National Science Foundation.

---

[1] National Science Foundation, "Awards Advanced Search," https://www.nsf.gov/awardsearch/advancedSearch.jsp.

F

Minimal Support Levels

The goal of this appendix is to estimate the minimum level of support needed for individual investigators, and more globally to maintain the flow of young people into the field of biological physics. The analysis starts with the basic ingredients and then addresses how these come together to define minimal effective grant sizes for individual investigators and a minimum level of support for the field as a whole. All estimates are based on publicly available data, as indicated.

INGREDIENTS

Faculty. At all but a handful of U.S. institutions, faculty receive a salary only for the academic year. Being active in research, and mentoring students and postdoctoral fellows, means collecting a summer salary from research grants. Faculty compensation varies widely, but the American Association of University Professors (AAUP) surveys salaries and parcels out the results by institutional categories.[1] What is most relevant to this discussion are the 227 doctoral (AAUP Category I) institutions; analysis is focused on the professorial ranks, since lecturers and instructors seldom serve as PhD advisers. A summary of these data can be found in Table F.1. Support for faculty includes fringe benefits, which again vary widely across institutions. The AAUP reports that Category I institutions' spending on retirement and

---

[1] See Tables 1 and 7 in American Association of University Professors, 2020, *The Annual Report on the Economic Status of the Profession*, 2019-20, https://www.aaup.org/2019-20-faculty-compensation-survey-results.

medical benefits is an average of 21.8 percent on top of salaries,[2] and this provides an estimate of fringe benefit costs to grants. The result is that 2 months of summer salary for an average faculty member is a direct cost of $34,100 per year. This likely is an underestimate, since science faculty have above-average salaries at most institutions, and because research in biological physics is not distributed uniformly across the Category I institutions.

Students. Supporting a PhD student means paying a modest salary (stipend) so that the student can be free to focus on their research, and compensating their host institution for the cost of the student's education. Making estimates here is especially challenging because stipends and tuition vary widely across institutions. A benchmark is provided by the National Science Foundation (NSF) Graduate Research Fellowship Program,[3] which sets a stipend of $34,000 per year and a "cost of education allowance" of $12,000 per year.

Travel. Doing research means engaging with the community. Although there are many ways to accomplish this, one important component is attendance at one scientific conference per year. For concreteness this is assumed to be the American Physical Society (APS) March Meeting. This cost has three components: (1) lodging, meals, and incidental expenses for 5 full days, taken from the per diem rates set by the General Services Administration[4] and averaged over the locations of the March Meeting in the decade 2011–2020; (2) meeting registration fees for students and faculty, taken from the APS (pre-pandemic); (3) travel, set very roughly at $500 per person round trip. A summary is in Table F.2.

TABLE F.1 Average Academic Year Salaries at PhD-Granting Institutions, by Rank

Rank	Mean Academic Year Salary	Fraction
Professor	$160,080	0.41
Associate Professor	$104,408	0.31
Assistant Professor	$90,764	0.28

SOURCE: Data from the AAUP faculty compensation survey.

---

[2] See Tables 8 and 9 in American Association of University Professors, 2020, *The Annual Report on the Economic Status of the Profession*, 2019-20, https://www.aaup.org/2019-20-faculty-compensation-survey-results.

[3] National Science Foundation, "Graduate Research Fellowship," https://www.nsfgrfp.org.

[4] General Services Administration, "Per Diem Rates," https://www.gsa.gov/travel/plan-book/per-diem-rates.

TABLE F.2 Costs of Attending American Physical Society March Meeting

Expense	Cost
Full member registration	$495
Early career registration	$300
Student registration	$195
5 days lodging	$903
Meals and incidentals	$340
Travel	$500

SOURCE: Registration fees from APS; lodging, meals and incidentals from GSA per diem, 5 days averaged over meeting locations 2011–2020.

Indirect costs. Grants in support of research provide support for facilities and administrative costs, also known as indirect costs or "overhead." This is added as a percentage of (qualifying) direct costs to cover infrastructure and other costs in the background of the research, and the percentage is negotiated by each institution with the federal government. These rates are available publicly by searching for "F&A rates" with the university's name. The committee assessed these rates for institutions listed in the top 100 doctoral programs in physics, focusing on those with identifiable biological physics groups on their departmental websites; results are summarized in Figure F.1. The average indirect cost rate is 56.1 percent.

Postdoctoral fellows. In physics, half of PhD students go on to postdoctoral research positions, and in 2015–2016 their median starting salary was $50,000 per year at universities and $65,000 per year at government laboratories; the split between these group is 75/25.[5] Since 2015–2016 there has been 10.8 percent inflation,[6] so an estimate of current median postdoctoral salaries is $60,000. As with faculty, fringe benefits are added on top of salaries; the same 21.8 percent is taken as an estimate.

A MINIMAL GRANT

Table F.3 is the budget for a minimal effective grant: 2 months of summer salary for a faculty member, stipend and cost of education for one student, travel to attend the APS meeting, and indirect costs. This is a very conservative estimate; notably, this analysis does not include any costs of actually carrying out the research, which for experimental efforts can be substantial, nor does it include the costs of

---

[5] P. Mulvey and J. Pold, 2019, *Physics Doctorates: One Year After Degree*, American Institute of Physics, https://www.aip.org/statistics/reports/physics-doctorates-one-year-after-degree-2016.

[6] See the "US Inflation Calculator," with data from the Bureau of Labor Statistics, https://www.usinflationcalculator.com.

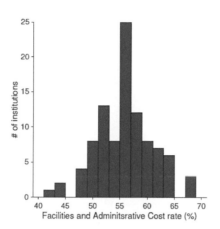

FIGURE F.1 Indirect costs at 97 PhD-granting institutions with biological physics groups identifiable from the physics department website.

infrastructure, which might range from computing to specially built experimental apparatus. Obviously many research projects also need the intellectual effort of more than just one student.

Despite this conservatism, this minimal grant ($123,736 per year) is essentially equal to the median annual award from NSF ($122,600 per year, see Appendix E). If one replaces the average faculty member with an assistant professor, the total drops slightly but remains above the level of the CAREER award.

SUPPORTING THE FIELD AS A WHOLE

As explained in Chapter 8, there has been tremendous growth in the number of physics students who do their thesis research on problems in biological physics; this number now is more than 150 per year. The goal of this analysis is to provide an estimate of the minimum support required to nourish this flow of young people into the field.

To begin, physics PhD programs typically take 6 years to complete, divided roughly into 2 years of coursework and 3 to 4 years of focused research activity. Producing 150 new PhDs each year thus means supporting ~500 students in steady state. As noted above, half of the PhD students in physics go on to postdoctoral research positions. While postdoctoral positions once were 2 or 3 years in duration, 4 to 6 years is now more common. This lengthening of postdoctoral periods has been led by the biomedical research groups, and many young biological physicists are competing with people from these groups as they look for their next positions, so this creates pressure on the community. Assuming that postdoctoral terms average 5 years, and with 150/2 = 75 new people each year, this gives a steady state

TABLE F.3 A Minimal Research Grant Budget, Supporting Faculty Summer Salary and One PhD Student

Expense	Cost
Faculty summer salary (2 months)	$27,425
Fringe benefits at 21.7%	$5,979
PhD student stipend	$34,000
Travel and attendance at APS meeting	$4,176
Indirect costs at 56.1%	$40,156
Cost of education	$12,000
Total per year	$123,736

TABLE F.4 Components of a Minimal Annual Budget to Support the Entire Biological Physics Community, Focusing on the Flow of Young People into the Field

Category	Unit Cost	Number	Total
PhD student stipend	$34,000	500	$17,000,000
Travel and attendance at APS meeting	$1,938	500	$969,000
Indirect costs at 8% (training grant)	$2,875	500	$1,437,520
Indirect costs at 56.1% (research grant)	$20,161	500	$10,080,500
Cost of education	$12,000	500	$6,000,000
Total student support via training grants	$53,230	500	$25,406,520
Total student support via research grants	$71,593	500	$34,049,500
Postdoctoral salary	$60,000	375	$22,500,000
Fringe benefits at 21.7%	$13,020	375	$4,882,500
Travel and attendance at APS meeting	$2,043	375	$766,125
Indirect costs at 56.1% (research grant)	$42,110	375	$15,791,250
Total postdoctoral support via research grants	$117,173	375	$43,939,875
Faculty summer salary (2 months)	$27,425	250	$6,856,250
Fringe benefits at 21.7%	$5,951	250	$1,487,806
Travel and attendance at APS meeting	$2,238	250	$559,500
Indirect costs at 56.1%	$19,980	250	$4,945,000
Total faculty support via research grants	$55,593	250	$13,898,556

NOTE: APS = American Physical Society.

community of 375 postdoctoral fellows. The committee estimates that these PhD students and postdoctoral fellows are being mentored by ~250 faculty members, which is roughly consistent with the number of distinct principal investigators represented in the Physics of Living Systems program over the decade 2010–2019. If these estimates are correct, then the core U.S. research community in biological physics consists (very) roughly of 500 PhD students, 375 postdoctoral fellows, and 250 faculty, or nearly 1,200 people. This is roughly consistent, for example, with the number of presentations on biological physics at the APS March Meeting.

These estimates of community size can be combined with the data above on the costs for supporting different components of the community, and results are summarized in Table F.4. The total depends on how support for PhD students is divided between training grants and research grants, but is in the range of $83 million to $92 million per year. It is emphasized again that this would support only the primary research personnel. Not included are the costs of research facilities, equipment and supplies, technical or administrative support staff, travel for collaboration, and so forth.

It also is possible that this discussion substantially underestimates the size of the field. As noted in Figure 8.2, if one includes both biological physics as a field of physics and biophysics as a field of the biological sciences, the enterprise is twice as large.

G

Committee Biographies

WILLIAM BIALEK, NAS, *Chair*, is the John Archibald Wheeler/Battelle Professor in Physics at Princeton University. He also is a member of the multidisciplinary Lewis-Sigler Institute. In addition to his responsibilities at Princeton, he has served as a visiting faculty member at the Graduate Center of the City University of New York, where he helped to launch an Initiative for the Theoretical Sciences. Educated in the San Francisco public schools, he attended the University of California, Berkeley, receiving an AB (1979) and a PhD (1983) in biophysics. After postdoctoral appointments at the Rijksuniversiteit Groningen in the Netherlands and the Institute for Theoretical Physics in Santa Barbara, he returned to Berkeley to join the faculty in 1986. In late 1990 he moved to the newly formed NEC Research Institute (now the NEC Laboratories) in Princeton, where he eventually became an institute fellow. Dr. Bialek's research interests have ranged over a wide variety of theoretical problems at the interface of physics and biology, from the dynamics of individual biological molecules to learning and cognition. Best known for contributions to our understanding of coding and computation in the brain, Dr. Bialek and collaborators have shown that aspects of brain function can be described as essentially optimal strategies for adapting to the complex dynamics of the world, making the most of the available signals in the face of fundamental physical constraints and limitations. More recently, he has followed these ideas of optimization into the early events of embryonic development and the processes by which all cells make decisions about when to read out the information stored in their genes. Throughout his career Dr. Bialek has been involved both in helping to establish biological physics as a discipline within physics and in helping biology to absorb the quantitative intellectual tradition of the physical sciences. For 25 years Dr. Bialek participated in summer courses at the Marine Biological Laboratory in Woods Hole, Massachusetts, serving as the co-director of the computational neuroscience course in the summers of 1998 through 2002. He also helped lead a major educational experiment at Princeton to create a truly integrated and mathematically sophisticated introduction to the natural sciences for first-year college students. He has received the Max Delbrück Prize in Biological Physics from the American Physical Society and the Swartz Prize in Theoretical and Computational Neuroscience, among other honors.

BRIDGET CARRAGHER is the co-director of the Simons Electron Microscopy Center at the New York Structural Biology Center and an adjunct professor of biochemistry and molecular biophysics at Columbia University. She received her PhD in biophysics from the University of Chicago in 1987. She worked in a variety of positions, both in industry and academia, until moving to the Scripps Research Institute in 2001. Since 2002 she has served, together with Clint Potter, as the director of the National Resource for Automated Molecular Microscopy (NRAMM), a National Institutes of Health (NIH)-funded national biotechnology research resource. The focus of NRAMM is the development of automated imaging techniques for solving three-dimensional structures of macromolecular complexes using cryo-transmission electron microscopy (cryoEM). The overall goal is to develop new methods to improve the entire process, from specimen preparation to the generation of the final three-dimensional map. In 2007 Dr. Carragher co-founded a new company, NanoImaging Services, Inc., whose goal is to provide cryoEM and other microscopy services to the biopharmaceutical and biotechnology industry. She serves as the chief technology officer of NanoImaging Services. In 2015, Drs. Carragher and Potter moved their academic laboratory from the Scripps Research Institute to the New York Structural Biology Center where they serve as the co-directors of the Simons Electron Microscopy Center. There they have established two additional NIH-funded national centers, the National Center for CryoEM Access and Training and the National Center for In-situ Tomographic Ultramicroscopy, as well as the Simons Machine Learning Center funded by the Simons Foundation.

IBRAHIM I. CISSÉ is currently the director of Max Planck Gesellschaft, heading the Department of Biological Physics at the Max Planck Institute of Immunobiology and Epigenetics in Freiburg, Germany. Prior to this, he was a professor of physics at the California Institute of Technology, and before, an associate professor with tenure in physics (and biology by courtesy) at the Massachusetts Institute of Technology (MIT). He received his bachelor in physics in 2004 from North Carolina Central University, and his PhD in physics from the University of Illinois at Urbana-Champaign, in 2009. He moved to Paris from 2010 to 2012, where he was a postdoctoral fellow at Ecole Normale Supérieure. He moved back to the United States in 2013, as a research specialist at the Howard Hughes Medical Institute's Janelia Research Campus before joining MIT in 2014 as a junior faculty. His research on single molecule and super-resolution imaging has been recognized through many honors, including being named a Pew Biomedical Scholar, a National Institutes of Health Director's New Innovator awardee, Science News "SN10 Scientists to Watch," a Vilcek Prize for Creative Promise in Biomedicine, and a MacArthur fellow.

MICHAEL M. DESAI is a professor of organismic and evolutionary biology and physics at Harvard University. Prior to this, Dr. Desai received a BA in physics from Princeton University and a PhD in physics from Harvard University. He then worked as a fellow at the Lewis-Sigler Institute for Integrative Genomics and Princeton University. He currently studies evolutionary dynamics and population genetics, primarily in microbial and viral systems. His group uses a combination of theoretical and experimental approaches to study how genetic variation is created and maintained. They also develop methods to infer the evolutionary history of populations from the variation observed in sequence data. Their focus is primarily on the dynamics and population genetics of natural selection in asexual populations such as microbes and viruses, which are often dominated by the random fluctuations in when and where rare mutational events occur. They are developing new approaches to population genetic theory to better understand the structure of genetic variation in these populations. To complement this theoretical work, the lab has developed high-throughput techniques that allow them to directly observe the evolution of thousands of experimental budding yeast populations simultaneously, tracking changes in fitness and other phenotypic characteristics and correlating these with the evolution of genetic variation within and between populations.

OLGA K. DUDKO is a professor in the Department of Physics at the University of California, San Diego. She received her PhD in theoretical physics at the B. Verkin Institute for Low Temperature Physics and Engineering in Kharkov, Ukraine, where she worked in condensed matter physics. She had postdoctoral appointments at Tel Aviv University in Israel and at the U.S. National Institutes of Health. The theory of single-molecule force spectroscopy developed by Dudko and collaborators has been widely adopted as a quantitative framework for extracting activation energies and rate constants of conformational transitions in macromolecules. Dr. Dudko's current research covers a range of problems in theoretical biological physics and is motivated by the notion that deep physics-based conceptual approaches can encompass living-systems complexity. Her research group is interested in physical principles of the spatiotemporal organization of the genome, and of the membrane fusion-mediated processes ranging from viral infection to neuronal communication. Their approach is to capture these principles in the form of analytically tractable physical theories that are reasonably simple and abstract yet generate concrete experimentally testable predictions. Dr. Dudko also serves as associate editor at *Physical Review Letters*. She was recently named a Simons Investigator and a fellow of the American Physical Society.

DANIEL I. GOLDMAN is a Dunn Family Professor in the School of Physics at the Georgia Institute of Technology. He received his BS in physics at the Massachusetts Institute of Technology in 1994 and his PhD in 2002 from The University of Texas at Austin, studying nonlinear dynamics and granular media (working with Harry Swinney). He did postdoctoral work in locomotion biomechanics at the University of California, Berkeley (working with Robert J. Full). Dr. Goldman's group focuses on discovery of principles individuals and collectives of organisms use to effectively interact with their environments, largely focusing on self-propulsion (locomotion). He compares bio and neuromechanical measurements of behavior to computational and theoretical models and has developed the "robophysics" approach to use robots (and robot collectives) as physical models of living systems. His group also studies the soft matter physics relevant to organism-environment interactions. Dr. Goldman is the lead of the Georgia Tech node in the National Science Foundation (NSF) Physics of Living Systems Student Research network and has been an instructor in the International Hands-On Research in Complex Systems School (now at the International Centre for Theoretical Physics). Dr. Goldman has received numerous awards, including a Burroughs Wellcome Career Award at the Scientific Interface and an NSF Presidential Early Career Award for Scientists and Engineers and is a fellow of the American Physical Society.

JANÉ KONDEV is the William R. Kenan, Jr. Professor of Physics at Brandeis University. He works primarily on problems in molecular and cell biology. He earned his PhD in physics from Cornell University in 1995. Dr. Kondev's research in the physical biology of the cell focuses on three distinct areas: regulation of gene expression, structure of chromosomes and their function, and dynamics of the cytoskeleton. He employs a combination of theory and experimentation on single molecules and single cells. Dr. Kondev established the Quantitative Biology Research Community (QBReC) that consists of undergraduate researchers who, while majoring in different fields of science, conduct collaborative research on specific biological problems and function as a single interdisciplinary research group. The QBReC program includes a freshman year laboratory and lecture course, which introduces students to science in an integrated fashion, combining physics, chemistry, biology, and mathematics, and collaborative research and mentorship opportunities.

PETER B. LITTLEWOOD is a professor of physics at the University of Chicago. Dr. Littlewood previously served as the director of Argonne National Laboratory, and before that was a professor of physics at the University of Cambridge and head of the Cavendish Laboratory. He is the founding executive chair of the Faraday Institution, the United Kingdom's independent center for electrochemical energy storage science and technology supporting research, training, and analysis. He began his career with almost 20 years at Bell Laboratories, ultimately serving for 5 years as the head of Theoretical Physics Research. His research interests include superconductivity and superfluids, strongly correlated electronic materials, collective dynamics of glasses, density waves in solids, neuroscience, and applications of materials for energy and sustainability. He is a fellow of the Royal Society of London, the Institute of Physics, the American Physical Society, and TWAS (The World Academy of Sciences). He serves on advisory boards of research and education institutions and other scientific organizations worldwide. He holds a bachelor's degree in natural sciences (physics) and a doctorate in physics, both from the University of Cambridge.

ANDREA J. LIU, NAS, is the Hepburn Professor of Physics in the University of Pennsylvania Department of Physics and Astronomy. Prior to becoming a professor, she received her PhD in physics from Cornell University, followed by being a postdoctoral fellow at the Exxon Research and Engineering Company, and then a postdoctoral appointment at the University of California, Santa Barbara. She then worked as a faculty member at the University of California, Los Angeles, before moving to the University of Pennsylvania. Her research group uses a combination of analytical theory and numerical simulation to study problems in soft matter physics ranging from jamming in glass-forming liquids, foams, and granular materials, to biophysical self-assembly in actin structures and other systems. The idea of jamming is that slow relaxations in many different systems, ranging from glass-forming liquids to foams and granular materials, can be viewed in a common framework. For example, one can define jamming to occur when a system develops a yield stress or extremely long stress relaxation time in a disordered state. According to this definition, many systems jam. Colloidal suspensions of particles are fluid but jam when the pressure or density is raised. Foams and emulsions (concentrated suspensions of deformable bubbles or droplets) flow when a large shear stress is applied, but jam when the shear stress is lowered below the yield stress. Even molecular liquids jam as temperature is lowered or density is increased; this is the glass transition. They have been testing the speculation that jamming has a common origin in these different systems, independent of the control parameter varied. On the biophysical side, her research has been motivated recently by the phenomenon of cell crawling. The morphology of the resulting structure is of special interest because it determines the mechanical properties of the network. Her group is developing dynamical descriptions that capture morphology. In addition, they are exploring models for how actin polymerization gives rise to force generation at the leading edge.

MARY E. MAXON is the associate laboratory director for biosciences at Lawrence Berkeley National Laboratory (LBNL). Dr. Maxon oversees LBNL's Biological Systems and Engineering, Environmental Genomics and Systems Biology, and Molecular Biophysics and Integrated Bioimaging Divisions and the Department of Energy Joint Genome Institute. She earned her BS in biology and chemistry from the State University of New York, Albany, and her PhD in molecular cell biology from the University of California, Berkeley. Dr. Maxon has worked in the private sector, both in the biotechnology and pharmaceutical industries, as well as the public sector, highlighted by her tenure as the assistant director for biological research at the White House Office of Science and Technology Policy (OSTP) in the Executive Office of the President.

JOSÉ N. ONUCHIC, NAS, is the Harry C. and Olga K. Wiess Professor of Physics and Astronomy, Chemistry and Biosciences at Rice University and the co-director of the National Science Foundation–sponsored Center for Theoretical Biological Physics. His research looks at theoretical methods for molecular biophysics and gene networks. He has introduced the concept of protein folding funnels. Energy landscape theory and the funnel concept provide the framework needed to pose and to address the questions of protein folding and function mechanisms. He developed the tunneling pathways concept for electron transfer in proteins. He is also interested in stochastic effects in genetic networks with applications to bacteria decision-making and cancer. Further expanding his ideas coming from energy landscapes for protein folding, his group is now exploring chromatin folding and function and therefore modeling the three-dimensional structure of the genome. He has received much recognition for his scientific achievements. He was elected to the National Academy of Sciences in 2006 for his contributions to understanding of protein folding and electron tunneling inside proteins. He received the International Centre for Theoretical Physics Prize in honor of Werner Heisenberg in Trieste, Italy (1989), and the Beckman Young Investigator Award (1992). He is a fellow of the American Physical Society (APS) (1995), the American Academy of Arts and Sciences (2009), the Brazilian Academy of Sciences (2009), and the Biophysical Society (2012). He was awarded the Einstein Professorship by the Chinese Academy of Sciences (2011). In 2014 he received the Diaspora Prize from the Ministry of Foreign Affairs and the Ministry of Industrial Development and Foreign Trade from Brazil. In 2015 he was awarded The International Union of

Biochemistry and Molecular Biology Medal. In 2017 he was elected fellow of the American Association for the Advancement of Science and in 2018 he was admitted to the Grã-Cruz class of the Ordem Nacional do Mérito Científico by the Brazilian government. In 2019 he received the Max Delbrück prize in Biological Physics of APS, and he received the title of Honorary Professor from his alma mater, Instituto de Física de São Carlos. In 2020 he was appointed by Pope Francis as an academician at the Pontifical Academy of Sciences.

MARK J. SCHNITZER is an investigator of the Howard Hughes Medical Institute and a professor at Stanford University with a joint appointment in the Department of Biology and the Department of Applied Physics. He is the co-director of the Cracking the Neural Code Program at Stanford University and a faculty member of the Neuroscience, Biophysics, and Molecular Imaging Programs in the Stanford School of Medicine. Dr. Schnitzer received his PhD from Princeton University in physics prior to his appointment at Stanford University. His research concerns the innovation of novel optical imaging technologies and their use in the pursuit of understanding neural circuits. The Schnitzer Lab has invented two forms of fiber-optic imaging, one- and two-photon fluorescence microendoscopy, which enable minimally invasive imaging of cells in deep brain tissues. The lab is further developing microendoscopy technology, studying how experience or environment alters neuronal properties, and exploring two different clinical applications. The group has also developed two complementary approaches to imaging neuronal and astrocytic dynamics in awake behaving animals. Much research focuses on cerebellum-dependent forms of motor learning. By combining imaging, electrophysiological, behavioral, and computational approaches, the lab seeks to understand cerebellar dynamics underlying learning, memory, and forgetting. Further work in the lab concerns neural circuitry in other mammalian brain areas such as the hippocampus and neocortex, as well as the neural circuitry of *Drosophila*.

CLARE M. WATERMAN, NAS, is a distinguished investigator, the chief of the Laboratory of Cell and Tissue Morphodynamics, and the director of the Cell Biology and Physiology Center at the National Heart, Lung, and Blood Institute (NHLBI) of the National Institutes of Health. Dr. Waterman received her bachelor's degree in biochemistry in 1989 from Mount Holyoke College and her MS in exercise science from the University of Massachusetts Amherst, prior to obtaining her PhD in cell biology from the University of Pennsylvania in 1995. After completing postdoctoral training at the University of North Carolina at Chapel Hill, in 1999, she joined the Department of Cell Biology at the Scripps Research Institute in La Jolla, California. After obtaining tenure at Scripps as an associate professor, Dr. Waterman joined the NHLBI in 2007. She has also trained hundreds of PhD candidates and postdoctoral scholars through her teaching in the physiology course at the Marine

Biological Laboratory in Woods Hole, Massachusetts, where she served as faculty from 2000–2009, and as its first female director from 2009–2014. The physiology course is an intensive 7-week laboratory summer course that has run for over 125 years. It is designed to bring together senior PhD candidates and early postdoctoral researchers to work on cutting-edge questions in cell physiology. Her research program is focused on understanding how proteins self-organize into cell-scale macromolecular ensembles that mediate the dynamic morphological and physical processes driving cell migration. The ability of cells to directionally move is critical to embryogenesis, development of the vascular and nervous systems, immune response and wound healing, and its regulation is compromised in vascular disease, immune disease, and cancer. Dr. Waterman invented the method of fluorescent speckle microscopy and used this and other state-of-the art light microscopy methods to elucidate how macromolecular protein complexes self-organize at the cell-scale to mediate directed physical outputs that drive specific cell shape change and movement. She has pioneered an integrated approach that demonstrates how cellular structures composed of the microtubule, filamentous actin, and integrin adhesion proteins are dynamically built and maintained; how they physically interact with one another; and how cell signaling coordinates their structure and dynamics to specifically mediate cell migration. Her work has shown that specific transient protein-protein interactions in a "molecular clutch" generate organized and directed forces in the cytoskeleton and transmit them through integrin-based focal adhesions to the extracellular environment to drive cell motility and morphogenesis of the vasculature.